OF ONE MIND:
THE
COLLECTIVIZATION
OF SCIENCE

Masters of Modern Physics

Published Volumes

The Road from Los Alamos by Hans A. Bethe
The Charm of Physics by Sheldon L. Glashow
Citizen Scientist by Frank von Hippel
Visit to a Small Universe by Virginia Trimble
Nuclear Reactions: Science and Trans-Science by Alvin M. Weinberg
In the Shadow of the Bomb: Physics and Arms Control
    by Sydney D. Drell
The Eye of Heaven: Ptolemy, Copernicus, and Kepler
    by Owen Gingerich
Particles and Policy by Wolfgang K.H. Panofsky
At Home in the Universe by John A. Wheeler
Cosmic Enigmas by Joseph Silk
Nothing Is Too Wonderful to Be True by Philip Morrison
Arms and the Physicist by Herbert F. York
Confessions of a Technophile by Lewis M. Branscomb
Making Waves by Charles H. Townes
Of One Mind: The Collectivization of Science by John Ziman

OF ONE MIND: THE COLLECTIVIZATION OF SCIENCE

JOHN ZIMAN

American Institute of Physics

AIP Press
American Institute of Physics
500 Sunnyside Boulevard
Woodbury, NY 11797-2999

Library of Congress Cataloging-in-Publication Data

Ziman, J. M. (John M.), 1925–
    Of one mind: the collectivization of science / John Ziman.
        p.   cm.—(Masters of modern physics; v. 16)
    Includes index.
    ISBN 1-56396-065-6
    1.   Ziman, J. M. (John M.) , 1925– .   2.   Physics—History.
    3.   Physics—Philosophy.   4.   Physicists—Biography.
    5.   Physicists—Great Britian—Biography.
    I.   Title.   II.   Series.
    QC16.Z56A3   1994   94–34902
    306.4'5—dc20   CIP

This book is volume fifteen in the Masters of Modern Physics series.

Contents

SUBDIVIDING THE REALM OF KNOWLEDGE

GUARDING THE FRONTIERS

ENLARGING THE SCIENTIFIC DOMAIN

THE WORLD BEYOND SCIENCE

INTO THE SOCIAL DIMENSION

About the Series

Masters of Modern Physics introduces the work and thought of some of the most celebrated physicists of our day. These collected essays offer a panoramic tour of the way science works, how it affects our lives, and what it means to those who practice it. Authors report from the horizons of modern research, provide engaging sketches of friends and colleagues, and reflect on the social, economic, and political consequences of the scientific and technical enterprise.

Authors have been selected for their contributions to science and for their keen ability to communicate to the general reader—often with wit, frequently in fine literary style. All have been honored by their peers and most have been prominent in shaping debates in science, technology, and public policy. Some have achieved distinction in social and cultural spheres outside the laboratory.

Many essays are drawn from popular and scientific magazines, newspapers, and journals. Still others—written for the series or drawn from notes for other occasions—appear for the first time. Authors have provided introductions and, where appropriate, annotations. Once selected for inclusion, the essays are carefully edited and updated so that each volume emerges as a finely shaped work.

Masters of Modern Physics is edited by Robert N. Ubell and overseen by an advisory panel of distinguished physicists. Sponsored by the American Institute of Physics, a consortium of major physics societies, the series serves as an authoritative survey of the people and ideas that have shaped twentieth-century science and society.

Preface

This book was conceived some fifty years ago. If my Good Fairy had then told me, a schoolboy in New Zealand, that she would eventually make me a Professor of Theoretical Physics at a distinguished British University, I could not have wished myself greater fortune. Science was my obsession, physics was my vocation, mathematics was my craft. My sole ambition was to become a serious player in the great research game.

By the time I was forty, the Good Fairy had redeemed her promise. She had protected me as I moved, step by step, along a very conventional curriculum vitae: undergraduate studies in New Zealand and Oxford, graduate studies and a post-doctoral fellowship at the Clarendon Laboratory at Oxford, ten years as a university lecturer at the Cavendish Laboratory at Cambridge, and then a professorial chair at Bristol. Examinations had been passed, a D.Phil. dissertation approved, a number of research papers and two books published. It was not easy work, I suppose, but from the beginning always enjoyable.

Early scholastic success must have germinated the original question: how was it that I could get paid and esteemed for doing just what I most enjoyed? I began to reflect on the nature of science, and on the scientific life that I was experiencing personally with such gusto.

By the time I got to Bristol, this embryonic thought had developed in an unexpected direction. I had played a small part in the slow politics of opening up the college system at Cambridge to women. I was directly involved in the whole business of scientific communication at all stages, including a primary physics journal, the international network of abstract journals and databases, a major physics review journal, a monograph series, and a general review journal. Organizing study programs at the International Center for Theoretical Physics at Trieste gave me unique oppor-

tunities to observe the growth and spread of science outside its traditional
centers in Europe and North America. Above all, I had become one of the
highly cosmopolitan inner group of the "Invisible College" of Condensed
Matter Physics, and was traveling here and there around the world to plan
and take part in scientific conferences, summer schools, workshops, colloquia and other meetings.

It was all an exciting whirl of activity, facilitating the general advancement of knowledge but taking its toll of my own personal contributions.' I
began to realize that science is generated out of a paradox. It is clearly the
product of individual men and women: it is equally clearly the collective
product of a community. On some occasions we celebrate the phenomenal
independence of mind and spirit required for discovery: on other occasions we emphasize the astonishing unity and coherence of the scientific
enterprise. The question had taken on a new form: how can it be that a
wild mob of rabid individualists can combine voluntarily to create the
most elaborate and exquisite artifacts, material and abstract, ever known
to humankind?

Ever since then, I have been fascinated by this paradox. It is obviously
far more than a very difficult many-body problem that might eventually
be solved by, say, some esoteric statistical psycho-mechanical formalism
validating seven new Laws of Socio-Dynamics. Physics is a useless guide
into the enigmas of social life. The ancient and modern masters of the human sciences, from Plato onwards, have always understood that our sharing in the common dream of being alive together is quite as ineffable as
the perennial mystery of being each our own conscious selves. Indeed, the
two mysteries are one: how we are to ourselves is inseparable from how
we are to others.

In recent years, I have been caught up in the somewhat dreary business
of science policy essentially as an independent analyst or constructive
critic of the decisions that have to be made by scientific leaders, administrative officials, politicians and others at the interfaces of science and government. At the same time, as Chairman of the Council for Science and
Society, and also as coordinator of a major program of research on public
understanding of science, I have been forced to see science and technology, scientists and technologists, through the eyes of other people, some
of them highly qualified humanists, lawyers, or political activists, others
more representative of citizens in general.

What I have come to see is that it is very misleading to dissect the personal and social components out of science, analyze them separately, and
then put them together again to see how they "interact." That "interaction" is precisely where the *action* is. The most creative individual is mo-

tivated by the social group in which he or she is embedded: the most imperious institution is a medium through which its members carry on their personal lives. Occasionally—I have in mind Michael Faraday at the Royal Institution, or Robert Oppenheimer at Los Alamos—the biographical and institutional images merge. In general, however, we put the psychological and sociological pictures into different pigeonholes, and take them out for study on different days. I realized that a conscious effort was needed to focus stereoscopically on the vital area where they overlap.

With this intent, I retired voluntarily from professional physics in 1982. But even after such a long period of gestation, this book is more descriptive than analytical. It does not pretend to be a psycho-social microscope. It is more like a bundle of optical fibers gathering images and insights from a wide range of localities in that many-dimensional focal area. These images are authentic, in that the various chapters are reproduced, at full length as they originally appeared, whether as popular articles, book reviews or invited lectures. And yet, although they were not generated according to a pre-conceived plan, it turns out that they can be arranged to combine into a larger picture. Even where they overlap, they are repetitive rather than contradictory.

This picture is not underpinned by a deep, wide-ranging theory. From exploratory treks into the social sciences, I gather that generalized models of social action are out of fashion, anyway. Current developments in "metascience"—that is, the history, philosophy, sociology, psychology, politics, economics, etc. of science and technology—are not in that direction. The post-modern temper is for revealing conflicts of interests rather than for establishing coherent general representations. The "folk" language in which these pieces were written is belittled by some pretentious academics. I believe that the theoretical models implicit in this language are still our most sensitive and convincing guide to the bewitching subtleties of human behavior.

From time to time, I have drafted grandiose schemes for formal research on this whole topic. But this would have to start with a scholarly survey of innumerable published accounts of scientists at work and play, often at second or third hand, from a very mixed bag of observers. These accounts are extraordinarily diverse. Many different conceptions of the scientific life are to be found in the various publications that are reviewed in the following chapters.

The unifying factor in this book is the instrument through which these heterogeneous images were all transformed and transmitted—my own personal interpretation. This has acted as a filter, imparting a characteristic tint to every scene. Wherever I have happened to look, I have tended to pick out and remark on the features that relate the individual to the group, and the group to the individual. As I have explained, the overall picture that emerges is not one that I set out with, but it does emphasize these features, to the exclusion of much else that I might have commented on. This mode of selection and presentation is unavoidably highly subjective. These were the topics and localities that happened to come under my eye, these were how I saw them at the time, and this is how I have now arranged and interpreted what I then reported.

Where do we start? Obviously, with the individual, in the portrait mode. The zero'th order history of physics, as of any advanced science, is often presented in the form of a family photograph album. Figures in old-fashioned dress are labeled with resounding names and complicated affiliations. So here is Great Great Grandfather Isaac, and his Dutch cousin Christian. The French branch of their family business was later expanded by Pierre, until James came down from Scotland in the 19th Century and organized a big merger with another firm—and so on.

These paste-board images are rounded into flesh and blood by scholarly biographies and personal memoirs. The *Masters of Modern Physics* series, and similar volumes of autobiographical material, are of notable service in turning scientists from plaster saints into living people. But the more conscientious the biographer, or the more honest the autobiographer, the more difficult they find it to efface their subject's self. From this abundant evidence, as well as from my own experience, I still believe that there is much psychological truth in the cliche of the "Lonely Seeker after Truth," even if it applies to only one of the many dimensions of personal being. This is the heading under which I have placed brief sketches of a few of the innumerable portraits in the physics family gallery.

The other tradition in writing about high science is philosophical. The story is told in terms of anomalous observations and unexpected discoveries, conjectures and refutations, crucial experiments and unifying theories. The mission of physics, in particular, is encapsulated in its old academic title—"Natural Philosophy"—and this is the spirit in which it has always defined itself as a collective enterprise. As we all know, the philosophy of nature is not revealed directly to any one human mind, however diligent in enquiry or thoughtful in contemplation. Physicists make sense of what they are doing as individuals by reference to this common mission and to the world picture that they are thus helping to put together.

"What the Seekers Find" gives some idea of the type of world depicted by modern physics, albeit with only glimpses of the mathematical structures around which this is built. These chapters also exemplify some of the ways in which physicists make discoveries—and put to the test each other's discovery claims. But these practices need to be located in a broader philosophical frame. The bare fact that our present understanding of the nature of the "physical world" is the product of the collective labors of a whole community does not tell us whether or not this is a true representation of objective reality. The basic principle of this whole book—that you and I and the rest of us act on the understanding that we are all living in the same world—has profound implications for any general theory of scientific knowledge. This section concludes with an outline of the epistemological rules governing the physics game, along with all other branches of science.

How is it that many thousands of proud and independent scientists are able to cooperate voluntarily in this vast enterprise? The simple answer is: division of labor. Adam Smith's recipe for economic productivity is the organizing principle of modern science. Physicists, like all scientists, work efficiently as individuals by intensively cultivating their own research plots. The process of "Subdividing the Realm of Knowledge" is not simply a procedure for cataloging scientific information or for administering academic institutions. It is a dynamic pattern of intellectual and social boundaries, continually changing in response to the growth of knowledge. Every scientist defines himself or herself as a *specialist*, active and expert in a particular research field or problem area. This is more than a statement of personal *interest*, like saying that one plays water polo, or enjoys baroque church music: it is the assertion of a *social* role, with rights and responsibilities in relation to other scientists in the same or neighboring fields.

In talking about scientific specialties, we simply cannot avoid the use of geographical terms such as "field," "domain," "area," or "boundary." Although this metaphor has been thoroughly naturalized and almost completely absorbed into familiar usage, I cannot resist its appeal as a general mental model around which to arrange the whole of this book. Having considered the internal boundaries that minutely compartmentalize science, we inevitably ask about its external borders. "Guarding the Frontiers"—that is, taking thought and action to demarcate scientific knowledge from other forms of human understanding—is a communal responsibility that cannot be left entirely to the whims of individual scientists. Physicists seem both attracted to, and repelled by, the supernatural. As card-carrying materialists, they believe that their official truths are comprehensive (if not entirely comprehensible): as investigators of the infinitesimally microscopic and infinitely macroscopic, they are always on the look out for traces of invisible, metaphysical entities. Expert knowledge of a narrow research specialty is not much use when it comes to distinguishing between imaginative speculation and pseudo-science, between observational anomalies and experimental artifacts, between self-deception and intentional fraud.

The interplay between psychological and social factors along these disputed margins is not only particularly delicate. The same factors are in dynamic tension in all the processes by which the frontiers of knowledge are pushed forward and made firm. "Enlarging the Scientific Domain" is the principal communal mission of physics—a mission that has traditionally distinguished it from most other scientific disciplines. But the expansion of science "for its own sake" expresses only one of the values to which physicists now find that they are committed. Their outward urge is pulled in particular directions by societal demands, such as the satisfaction of manifest human needs, the greater glory of the state, the desire for commercial profit, the exercise of technical virtuosity and many other admirable or deplorable influences. The forces of expansion are still very undisciplined, but natural philosophy can no longer pretend to be in general command, and technology policy is claiming the authority to set new grand strategies.

Scientists are finding that they must adapt to the changing role of their profession. This means accepting personally the new missions of their community. Drawn into "The World beyond Science," they are being loaded with unforeseen obligations, as technical experts, as entrepreneurs, as managers, as military innovators, as political advisers, as publicists and educators, and as ordinary citizens. Research scientists are urged to be "socially responsible," to foster "public understanding of science," and generally to play a much more active and direct role in the "real" world.

It is not fair to suggest that these are matters about which they never previously cared. Even in the Victorian era, there was public concern about the place of science in national life. I have to confess, however, that like most scientists of my generation, I had been led to believe that the search for truth was an absolute good, which would eventually contribute more than a mite to the good causes that urgently troubled other people. We have learned from the involvement of science in war and environmental catastrophe, political tyranny and economic exploitation, that the advancement of knowledge does not take moral precedence over other callings. We scientists are having to educate ourselves and others about the rights and responsibilities, the strengths and weaknesses, of our chosen vocation.

The increasing incorporation of science into modern life affects institutions even more than individuals. The new conceptions of science that we observe as we go "Into the Social Dimension" not only link it much more closely with technology: they also indicate radical internal changes of structure. The relationships of scientists with one another, and with their institutions, have been quite transformed. This shows up very clearly in the considerations governing research plans and career paths. The core of scientific individualism—personal choice of research problems—is increasingly subject to peer review and other forms of social control. Scientific publications are no longer treated as personal contributions to the public stock of knowledge. They are systematically totaled as numerical indicators of individual and institutional productivity, or they are designated "intellectual property," to be traded competitively in commercial markets.

Above all, most scientific activity is now undertaken by groups of researchers, sharing the use of elaborate equipment and working closely together in organized teams. Scientists still explore new realms of knowledge, but in platoons, companies, or regiments, rather than as solitary scouts. In a nutshell, this highly individualistic profession is being "Collectivized," and the traditional stereotype with which I opened this analysis is rapidly going out of date. The seekers are far from lonely, and they are motivated by more than a desire to discover the truth.

Let me present the message of this book in personal terms. In the spring of 1954, I drove over from Oxford to Bristol to talk to Nevill Mott. He had invited me to apply for a lectureship at Cambridge, where he was about to take up the Cavendish Professorship. We met in his dark, panelled office in the H.H. Wills Physics Laboratory, a large, heavy, but hospitable building which Arthur Tyndall, Mott's predecessor as Head of the Physics Department, had charmed out of one of the local Bristol tobacco barons. Our conversation was, naturally, very cordial. As well as being a very great physicist, Nevill Mott is a warm and friendly man, with a gift for putting his juniors at ease.

What struck me, however, was that he seemed so much at ease with himself. In some ways, his position in life was like that of an 18th Century nobleman. In return for material security, social esteem, freedom from mundane tasks, and complete independence of thought, word and deed, he was expected to be fully committed to larger causes, and to devote himself to the public welfare as he saw it. As a member of an elite stratum of society he was under little pressure to perform or conform, so he could easily have been idle or irresponsible. But Nevill Mott obviously had a very clear conscience. He more than paid for his privileges by his contributions to knowledge and to national life.

That year, we both moved to Cambridge, where we had a good deal of fun successfully persuading the Cavendish to teach quantum mechanics properly and unsuccessfully trying to reform the Natural Sciences Tripos.

In due course, Nevill was awarded a Nobel Prize. In 1964, I recognized that it was time to move again, and eagerly accepted the offer of a professorial post at Bristol. Eventually, in 1975, I took my turn as Chairman of that outstanding Department. The Laboratory building had not been improved in beauty or comfort by an incongruous modern wing, but as Director I inherited the same dark, panelled office where I had talked with Nevill Mott twenty years before.

How much had really changed? People looked back to a time when the Lab was just one big family. But it had expanded enormously, in academic and support staff, in student numbers, and in research facilities. It had to be organized more systematically, and had inevitably crystallized into relatively autonomous research groups, each with its own intellectual and social life. Nevertheless, it seemed to me that the Director's job was still primarily to maintain an open and supportive environment for a bunch of self-winding, immensely talented physicists, following individual careers in the traditional style. By this time, it is true, the whole system was freezing up, through lack of new posts and research resources, but that only meant that one had to be more frugal, more selective and more aggressive on behalf of the high flyers. We had lost many of our privileges, but we were still an elite, running our own affairs and renewing ourselves in each generation through proven merit.

Nowadays, when I call in on my old friends at Bristol or Cambridge, I find a very different situation. In less than twenty years, the social role of the physics professor has changed out of all recognition. My own successor at Bristol, John Enderby, occupies that same old dark, panelled office, and is just as committed as we were to serving the public through the advancement of knowledge. But he is not free to perform this social duty in his own particular way, according to his own lights. He is expected to behave like a middle manager in an industrial firm, reporting regularly to the chief executive officer of the university, and competing with other senior professors to maximize the grant income and performance ratings of their departments and research groups. He too has been "collectivized," along with all his professional colleagues.

The task of understanding the relationship between personal and social factors in science has thus become more complex and more urgent. It has become more complex because the whole concept of a scientific career is undergoing radical change. Historical and biographical studies of the scientific life of a generation ago are of limited relevance to what is happening today. We have to observe and reflect on our own actions in real time—a notoriously difficult undertaking.

It has become more urgent because, as actors in this historical drama, we need to grasp the meaning of our own parts, and extemporize appropriate lines, as the unwritten plot unfolds. I don't have any settled notion of how things will turn out. All I can suggest is that perhaps it helps us all to adapt to unexpected events if one or two of us soliloquizes aloud as these events occur. That is the spirit in which I offer this work to the public.

LONELY SEEKERS
AFTER TRUTH

There are thousands of portraits in the scientific family album. And behind each of these portraits there may be a fat volume of biography or autobiography. But almost every text in this vast archive is written in two languages. According to one reading, it is an account of explorations and discoveries in the symbolic domain of knowledge. The text can also be read as the story of a personal journey through the life-world—the domain of everyday reality. The scientific reading is often very difficult to grasp in full, but is usually far more exciting than the life-world story. Indeed, it is customary to treat the latter as secondary to or dependent on the former, except as an inferred source of interest or inspiration.

To be honest, the great majority of scientists, even the most extraordinarily creative, lead very ordinary personal lives. They may not be driven as far as Lev Landau into pretending that autobiography is somehow entirely unrelated to theoretical physics, but they often interpose psychological barriers between the two domains. It takes a very perceptive biographer, such as Abraham Pais to penetrate these barriers, even for such a reflective genius as Albert Einstein. The best of such writing does throw some light on significant events in the scientific domain, but mostly in revealing the immense diversity of the sources of individual creativity from which such events spring.

In this book, we turn the conventional study on its head, and ask how the life-world paths of individuals depend on what is happening to them in their scientific lives. The standard account of what exactly Einstein dis-

covered is complemented by the story of how those discoveries transformed him into a public figure. He was exceptional in being able to resist
the social pressures of fame. Many scientists have found that outstanding
personal achievement in the knowledge domain attracts to them such a
swarm of life-world obligations that they are diverted from the path of individual discovery into advisory or supervisory roles in academia and
government, for which they may not be well fitted. And yet the career of
Freeman Dyson shows how difficult it is to sustain consciously the role of
the lonely seeker after truth, into which he had been projected by early
scientific success. Social institutions such as the Institute for Advanced
Study at Princeton, designed precisely to facilitate this stereotypical way
of life for a few chosen mortals, do not automatically produce immortal
contributions to science.

The perverse logic of history generated a strangely similar institutional
environment in Stalin's Russia. The Soviet Academy of Sciences provided
a materially impoverished but intellectually rich home for a number of
talented scientists—especially mathematicians and theoretical physicists—whose personal goal was to shine as individuals in the domain of
ideas. The life stories of Lev Landau, Mark Azbell, and Andrei Sakharov
all began in that spirit. They each followed their star through grave obstacles and perils to this Nirvana, where they were peculiarly protected and
privileged. They all performed their scientific roles superbly: nevertheless, in mid-career their life-world paths diverged spectacularly.

Landau successfully forged straight ahead, as if totally unaware of the
surrounding lies and injustices, until he was cut down by the very earthly
tragedy of a motor accident. Azbel, by contrast, risked a decent political
gesture, which compromised his further scientific career and faced him
with the insecurity of his position as a talented Jew in a society riddled
with anti-semitic prejudice. He bravely took the lead in the "refusenik"
group, stubbornly battling for permission to migrate to a country where
their individual rights would be respected, and where eventually he was
able to resume his personal vocation as a physicist. Sakharov, as we all
know, went much further. His own sense of honor drew him, step by step,
out of the scientific domain into the immense, dangerous, perplexing
world of ordinary humanity. By devoting himself heroically to this cause,
he transformed his intellectual individualism into a societal force, with an
incalculable moral impact. Indeed, the second half of his story really belongs in Part 6, as the supreme example of a fully committed scientist
crossing the bridge into "The World Beyond Science."

The Soviet era has gone down the drain of history. Nevertheless, the
forces that came to the surface there—and I don't mean just Marxist com-

munism—are still with us, and will surely manifest themselves in other ways. As it happens, my examples are all drawn from theoretical physics, since this is the field of my own scientific experience. We know that these effects were very much more serious for other branches of science, such as genetics, than for physics. This historical episode thus illustrates vividly the impossibility of detaching the idealized concept of the scientist as a "lonely seeker after truth" from the social environment in which that role is located.

The individualistic stereotype is inadequate, even for areas of knowledge which are apparently quite isolated from the life world. This observation can be verified in the personal life of every scientist, from those singled out as "Masters of Modern Physics" to the most humble members of the scientific community, at all times, in every country, in every field of study. That must be the base line from which this analysis sets out.

What Exactly Did He Discover?

I t is less than four decades since Albert Einstein died, yet many different ent personae have been supposed behind the familiar mild exterior. Nobody would impute any lack of psychic integrity in the man himself. True enough, he was a peculiarly self-contained person whose inner life was always opaque, even to his most intimate companions. But there was no harsh discontinuity or irreconcilable inconsistency in his temperament, and we have no reason to suppose that he was nervously guarding some guilty secret like Newton's heretical Unitarianism. His private and public activities are amply documented, and are seldom inexplicable to an intelligent and imaginative observer. Yet even in his scientific work, Einstein can be represented as playing several different roles, in several quite different dramas.

Abraham Pais is a distinguished theoretical physicist who knew Einstein well in his later years at Princeton. In Pais' *"Subtle Is the Lord:" The Science and Life of Albert Einstein*, the personal and political aspects of his life are dealt with at length in a sympathetic spirit, but this is an intellectual biography, and will long hold its place as the authoritative account of his scientific achievements. It is extraordinarily well done. For anyone who can read the standard language of physics, this is a fascinating book. Here are the famous formulae that we have had to learn or to teach, here are the subtle concepts and profound arguments that we have all had to master, seen in their moments of discovery, when all is glory. It is all familiar, and yet it is a pleasure to be told it all again.

Consider him, first, in the role of iconoclast, or revolutionary. The theory of relativity upset everybody's settled notions of the uniqueness and universality of space and time. Not, of course, that people took an entirely conservative attitude towards their everyday co-ordinates. Even in 1905, the passengers on a steamship to Australia did not try to carry with them, all the way from Britain, the exact moment of twelve noon, or the direction of Nor'Nor'East. In the end, that would have meant lunching at midnight, standing on their heads. Life on a rotating planet is more conveniently ordered according to local space-time conventions, which vary systematically from place to place. But the ship's navigator had his chronometer set to Greenwich time, and knew how to orient the ship in relation to the celestial sphere of the fixed stars. For what had always seemed convincing philosophical reasons, these were considered unambiguous, absolute co-ordinates. Any event, anywhere in the universe, could in principle be located at a particular point, at a particular moment, in this complete, unique framework.

It was very shocking, then, to have young Einstein point out that this was not a very convenient scheme for talking physics when one is dealing with objects that are travelling at very high speeds. Even as a schoolboy he had tried to imagine what an electromagnetic wave—that is, a beam of light—would seem like if one tried to catch up with it and observe it as if it were nearly at rest. This almost unthinkable conception was somehow unphysical: the observer on the high-speed space-ship would be seeing something that was not permitted by the laws of physics. The only way to make scientific sense of such a situation was to conjecture that the observer would automatically carry around with him his own personal framework of space and time, in which the light would still appear to be travelling at its usual enormous speed. Every scientist, on every moving space-ship or planet, would naturally prefer to plot the events of the universe on a slightly different map, according to a slightly different calendar. Every world-view was thus, in some degree, *relative* to the situation of the viewer.

At first this sounds reasonable enough; the perspective from Sydney is not by any means the same as it is from London. But when Einstein insisted that clocks on fast-moving vehicles would appear to be running fast or slow, depending on the point of view, this was clearly prejudicial to good scientific order and discipline. In fact, he was lucky in his radicalism. Other theoretical physicists, older and better-established, had been having similar dangerous thoughts. Einstein brought this revolution to a head, but he did not make it quite alone. Nor was it quite as anarchical as many people seem to think. Different observers certainly prefer to use different maps and clocks, but these are not arbitrary or idiosyncratic. They are all related to one another by strict mathematical formulae, just as a map on Mercator's projection is precisely related to a map of the same region on a stereographic projection. All that Einstein was saying was that there is no special, no unique point of view, corresponding, say, to a point at rest in the ether. Strange physical phenomena might be observed on bodies moving nearly at the speed of light, but most of the ordinary laws of physics remain unchanged. As in all the best revolutions, the classical achievements of the past were preserved and enhanced.

See Einstein as iconoclast if you like—but see him also as a creative artist. The General Theory of Relativity that he called into being in 1916 is a daring product of pure thought. It was built upon the Special Theory with the aid of only one further fact of nature—the familiar equivalence of inertial mass and gravitational weight. Accelerate away in your space-shuttle, and everyone in the cabin will become heavier by exactly the same factor: get into orbit, where acceleration and gravity are balanced, and you will find yourself as weightless as the cup of coffee in your hand. Every physicist since Galileo has known this principle as an "accidental" law of physics: Einstein reformulated it as a logical necessity, derived from the curved geometry of the space-time continuum. As Professor Pais shows in detail, this amazing idea was not the product of a moment of artistic inspiration. First there had to be some concentrated physical thinking. What sort of physics would you expect to find in the cabin of the space-shuttle, up there in orbit? How do you measure lengths and intervals of time, electric and magnetic fields, energy and momentum, paths of light rays and of charged particles, in a weightless environment? Can you do all the usual experiments, and get all the usual answers, as if the only effect of the motion were to reduce the force of gravity to zero? Or (here is the inspiration) can you now go back to Earth, and calculate the tiny effect of gravity on, say, the propagation of light as if this were due to an unsuspected acceleration of your whole laboratory? The principle of equivalence, alone, is a powerful tool in the hands of a supreme artist of physical theory.

But now you need a systematic way of thinking about all the different representations of space, time, matter, motion and electricity that might be used by different observers moving on the most complicated curved and accelerated trajectories relative to one another. This is like asking for a general mathematical formalism for mapping all possible lines and surfaces in a space of many dimensions. Einstein was fortunate to find a mathematician who knew about the progress that had been made on this daunting problem, and taught himself the language and concepts of the tensor calculus and differential geometry. However complicated your path through space and time, at each point it has well-defined local values of curvature and torsion. The surface of the sea, on a calm day, exhibits its intrinsic sphericity whenever a ship is seen to sink below the horizon. Out of more sophisticated generalizations of spatial curvature and sphericity,

Einstein constructed a formalism for the laws of physics as they might appear locally to any particular observer, and showed how these formulae were transformed as one took on the point of view of any other.

In this formalism, the generalized "curvature" of the local space-time framework can equally well be described as an unsuspected acceleration—in other words, it will look exactly like a local gravitational field. But what puts this warp into the space-time framework? Surely this must come from the well-known source of all gravity—the presence of matter itself. There is nothing in physics more beautiful and perfect than Einstein's formula of equating a highly abstracted measure of the local density of mass and energy to an equally abstracted measure of the "curvature of space and time" in that neighborhood.

But what does this formula tell us? Whole treatises have been written on the subject. For a start, take the Sun as your local concentration of matter, and calculate the path of a particle, step by step, curving gently as it follows the dictates of the warped geometry. You will find the orbit of a planet, almost exactly as Newton described it, apparently attracted by the "gravitational force" of the Sun. Carry this calculation to a higher order of accuracy, and you can predict an anomaly in the behavior of the orbit of Mercury which has long been observed and never explained. That was one of the most exciting moments in Einstein's life. Work out that a ray of light will be bent as it passes very close to the Sun, and an expedition will be sent out to Africa to observe the stars at the next eclipse. No wonder Einstein became world-famous, in 1919, when this prediction was precisely confirmed.

The products of creative scientific artistry are not purely idiosyncratic, however. Other theoretical physicists and mathematicians were snuffling round in the neighborhood of this theory, and would probably have unearthed the same formula some time in the next fifty years, even if Einstein had never lived. It was the outcome of profound thought. In its final abstract form, it is extraordinarily simple and compelling. It has been the inspiration of astrophysics and cosmology for half a century. We still think it is perfectly correct. And yet we know that it could still prove to be wrong. The role of the scientific genius is more narrowly defined than that of the creative artist, although no easier to emulate.

To the present generation, Einstein's name is inseparable from the formula that allows nuclear mass to be transformed into nuclear energy. The lifelong pacifist is cast incongruously in the role of the magician whose incantations unleash demonic powers. This formula does, in fact, come out of relativity theory—but it is only one example of the immense technological consequences of Einstein's theoretical research. The hypothesis that light itself is quantized, which he put forward in that same miraculous year, 1905, is the basis of all our understanding of the interaction of light and matter, and thus of every sort of photo-electric or electro-optical device. He can be blamed for the pretty picture on our TV screens, or for the optical fibers that will soon be replacing telephone cables, as much as for the Bomb.

The light-quantum hypothesis itself smelt of magic to most physicists of the time. The triumph of 19th-century physics had been to prove that light was a moving wave of linked electric and magnetic fields, spreading out continuously in space. It seemed quite nonsensical, if not downright reactionary, to go back to Newton's way of talking about a beam of light as if it were a stream of discrete particles. Yet Einstein sensed that there were some subtle inconsistencies in Max Planck's analysis of the emission of light from a red-hot or white-hot object. If the light could only be emitted in "quanta," then surely it must be absorbed in the same way. Let us look more closely at the photo-electric effect, he suggested. How is it that each electron that is ejected from a metal surface can pick up such a lot of energy from the light that is played on it? It is as if the waves breaking on an ocean beach occasionally picked up one-ton rocks, and hurled them a thousand feet into the air. Think of the light as a hail of "photons," dislodging the electrons by the sheer force of their separate impacts, and everything fits.

It is a very simple, direct hypothesis. It can be tested in a variety of ways, by very elementary apparatus. It works for the light entering the eye, for telescopes, for x-rays, and for the interactions of strange particles in billion-volt accelerators. Every physicist knows it, and uses it without a second thought. Yet it retains its magical quality. Allow me some advanced mathematical formalisms, and give me time to work it all through, and I am confident that I could show you that the wave and particle properties of light are complementary aspects of a logically coherent, self-consistent conceptual scheme. I know for sure that this scheme gets the right answers for all the experiments that have ever been devised to test it. But I am not sure that I could make you believe in it as the true nature of light.

Never a professional philosopher, Einstein cast himself into a more and more philosophical role in relation to the later developments of quantum physics. This is a paradoxical story, told by Professor Pais with great clarity and understanding. Until 1925, Einstein had been at the forefront of the scientific enterprise to make sense of atomic and nuclear phenomena. Then, when Heisenberg and Schrödinger produced the new quantum theory which cleared up almost all the old confusions, Einstein held back, and seemed to take a very negative, critical position. In a famous correspondence and series of debates, he argued and argued against the doctrines he had labored so long to bring to birth.

Einstein's critique of the new quantum theory was not directed against its technical achievements. He did not, himself, make use of the new formalisms to explain atomic and nuclear phenomena, but he had no objections to those who did so, and was perfectly sensible of their successes. It was just that he could not make himself believe that this was, indeed, the true nature of things. He was particularly concerned about the probabilistic interpretation of physical phenomena. He knew, of course, that many sub-microscopic events, such as the emission of a photon by an atom or the radioactive decay of a nucleus, occur as if at random. Was it sufficient to describe such events as if their randomness were fundamental and intrinsic to the situation? "God does not play at dice," he insisted, and asked for an even better theory which did not suffer from this defect. It is proper to call this a philosophical stance, because it did not result in any new physics. All that he could do was to suggest some very cunning "thought experiments" which might have overthrown the new orthodoxy. But these experiments could not be carried out with the apparatus then available, and most physicists brushed his objections aside. In recent years, the issue of "causality" in quantum phenomena has been reopened, in the very spirit that Einstein affirmed, but it has still not been resolved. An experiment along the lines he proposed has recently given results that are consistent with the standard quantum formalism—but what does that prove against a metaphysical dissenter?

Einstein as philosopher still fascinates. In *The Cosmic Code: Quantum Physics as the Language of Nature*, Heinz Pagels, another thoughtful physicist, travels for many chapters along "the road to quantum reality," insisting that for the general public this must be the most exciting question in physics. For a reader drawn along the same road, he can be recommended as a guide. But Jonathan Powers is a genuine philosopher. As his elegant little book *Philosophy and the New Physics* makes plain, almost everything that physicists have ever told us about "the true nature of things" is worthy of skeptical philosophical analysis. This is the deeper source of Einstein's lifelong interest in philosophy, and of his profound grasp of the intellectual strategies of scientific theorizing.

Einstein's personal life was not without its strains and tragedies. Neither of his marriages seems to have been very happy, and he was driven into exile by the advent of Hitler when he was already over 50. At the age of 40, he suddenly became one of the most celebrated people in the world, and was irresistibly plunged into public affairs. As the Centennial Symposium shows, much symbolic responsibility was loaded on him, to add to his own concern about the state of the world. He did his duty as a supporter of Zionism and of pacifism with beguiling modesty and simplicity. Yet it was only in his physics that he finally came to play a heroic role— not in his triumph, but in his patience and persistence.

He could never be satisfied with a general theory of space, time and matter that did not include the classical laws of electricity and magnetism. Gravity had been transformed into a geometrical convolution of the continuum. Relativity theory showed that magnetic forces were merely the effects produced by electricity in motion: was it possible to represent the electric field itself in the language of space-time geometry? This was the goal of almost all his scientific work for the last thirty years of his life. He searched for a "unified field" theory with all his physical understanding, with all his mathematical skills, with all his imagination, and with all his will. He worked with a number of gifted collaborators. He would go into the fifth dimension, if necessary. There were times when he thought he had the creature in his grasp, but it would twist and turn and get away.

The world of physics did not share his confidence that this was an attainable goal, and passed over his work in silence. The world of physics was probably right. The equations that he was studying were just too simple to represent all the phenomena that had been discovered since he began his quest. Experiments with elementary particles were finding several other fields of force that would have to be geometrized in the same way. There were quantum effects that he was ignoring. He was proving himself the last of the classical physicists, rather than the first, and foremost, of the 20th-century breed. Nevertheless, in his cheerful persistence he provides a model for any scientist. It is unjust to say that he was already so famous that his final failure cost him little. The message is, rather, that the scientific hero embarks upon whatever research enterprise he believes to be worthy of his steel regardless of the risks of defeat.

These are some of the faces that Albert Einstein presents to us now. But for his contemporaries he must also have played a more immediate role— the fellow-citizen of the republic of learning. Consider the third of those marvelous 1905 papers, the one on the Brownian motion. It takes a little knowledge of the earlier history of physics to appreciate its importance. For a hundred years the question was: were atoms and molecules real? There was strong circumstantial evidence for them from chemistry, but could one actually detect them as distinct particles? How big are they anyway? If you look down a microscope, you certainly cannot observe atoms directly, but you can see some signs of the granularity of matter. Watch a tiny speck, such as a pollen grain, floating in a liquid: it seems to move irregularly, as if under the impact of tiny hammer blows in random directions. Is it being knocked hither and thither by the molecules of the liquid, in their eternal, restless motion? Not precisely: each molecule is much too small to move the grain perceptibly on its own. But Einstein calculated the statistical fluctuations in the forces produced by the impacts of myriads of molecules, and linked the suspected properties of the sub-microscopic world to the observed phenomenon. That young man from the Patent Office had not only solved this famous problem: he had demonstrated an uncanny mastery of the tradition of statistical physics. Here was a new member to be welcomed into the guild, a brilliant craftsman in physical intuition and mathematical analysis. The outstanding virtue of Professor Pais' book is that it places Albert Einstein in the milieu that he himself had chosen for his life work, and judges him by the most demanding standards of that most rigorous profession, as the greatest of them all, since Newton.

Portrait of a Disappointed Scientific Soul

Freeman Dyson must have been born in 1924. At Winchester, and at Cambridge, he demonstrated prodigious mathematical intellect. After a disillusioning couple of years of Operational Research in Bomber Command, he went to the United States in 1947 and began research in theoretical physics at Cornell, under the great Hans Bethe. A brilliant piece of mathematical analysis, proving that the theories of Feynman, Schwinger and Tomonaga were all essentially equivalent, soon persuaded Robert Oppenheimer to appoint him to a permanent chair at the Institute of Advanced Study at Princeton, where he has remained ever since. What, if I may say so, could be more donnish and less exciting?

Dyson's *Disturbing the Universe* is the first of a series of books "by distinguished scientists throughout the world, setting down their own accounts of their lives in science." This enterprise, supported by the Alfred P. Sloan Foundation, has much to commend it, although I just wonder how enlightening it will prove to be: very few outstanding scholars have the wit and nerve to equal Jim Watson in *The Double Helix*.

Dyson, in fact, has both wits and nerves—perhaps too much of both to be comfortable in himself or with others. He writes with scrupulous charm and lucidity, exactly in keeping with his contributions to physics. These have been *theorems*—i.e. solutions to difficult mathematical problems—rather than *theories*—novel interpretations of phenomena—but always technically beautiful, and complete in themselves. There is a sureness and finality about a scientific paper by Dyson that puts him in the top ranks of living masters of this refined and sophisticated art. He does not in fact say very much about his real work as an academic scientist, over the years, presumably because it is much too difficult to explain to the lay public—but he has certainly not been idle.

Although this book does not significantly illuminate the interior of the Ivory Tower, it testifies with remarkable sincerity to the perils that a scholar can encounter when he sallies forth to do good deeds in a less abstract realm. In a letter to Oppenheimer, Dyson refers to "the mixture of technical wisdom and political innocence" in which the first Atomic Bombs were made at Los Alamos in 1943: for all his intelligence and sensitivity, many of his projects could be characterized by the same phrase. For example, he admits ruefully that his famous article in *Foreign Affairs* in April 1960, arguing against the nuclear test ban treaty, was "a desperate attempt to salvage an untenable position with spurious emotional claptrap." This fully deserved Oppenheimer's cryptic quotation of a Hungarian proverb: "It is not enough to be impolite; one must also be wrong"—although the folly was soon redeemed by the work he did for the Arms Control and Disarmament Agency in the summers of 1962 and 1963, getting that treaty signed.

Why is it that Dyson, with every good intention, has been so unlucky in his extracurricular ventures? There are enough clues scattered around in this very pleasant and readable book to piece together an explanation of which the author himself may not be quite aware. This is only a hypothesis, but is simple enough to be instructive, for it applies to others of our most admired and admirable Mandarins.

In the first place, notice the great value he places on technical virtuosity. He empathizes immediately with the enjoyment, reported at second hand from Los Alamos, of actually building the bombs. He appreciates the "technically sweet" solution that Teller and Ulam discovered for making a hydrogen bomb. Disillusioned by his wartime experiences, he argues that "in the end it is how you fight, as much as why you fight, that makes your cause good or bad." The whole nuclear power industry is in a mess, because "nobody any longer has any fun building reactors"—i.e. the honest work of a thousand engineers cannot be made redundant by one simple, scintillating notion.

Of course we must agree with him that there is no substitute for "quality" in all creative enterprises. We must also share his unstinting admiration for people with the true creative touch—Hans Bethe, Robert Oppenheimer, Richard Feynman, Edward Teller and others. There is a modicum of truth in his suggestion that "anyone who is transcendentally great as a scientist is likely to have personal qualities that ordinary people would consider in some sense superhuman." But Dyson goes further: he seems to have surrendered himself completely to a succession of charismatic leaders, and followed them enthusiastically into their "half serious and half preposterous" schemes. For 15 months, through 1959, for example, he worked on project Orion, which was to have sent a vehicle hurtling into space by firing a few atomic bombs behind it, like fire crackers. He is now associated with Professor Gerard O'Neill's plan to colonize space by mining the moon, although personally he would like to devise some nice, small (but beautiful!) and rather risky way of swanning around in the solar system on the cheap, independently of big science, big technology and big, bad governments.

Freeman Dyson, the middle-aged professor of the highly calculated science of theoretical physics, retains all the romanticism of the school boy who clearly saw that the answer to the problems of war and injustices was Cosmic Unity. He is still "obsessed with the future." He dreams of the "greening of the galaxy" by a series of enormous technological projects in which whole planets are eventually carved and redistributed as thin shells around their slowly waning suns. Meanwhile, on earth, mankind will live in separate little nation states, armed only for defense like the Swiss and the Swedes. In the shorter term, the good life is exemplified by the vigorous adventures of his own son, sailing and pioneering amongst the Canadian Pacific islands.

Should such boyish delights have been renounced in sober maturity? Not at all, especially if the fantasies had been his own invention. But they are not all that original in conception and they do deserve an ounce or so of skeptical comment. Space travel can (surely!) never be as cheap and improvised as 'messing about in boats' or hang gliding, because the extraterrestrial firmament is so utterly hostile to terrestrial forms of life. The notion of an ocean barge or "rock eating" desert crawler that can reproduce itself robotically by solar energy is vastly incompatible with some simple arithmetic on power densities and thermodynamic potentials: in any case, it is of no significance whether human beings or artificial intelligences are in control. Yes, these are delightful "thought experiments"; but I got the uncomfortable feeling that Dyson wanted to believe in them as future science fact rather than present science fiction.

What reveals itself, apparently unconsciously in this very self-conscious and superficially revealing book, is a profound dissociation between reality and ideals, between the technical act and the philosophical thought. It is a self-portrait of a man with an immense talent for the manipulation of logical structures, and their transformation by vigorous argument into perspicuous new forms. He also has the capacity to wonder and to dream, and to float away on silvery clouds of fancy. But he seems to keep those two aspects of his personality strictly apart. That is to say, his

mathematical physics is not quite imaginative enough to rise to the sub-
limest heights of scientific genius, and he doesn't seem to apply quite
enough rational commonsense to the wilder schemes in which he is involved. There seems to be no central region of practical inventiveness
and sound judgement, where the good solid work of this world has to
get done.

For those destined by nature or nurture for life apart in the ivory tower,
this dissociation may not be a grave defect: academia demands no more
of them than technical expertise, and they may comfort themselves as
they like with any irrelevant fantasy. But Freeman Dyson is quite clearly a
humble, humane, spiritually deep and intellectually autonomous man,
who feels deeply for the fate of mankind, for the love of life, for comedy,
for tragedy, and for his own place in the scheme of things. In worldly
ways, life has been easy and kind to him; and yet this is a sad book written bravely from a disappointed soul. It doesn't tell us very much about
"what makes a scientist tick." It is singularly unrevealing about the Institute where for more than 30 years he has had to profess to be doing Advanced Studies without any teaching duties. But it is a moving human
document on the way things have turned out, and the way things now
seem, for at least one distinguished scientist of our own generation.

Landau and His School

Name the greatest Russian physicist of this century. The public vote would go for Andrei Sakharov—but for moral stature rather than for contributions to knowledge. A generation ago, Pyotr Kapitza would have been supported by many, in the mistaken belief that he was the master mind behind the Russian Bomb. Among physicists, however, Lev Davidovitch Landau would stand preeminent. He ought, by rights, to be still with us, for he was born in 1908; but a ghastly car accident in 1962 destroyed his intellectual powers and in 1968 he died.

Perhaps he does not rank in the public mind with the top dozen theoretical physicists of our times because, as he himself admitted ruefully, "I was born a bit too late," thus missing the "Golden Age" of the late 1920s, when all physics was being rebuilt from the ground up. But then again, he always insisted that it was sheer vanity to tackle only the most "important" problems of science, so there is no saying whether he would have invented quantum mechanics and won that sort of fame.

Perhaps one has to accept that he was, above all, a paragon of the professionals. His Nobel Prize citation, in 1962, was "for pioneering theories for condensed matter, especially liquid helium." Anna Livanova devotes half of her book *Landau: A Great Physicist and Teacher* to a most lucid elementary account of the theory of superfluidity of liquid helium, bringing the general reader as near as words can convey to an appreciation of this magical phenomenon and the wizardry he displayed in explaining it. And this was only one of the "Ten Commandments of Landau"—a tablet presented to him on his 50th birthday, engraved with the famous formulae which he contributed to quantum theory, thermodynamics, magnetism, superconductivity, nuclear physics and the theory of elementary particles. His range within theoretical physics was unrestricted: his knowledge and creativity affected every branch of the subject. Very little of this work has come to the outer surface of science, in the form of useful hardware or disconcerting philosophy, but that is because theoretical physics is, by its very nature, an esoteric activity which is conducted in its own dimensions of the intellect, in its own particular language. Landau's scientific thrusts were always directed at the central core of subject, where only other theoretical physicists could fully appreciate their sharpness, precision and force.

The language of theoretical physics is, of course, mathematics, of which Landau was a master. Indeed, like every top professional, in performance he was oblivious to mere "technique," which he had already transcended. There is a deceptive simplicity about the arguments of his papers, as if, somehow, everything followed naturally and inevitably from the basic physical laws. This apparently intuitive approach concealed a deep understanding of the mathematical inwardness of the situation, which few other physicists could adequately grasp. What we all admired about his scientific work was its unity—of mathematical formulae with physical concepts, of theoretical hypothesis with experimental fact, of romantic imagination with classical rigor. It was eclectic—and yet all of a piece.

I wish I could communicate to a wider public the special *quality* of Landau's scientific achievements: but it is a bit like trying to express an appreciation of Mozart to the tone-deaf. One can get no closer than the mystical generalities by which Herman Hesse hinted at the nature of the glass bead game. Can one grasp more from his scientific personality, of which Anna Livanova gives such an evocative picture? Many great scientists, like many great artists and writers, appear, so to speak, quite inconsistent with their creations. This was not the case with Landau: he exercised his mind as brilliantly in discourse as in his own counsel, and was as influential as a teacher as in his published papers. The "School of Landau," which he started at the age of 25, and unsparingly sustained thereafter, may have been his greatest achievement.

It was certainly a remarkable institution. There was no public quantification for admission. One just had to satisfy Landau personally that one had reached his "theoretical minimum"—that is to say, that one knew all of theoretical physics sufficiently well to solve any problem he wanted to set! The informality of the arrangements—a telephone call to ask for a test on one of the topics, a quiet room in which to struggle with it—belied the rigor with which these examinations were relentlessly conducted. Even for a brilliant student, it took several years of intense effort to achieve the required standard: over the whole period from 1933 to 1961, only 43 candidates (in the whole of Russia) were successful. It is not surprising that at least seven are now members of the Soviet Academy of Sciences.

For the rest of us, all over the world, some idea of the scope of the theoretical minimum can only be got at second hand, from the encyclopedic *Course of Theoretical Physics*—another of Landau's unique contributions to science. In fact, like almost all his published work, these many volumes were not actually written by Landau: he merely contributed the ideas, the mathematical proofs, the new formulations, and continual deep criticism, to manuscripts written and rewritten by his selflessly devoted co-author Evgeny Lifshitz—an accomplished physicist in his own right. It seems that he never read anything for himself either, relying on his students and colleagues to tell him exactly what they had done or what was worth noticing in the literature.

What went on inside the "School of Landau"? The key institution was the weekly Seminar—every Thursday at 11 a.m. on the dot—where pupils or visitors explained their own or other new work to the group. A theoretical physics seminar is an oven, where bright ideas get properly baked by collective criticism—or they frizzle to nothing. This particular audience was trained to be uninhibitedly scathing of every apparent imperfection, whether of exposition or logic. The atmosphere was informal, yet there was never any doubt that "Dau" (as he was known to his intimates) was in complete command of the subject and the proceedings. Landau probably picked up this way of focusing individual opinions into a collective understanding from the year he spent attending Niels Bohr's Seminar in Copenhagen, in 1929, when he was just 20. But he was much tougher and more aggressive than Bohr, and everything that went on in his research school was more deeply marked with his personal scientific style and preferences.

The central principle of his teaching was respect for convincing argument. There could be no half-truths in physics: the results had to follow from the premises, the theory had to fit the experiments, there must be no hazy steps in the mathematical analysis: otherwise it could be dismissed at once as "pathology." Theoretical physics is the sort of science where these principles are prime virtues, provided that they are combined with wide knowledge and deep understanding of just what mathematical arguments and experimental facts might be brought to bear upon the point in question. Landau selected pupils with talents consistent with his own style of thought, and instilled these principles into them to the extent that they could be frank and free and democratic with him and one another. The morale of this elite scientific group can still be detected as a positive force in Russian science.

Nevertheless, he was damnably intolerant. A first impression of "pathology," in an idea or a person, might never be overcome. Only a genius, who seldom made serious mistakes, could be forgiven such absolutism in his preferences. Even Landau could be disastrously wrong. Anna Livanova does not mention the manuscript paper by I.S. Shapiro which Landau would not approve for publication in 1956. A few months later, Lee and Yang arrived at the same idea about the conservation of parity, for which they were soon awarded a Nobel Prize. And for lesser minds the completeness and coherence claimed for his *Course of Theoretical Physics* opens doors towards academicism. Landau's scorn for pure mathematics tended to close the minds of his pupils to novel methods and concepts which were not already within the "minimum" of theoretical physics.

This absolute confidence in his own judgment was only made tolerable by his absolute integrity and honesty. Landau was never an "easy" man, in himself or with others. Although there is ample evidence of his sympathy with the problems of others—especially schoolchildren and aspiring students—he evidently had no capacity for empathy except in purely scientific matters. Despite his outward openness and directness, personal affairs were taboo in conversation with his pupils. "This is an item in your biography," he would say. Anna Livanova's biographical memoir of him lacks many such items. He practiced a superficial unpretentiousness in clothes and social manners, but never concealed from lesser mortals his disdain for their follies or weaknesses. He was driven from Leningrad and Kharkhov by irate seniors whose vanity he had pricked. It was the brave and wise Kapitza who took him into his Institute in Moscow, and provided a protected environment in which he could thrive. Although we talk of the "School of Landau," it was not until after his death that an independent L. D. Landau Institute of Theoretical Physics was actually created.

And in spite of the fact that he was entirely dedicated to physics, to the pursuit of knowledge in its most refined and superhuman form, he cannot be separated from the background of his country, his people and his era. Anna Livanova refers to "one very burdensome year" in which Landau reconstructed for himself, without paper and pencil, the theory of shock waves. I had to turn once more to the valuable little memoir by F. Janouch (published in March 1979 by CERN) to check that this was 1938, when Landau was thrown into prison, and would have been yet another innocent and irreplaceable victim of Stalin's purges had not Kapitza, with the utmost courage, and Bohr, with infinite tact, interceded for him. It should not be altogether ignored, however, that in 1954 he was made a Hero of Socialist Labor—presumably for secret work, during and after the war, not unconnected with nuclear weapons and other explosive devices. To our eyes, now, he belongs unmistakably to world science, but he had little direct experience of life outside Russia, and was no "cosmopolitan" in his attitudes.

In the circumstances of Soviet life, it is not surprising that he kept his own counsel on political issues. But what did he really think about the human condition? The deficiency of evidence is tantalizing in someone with such a powerful, alert understanding, so confident in his opinions. He was, apparently, interested in the private lives of his friends and always ready with advice on how to solve their problems. But that was precisely how he saw the vicissitudes of life—as "problems" to be scrupulously analyzed and solved. Like some (but not all) very able scientists, he was a scientific fundamentalist: he derived so much confidence from the power of scientific argument in its own sphere that he was convinced that it could be applied to every other.

Or was this extreme rationalism, with its denigration of all emotional considerations, simply the attempt of a very determined person to keep control of his own feelings? He believed that "every person should and must be happy. That is his duty to himself, to life and, if you will, to society." In his own work, it is clear enough what he meant—and although his biographers refer only perfunctorily to his marriage there is ample independent gossip about his "girls." It seems, however, that he did not appreciate for himself many of the things that make other people happy. It would be absurd to suggest that a person of such transcendental genius dedicated himself to science because of this fundamental deficiency in his own personality: yet I have noticed the same trait in other theoretical physicists and mathematicians of lesser stature. He was the most perfect, most complete, and yet most sincerely admired and loved, example of a not unfamiliar human type.

Most of this comes out very clearly and sympathetically in this little book. But one major factor is missing. Look at his name. Lev Davidovitch Landau was not just a Russian: like many of his brightest pupils, he was also a Jew. Was he really "born too late"? Had he been a student in the 1950s, he might not have found a patron such as Kapitza to protect his acerbic intelligence from jealous enemies. Were he an applicant for admission to a Russian university now, he would have to surmount formidable barriers of officially sponsored anti-semitic prejudice. I am told by a member of the Landau Institute that they have recently managed to find a post for another such brilliant young Jewish theoretician—but who knows whether Landau himself might not have ended up as a factory bookkeeper or an immigrant in Israel. Like many Russian Jews of his generation, he cannot have been unaware that science was one of the few careers that were then open to their talents. Does this explain their exaggerated attachment to pure science, as if "physics" were an end in itself, less a philosophical exercise than a self-justifying religion? This question is worth asking, because the life and work of L.D. Landau exemplifies the traditional ethic of pure science to the highest, despite the morally bankrupt, politically vicious and socially corrupt environment in which history had placed him. He did not, like Sakharov, stand up against infamy, but he stood uncompromisingly for truth.

Separation: On the Refusenik Mark Azbel

I first came across the name M. Ya. Azbel in about 1956. He was one of the three authors of a very remarkable paper, published in the Russian *Journal of Experimental and Theoretical Physics*, showing how the electrical resistance of a very pure and prefect crystal of a metal might be expected to vary with direction in a high magnetic field at a very low temperature. This paper was a decisive breakthrough in the electron theory of metals, which was my own scientific specialty. It was not surprising to see the same name attached to other papers of similar brilliance, or to hear, later, that Azbel had moved from Kharkhov to Moscow. Some of my scientific colleagues who visited Moscow in the sixties mentioned him as one of the most stimulating members of the Landau Institute for Theoretical Physics, where he was chairman of a department; he was also a professor at Moscow University. In 1973, I heard he had applied for a visa to go to Israel. The plight of Jewish "refuseniks" in the Soviet Union was becoming a serious human rights issue at that time, so it was natural enough for me to join the campaign on behalf of this Russian "opposite number." We sent letters and telegrams to various Soviet dignitaries, and I even spoke to Azbel on the telephone, direct from Bristol to Moscow, when a group of refuseniks were on a fortnight's hunger strike to draw attention to their situation.

Most of the refuseniks had lost their regular scientific jobs, and were deliberately excluded from institutes, universities and libraries. As time went on, they found it more and more difficult to keep up their scientific expertise and interests, so Alexander Voronel, another physicist, set up a regular "Sunday Seminar" in his apartment. There, a score or so of scientific refuseniks would meet and give talks about their scientific work. Occasionally they would be visited by sympathetic Western scientists who happened to be in Moscow on other business. In the summer of 1974, they had tried to expand the seminar into an International Scientific Symposium, but this was frustrated by the authorities. The foreign participants were not admitted to the country and we heard with dismay that Voronel, Azbel and other members of the group had been arrested. Perhaps it was the international outcry that won their release after a couple of weeks. Voronel got his visa at the end of 1974, and Azbel took over the leadership of the seminar. Many other refusenik scientists were able to leave Russia during the next two years, but not Mark Azbel. Agitation on his behalf was kept up in Britain, although we never had the organizational effectiveness to match the Americans. It was particularly disheartening that the Royal Society could not be persuaded to lift a finger in support of scientists in other countries.

By 1977, it began to seem as if Azbel was trapped in the Soviet Union for the rest of his life. I felt that I had failed him when I found it impossible to get a paper of his published in *Nature* or in the *Proceedings of the Royal Society*. It had been carefully smuggled out of Russia, but was so cryptically written that I could not puzzle it out. In fact, it had to do with the physical properties of DNA, but I could not find a molecular biologist who could understand that aspect of it and help to rewrite it. It was impossible at that time to discuss it with Azbel himself, so that even this little gesture of scientific solidarity was frustrated. Azbel and his companions were not to be deterred, however. In May 1977 they managed to get a number of foreign scientists to Moscow for a special jubilee meeting of the seminar. Fifty or sixty people crowded into his apartment, to celebrate the fifth anniversary of this unique scientific institution, and to emphasize the support of the world scientific community. Was it the public impact of this meeting that won Azbel his exit visa a couple of months later? It was a surprise and a delight for us that he was free after five years of waiting. He took up the Chair he had been offered at Tel Aviv University, and visited England a few months later, so that he and I got to know each other at last.

I have told this story from the viewpoint of a very minor participant, far from the scene of action, because I wanted to try to show how little we knew of the background of the lives of the people we were trying to help and of what was actually happening to them. This ignorance obviously affected the tactics of our support. As letters failed to arrive, and telephone links were cut, one had to rely on foreign visitors to Moscow to find out what was going on and what ought to be done next. It was difficult to determine who might be a friend and who had influence in that corrupt and convoluted society. Above all, there was an unanswered question at the strategic level: why should so many "successful" Russians have voluntarily entered the limbo of "refusal," from which there was no certainty they would ever escape? They were obviously in trouble and needed our help, but how did their personal situation relate to their previous careers as Soviet scientists and Soviet citizens?

Through an evocative and gripping account of his own personal experiences in his book, *Refusenik*, Mark Azbel describes the condition of the Jewish intellectual in the Soviet Union, and thus illuminates the cruel dilemmas of many hundreds of thousands of talented and self-aware people, trapped in the unhappy country from which he, fortunately, managed to escape. His scientific career was outstanding, but not unconventional, even by Western standards. The clever Jewish boy, precociously brilliant at mathematics, is a familiar figure in our schools and universities. Azbel's parents were both doctors, intensely hard-working and conscientious. Born in Kharkhov in 1932, he survived a terrible railway journey to the Far East at the beginning of the war, and returned in 1944 to an excellent school. Having won first prizes in several mathematical olympiads, he managed to get into the physics department at Kharkhov University. This, also, was a first-rate scientific institution, where he soon found his way to the frontiers of research. At the age of 24 his "candidate" dissertation—our equivalent of a PhD thesis—won him an appointment on the faculty and earned him the national and international attention to which I have already referred. There are delightful descriptions of moments of justifiable pride: the day when he argued boldly with and convinced the great Landau; the day when he spoke for the first time at Landau's seminar at the Kapitza Institute in Moscow, with all the demi-gods of Russian theoretical physics in the audience. For Kharkhov, read Bristol; for Moscow, read Cambridge; for Landau, read Dirac or Mott: that sort of academic glory might come to any brilliant student in any great scientific community.

Soon the young Mark Yakovlevitch was a "visiting member" of the group around Landau, one of the finest theoretical physicists in the world. The eventual move from Kharkhov to the Chair in Moscow in 1964 was not quite as easy as it seemed from the outside, but the Rector of the University, Ivan Petrovsky (one of Azbel's heroes), made exceptional efforts on his behalf. In 1966, his name was on the public list for a Lenin Prize, and it was likely that he would soon become a Corresponding Member of the Academy of Sciences, and, in time, a full Academician. At the very last moment, his name was crossed off the list. What had happened was very simple. Some years before, Mark Azbel and his friend Sasha Voronel had become very friendly with the writer Yuli Daniel. The families would meet at each others' apartments, in Moscow or Kharkhov, for the standard Russian pastime of talk and talk and talk. When Daniel and Sinyavsky were put on trial for "slandering" the Soviet state, Azbel and Voronel were submitted to intense interrogation. They did not betray their friend and were not directly charged themselves, but Voronel was demoted and Azbel only just managed to retain his post. The system had to punish them for the merest symptoms of dissent.

Was Mark Azbel already a dissenter? As a child he had begun to think things out for himself. In his early teens, he accidentally had access to a random collection of "forbidden" books in many languages, and "by the time I entered the university," he writes, "my understanding of what was really going on in the USSR was practically the same as it is now." Like every sensible Soviet citizen, he had learnt to keep his mouth shut, except to the most intimate and trusted friends. It is true that he had once refused to become an informer for the KGB, but many promising young scientists and responsible people must have done the same. He had also refused to join the Communist Party, which spoilt his chances of travelling abroad. But in spite of his eccentric interest in literature and the cinema, he was not regarded by his colleagues as a "political person," and by 1968 his scientific career was once more "moving ahead in a most satisfactory way."

Nevertheless, beneath the successful surface of his life, there was deep discontent. From childhood, he had encountered, and had had to overcome, the virulent anti-semitism that is ingrained in Russian culture. At school and on entry to university, in getting an academic job and in moving to Moscow, he had just managed to squeeze past the most bitter prejudice and open discrimination. Because Jewish students, for example, as a heritage of their peculiarly intellectual culture, seem to excel in the mathematical and physical sciences, the examination questions are simply rigged against them, so that only an infinitesimal number have access to the education and scientific careers their talents merit. This had nothing to do with religion; every child of Jewish parents carries an internal passport in which his or her "race" is marked as "Jewish"; from this label, every act of discrimination then follows. Anti-semitism is not, of course, official policy, but informal government and party pressures reinforce popular prejudice.

Azbel records the changes in the "climate" of prejudice following such traumatic public events as the "Doctors' Plot," the Israeli victory in the Six-Day War, and the Leningrad trials of the unsuccessful hijackers. Early in 1971, thousands of Jews openly applied for exit visas and—"it was unbelievable"—were not all forcibly rebuffed: "The thought of ever being able to escape the Soviet Union had never before entered my own mind.

But when this hope arose, I had no doubt that I would be among those who made a try for it. Overnight, it became unbearable to look forward to a lifetime as a Jew in Russia, tolerated only for his talents."

The road to Vienna and Jerusalem was longer and harder than he could have expected. The second half of his book recounts the wearisome practicalities of getting together financial resources and documentation, bluffing through red tape to establish a valid application, resigning from the Institute and finding a semblance of employment to avoid being charged with parasitism. "I moved into a second life, outside the Soviet system I had always known." There are poignant accounts of the numerous other refuseniks who came to the distinguished professor for advice and assistance. There were the public demonstrations, the meetings with foreign scientists, the grave physical hardships of the hunger strike, the lecture transmitted by telephone to Israel, the seminar to be organized and led— all with their inevitable consequences of KGB surveillance, arrest and interrogation. There was a wife and young daughter to be concerned about, and friends to celebrate with, or to console in adversity. And all the time, there was the effort to complete some research in an entirely new field, with no opportunity to discuss and revise his papers in consultation with foreign experts. It was "a harried and nervous life wherein day after day we trod the minefield laid out for those betrayers who proclaimed their intention to transfer allegiance from the Soviet motherland to malign and sinister Zion."

The attitude of his former scientific colleagues (many of whom were also Jewish) is revealing. By resigning from the Landau Institute before actually applying for a visa, he had sacrificed himself to save them embarrassment. Yet the director, Isaak Khalatnikov, would have nothing more to do with him and actively impeded his departure from the country. Another old colleague, Lev Gorkov, never spoke to him again. At first, the others also shunned him, but in time they relented. Towards the end of his long wait, he found that the "hostility and contempt I had faced when I was first in refusal had reversed itself: I was conscious of a real concern and respect. I was quite stunned. It was not only my science for which they respected me, but for my activity in dissent and Zionism." The way the refuseniks were treated was not only an infringement of their human rights, it was also a gross violation of the traditions of scientific practice. The majority of Russian scientists probably understood this well enough and did what they could to help the victims, but the Khalatnikovs and Gorkovs who participated actively in these offenses should no longer be considered members of the international scientific community.

In 1981 the meetings were blockaded by the KGB, and Victor

Brailovsky, who took over when Azbel left, was arrested, and sent into internal exile. Mark Azbel and his companions insist that they were not "dissidents": that they were not attempting to alter the Soviet political system, but were merely separating themselves from it. But of course that very act of "refusal" was an attack upon the system at a fundamental point. In a similar spirit of naive pragmatism, they created their seminar, to help each other in trouble. It was only afterwards that this could be seen as a return to the "real" science of free and independent discussion, challenging the whole corrupt apparatus of the state technical machine. Through these decent, stubborn, courageous men and women, "science," for once, found that it had a conscience.

The Dark Side of Science

Why should the case of Andrei Dmitrievich Sakharov be given such special attention? There are hundreds of millions of people in this world who are materially much worse off. There are millions whose political rights are as severely constrained. There are many thousands who are unjustly suffering more severely for the sake of their principles. I shall argue, nevertheless, that it is for the good of all mankind that his particular case should be given the widest publicity and support.

We are considering the case of a man whom few of us have seen, a man who lives in a far country, under the sway of an alien flag. And yet he is no foreigner to us. We have come to regard him as one who works with us as a colleague, who is on our side, as a comrade—indeed who belongs to our family as a brother. Such a relationship can be expressed only in the language of personal affections: it can only be defined with respect to each one of us, in our own particular terms. Let me try to explain my own feeling about this great man, in the hope that it may be typical of the attitude of many other people.

The question I might ask of Andrei Dmitrievich is—"What do we have in common? What interests do we share? To what human groups do we both belong?" He cannot answer these questions in person, so I must search in his life and works for answers.

In the beginning we were, one might say, fellow students of physics and mathematics. We shared in the enjoyment of one of the greatest cultural creations of mankind. He must have discovered beauty in the masterworks of our Newton and Faraday, just as I did in those of Lobachevsky and Mendeleev—and as we both did through the genius of Albert Einstein, transcending national frontiers and languages.

Having mastered the science of today, it is natural to move on to the

creation of new knowledge, for tomorrow. Research is a calling with a famous tradition of personal solidarity. Science, we insist, is transnational;
we celebrate its universality. Surely, that must be a bond that strongly
unites us, as scientists, with our eminent Russian colleague.

Alas, Andrei Sakharov never fully entered the world community of science. His life trail led into quite a different realm of human activity, into
secret research on nuclear weapons. In this "fantastically terrifying
world" as he later called it, he was quite cut off from the rest of us.
Should I call someone a fellow-scientist, when I am unable to share with
him the fruits of our labors? It is clear that he excelled vastly at what we
might call "Black Science"—like "Black Magic," the application of
Knowledge in the power of evil. I don't doubt that he well deserves his
reputation for the superb artistry of his solutions to a number of technical
problems; but I think the world would be better off without those regiments of talented people in all countries devoting themselves to their
"jobs" in just that way, and I do not regard them as "colleagues." Of
course, this is an opinion on which Andrei Dmitrievich would now entirely agree.

Nevertheless science is not all technique: in its deepest philosophy it
demands private integrity and public rationality. Even "Black Science"
cannot thrive when these virtues have been crushed. Like many other Soviet scientists, Andrei Sakharov was sickened by the corruption and folly
in all the business of society around him. He felt the necessity of taking a
much more "scientific"—that is rational—approach to economic, educational and military affairs. Since corruption and folly are to be found in all
countries, this is a spirit that can be shared by all enlightened persons:
there are few with his courage in making his views known directly to
those with real power.

Unfortunately, as he began slowly to understand, the well-intentioned
technocrat is impotent within the closed political system. His attempts to
influence atmospheric tests of nuclear weapons were disdainfully brushed
aside. His moving description of his disillusionment has its message for
us all. There can be no compromise with tyranny—only acquiescence, or
open defiance at terrible personal cost.

It was the publication of his pamphlet on "Progress, Co-existence and
Intellectual Freedom" in the summer of 1968 that changed him for us
from a scientific colleague to a political comrade. Not only did he ex-
pound with steely eloquence a world view that resonated with liberal
democratic opinion throughout the world; he also gave a lead to the liberal intelligentsia of his own country, far outside the scientific-technical
sphere. As he tells us "the complexity and diversity of all the phenomena

of modern life" demands "deep analysis of facts, theories and views, pre-supposing unprejudiced, unfearing open discussion and conclusions." The details of his theses on "the destruction threatened by the division of mankind" and on "the need for intellectual freedom" are not now of great significance. But by the attempt to initiate honest public discussion on these great themes over the whole political and cultural spectrum, he bound himself to us as a fellow intellectual, reacting with full personal responsibility to our life and times.

Unlike many of us "Western liberal intellectuals," however, he had "an adequate notion of the tragic complexity of real life" (in particular, life in the socialist countries). His own courageous defence of persecuted dissidents taught him the importance of freedom under the rule of law—not merely as a general condition for political, economic and cultural progress, but as a matter of supreme concern in itself. He began to speak, for example, of the freedom to emigrate as "an essential condition of spiritual freedom." In his Memorandum addressed to Brezhnev in 1971 he gave the defense of human rights as "the loftiest of all aims" of the state.

We are all well aware of Andrei Sakharov as one of the world's leading defenders of the international code of human rights. With just a few incredibly brave supporters, he kept alive the ideal of justice within the glacial cruelty of a totalitarian society. The struggle for human rights—freedom of expression, freedom of conscience, liberty of person and of movement, freedom from discrimination and exploitation—is worldwide. For many of us, it transcends the traditional political doctrines and economic utilities: it could be the decisive moral and spiritual issue of our days. We are bound to Andrei Dmitrievich as to one of the noblest of fellow citizens on this our Earth.

In his earlier years, Sakharov spoke as a member of the elite of his country, already concerned with large affairs such as economic policy and war. His pilgrimage has evidently shown him more of the humbler and simpler issues of life, the everyday concerns and values of us all. In concluding a horrifying account of contemporary Soviet society he says "…it is not a matter of accident that religion and life-affirming systems of ethical philosophy…put their emphasis on the human being and not on the nation. It is specifically the human being who is called upon to acknowledge guilt and succor his neighbor." I salute Andrei Dmitrievich above all as a fellow human being, as a member of our family, as a brother. That is why his fate is so very much the concern of humanity as a whole.

And that is also the reason why as a scientist, I am particularly concerned about his case. For we must honestly recognize that modern science is not entirely on the side of sweetness and light. "Black Science" is stronger than ever, in the service of every form of cruelty, greed and hate. We are hard put to it these days to defend our traditional norms—the universality, the disinterestedness, the openness of a transnational community bringing beneficial knowledge to all. Andrei Sakharov was scarcely allowed any opportunity to give his immense talents to that community, and to assume his natural place as one of its wisest and most powerful leaders.

He leads it spiritually, all the same. We are coming to realize that science cannot exist solely by its own standards, for its own sake. It is a part of our total culture, growing well or ill as the culture grows well or ill, given over to the forces of darkness or of light as those forces hold sway around us. Respect for fundamental human rights is the only firm foundation for an honest, straight-growing "White Science," applying itself to human benefit and need. It is only in a society whose loftiest aim is the protection of these basic rights for all its citizens that science can be an honorable pursuit.

The case of Andrei Dmitrievich Sakharov is thus of unique importance. He is more than a brave and distinguished man suffering a grave injustice. He is more than the boldest champion of freedom and justice for the people of a great nation. Amongst all who care for the future of science itself, he stands the foremost against the enemies of humanity and truth.

WHAT THE
SEEKERS FIND

Physicists and philosophers alike have a very cool, black and white image of the domain of scientific knowledge. Although it is supposed to represent buzzing, many-splendored, everpresent reality, it is distant and unchanging. Standing on the theoretical shoulders of their predecessors, peering through the intricate instruments constructed by their colleagues. manipulating the elegant algorithms of the mathematicians, each generation of physicists glimpses a little more of this sublime landscape. The communal mission of physics is to map this domain accurately. The mission of the individual physicist is to overcome the difficulties that are continually encountered in this collective enterprise.

The orderliness of the physicists' world picture, and of the elaborate equipment needed to survey it, would seem to indicate that the path of the professional researcher should be equally ordered. But this path is actually followed in the life-world, where uncertainty is the ruling principle. Discoveries are unexpectedly made, or inexplicably missed, by sheer chance. The sharp, clear image of a distinct scientific domain is blurred by the phenomenon of "serendipity" which plays such a very important part in the scientific life.

Indeed, the focal sharpness of the image may at times be very misleading. We suddenly realize that our vision is fading peripherally into misty speculation, and we no longer know quite where we are. For more than a century, physicists have been trying to construct an unambiguous account of the phenomenon of irreversibility. Perhaps this lack of success is a sign that we have quite lost our way, and will have to completely revise our maps of this area. What now should we believe? Which path should we take through the wilderness? This is another typical hazard in the personal pursuit of scientific knowledge.

Philosophy has favored fundamental physics for its conceptual clarity and precision. But the whole scientific domain cannot be mapped out in that ideal mathematical form. Reality has its own rationale, which scientists must accept and report as it reveals itself through their research. The computer models that they manipulate so exuberantly are mere simulations of genuine life-world systems. One of the moral hazards of the scientific life is to fall under the illusion that these highly schematic maps are equivalent to things as they are and as we might experience them.

The objectivity that is such a virtue of scientific knowledge does not reside in the ability to operate in a non-human domain abstracted from personal concerns. It arises out of the possibility of communication between conscious beings, and the plain fact that there are many features of our individual experiences that we share with our fellows. Objective knowledge can be achieved in this way, but only by unambiguous communication.

The scientific life is thus shaped by the means we use to convey observations and inferences to one another. Philosophers have almost always assumed that this is done most efficiently through the careful use of words, spoken or written. That is why the word-processor is not just a convenient tool: it provides an interactive link between the inner thoughts and the textual output of the scientific author. But many of our most significant observations and inferences come to us individually as visual images and mental models. Scientists have long been accustomed to conveying this information directly and unambiguously to one another by diagrams and pictures, which are thus much more fundamental to scientific practice than the philosophers seem to realize.

At the research frontier, the image of the scientist exploring and mapping a pre-existent domain is psychologically compelling. But this metaphor ignores the social element in all that we do as individuals. Scientists are not only members of a community in an everyday sense: the scientific domain that they inhabit mentally is itself a "social institution." What the individual researcher observes, or infers about the life world is not purely personal. It is framed and transformed by communication and public debate into the type of knowledge that we call scientific.

To note that this enormous body of facts and theories is built up collectively does not prove the fashionable sociological doctrine that scientific knowledge is purely arbitrary, or unassailably relative to its cultural context, or merely expresses the interests of a politically powerful group. Our overwhelming belief that we all live in the same world cannot be brushed aside so irresponsibly. But a thorough-going social model of science does suggest that the way in which scientists relate to one another must have a significant influence on their agreed accounts of their shared experiences. The subject of this book—the interface between the individual and communal aspects of science—is thus the site of a deep connection between scientific epistemology and science policy. As I have argued at much greater length in two earlier books (*Public Knowledge* [1968] and *Reliable Knowledge* [1978], both published by the Cambridge University Press) the philosophy of science cannot be separated from the practices, procedures and priorities of scientific life.

Resolving Little Local Difficulties

O n November 8, 1895, Wilhelm Roentgen inadvertently discovered x rays. Less than four months later, more purposefully, Henri Becquerel discovered radioactivity. The physicists now had the means for breaking into the atom. The era of "modern physics" had begun.

Forty years later, the revolution seemed to have run its course. Matter had been analyzed into its primary constituents: the protons and neutrons that combined to make atomic nuclei; and the electrons that circled around them in wave-mechanical clouds. The enigmatic theory of quantum mechanics had been invented, to explain their actions and reactions. The infinitely varied characteristics of chemical compounds, the electrical, magnetic and optical properties of solids, liquids and gases, could at last be understood.

But, as always in science, there remained certain little local difficulties. What were the forces binding protons and neutrons into compact, remarkably stable nuclei? Why didn't beta rays—the electrons emitted by certain unstable nuclei—all come out with the same energy? How should electromagnetism be described quantum-mechanically? Where did relativity fit into the picture? The revolution was not over, after all.

The physicists came back from World War II into a new epoch—"postmodern physics" one might call it. Armed experimentally with Ernest Lawrence's cyclotron, stimulated by Cecil Powell's discovery of a new unstable particle, the pion, in cosmic rays, intrigued by Wolfgang Pauli's invention of practically unobservable neutrinos to carry off the missing energy of beta rays, flummoxed by nonsensical infinities in all their calculations of the interaction of an electron with its own electric field, they pulled down the house they had built and set to work on deeper foundations for a taller, more robust structure.

A generation later, in 1970 say, the new building had many fine apartments—but it was an architectural monstrosity. The accelerators had grown several thousand fold in energy—and cost—and they had produced a whole zoo of exotic particles, real or fleetingly virtual. But the fairly simple rules by which they transformed into one another seemed to lack an underlying rationale. Neutrinos had not only been observed experimentally—they could even be used as instrumental probes. But what was the source of the "weak" force by which they interacted? The infinities of quantum electrodynamics could be "renormalized" away by highly suspect but effective mathematical trickery, leaving answers that agreed with experiment to ten places of decimals. But the same tricks apparently produced the same old nonsensical results when they were applied to the "strong" nuclear forces.

It was in 1974—the year of the "November revolution"—that the entirely unexpected experimental discovery of "charmed" particles clinched the argument for the existence of "quarks"—hypothetical constituents of protons and neutrons. Imagine half a dozen or so families of these, in various "colors" and "flavors," and put them together in twos or threes to make all the hundreds of animals in the particle zoo. Bounce electrons off a proton or neutron and you can even see the quarks as "partons," buzzing around like flies in a bottle. Disconcertingly, the quarks can never get out and live apart from one another for a while. But why should any of the characteristics of mundane entities persist in that infinitesimal domain?

The theory behind "the new physics" (as they call it, with typical lack of historical perspective) is far too sophisticated for words. Even when expressed in symbols, it is immensely complicated and opaque. It describes the nuclear forces satisfactorily. It is (thank goodness) "renormalizable." It combines the "weak" and the electro-magnetic forces beautifully. It predicts yet more particles, which have already been brilliantly observed. In spite of its extreme mathematical complexity, the underlying theory of "the new physics" is primarily an elaboration of quantum mechanics and does not pose any more paradoxes than the ones that Albert Einstein could not stomach about quantum mechanics itself.

For physical fundamentalists, the past decade has been a re-run of the triumphant epoch from 1925 to 1935. The ultimate constituents of matter have almost been identified, the forces between them have almost been unified. Almost—but not quite. As always, there are those little local difficulties. Why all those different families of quarks? What is the relationship between quarks and electrons? What determines the masses of the particles and the strengths of the forces? What would happen if we could bang them into each other with yet more energy? Oh, yes—and where does gravity fit into the picture? Defying monetarist parsimony, the show goes on: the next generation of revolutionaries are already harnessing up their tumbrils.

Such might be the "Kuhnian" account of 100 years of fundamental physics. Abraham Pais' 600-page chronicle, *Inward Bound: Of Matter and Forces in the Physical World*, shows that the real story was much more incoherent. Various strands develop independently for a while and then intertwine. Revolutionary ideas are born long before they come into their own, for, as Pais points out, "incomplete but profound ideas seep barely noticed into the body of physics." In detail, as in the large, "progress leads to confusion leads to progress and on and on without respite." The history of every separate "discovery"—Niels Bohr's theory of the hydrogen atom in 1913, say, or the observation of neutral currents in 1973—turns out to be almost as convoluted as the history of the whole of physics. Each page of Pais' survey could generate a separate book of 600 pages to give a fair account of what actually happened.

Pais understands the irreducible complexity of the history of science:

> I do believe that one can discern general themes in the history of discovery in science, and I shall even venture to mention some, but mainly for the purpose of emphasizing variety over uniformity. After decades spent in the midst of the fray I am more than ever convinced, however, that a search for all-embracing principles of discovery makes about as much sense as looking for the crystal structure of muddied waters.

That is why he pauses to recount some tiny incident illustrative of a period or a person—W. H. Bragg spending 17 years as professor of physics in Adelaide without a thought of doing research, before going on to win a Nobel Prize; Niels Bohr, in 1925, travelling by train from Copenhagen to Leiden and back, stopping on the way at Hamburg, Göttingen and Berlin, meeting other physicists at each station and coming back "a prophet of the electron magnet gospel." The scientific enterprise exists in the personal and conceptual realms simultaneously and is only meaningful when depicted in both those dimensions.

For the "modern" era, Pais does this beautifully. He gives a very clear account of the intellectual plot of the drama. At the same time he draws on the reminiscences of the actors, many of whom he knew personally in their old age, or selects aptly from the extensive biographical and anecdotal literature. His account of "post-modern" physics is, by his own admission, much more schematic. Apart from his own personal recollections of significant episodes—sitting in on an "illuminating evening's poker game" with Robert Oppenheimer, John von Neumann and Edward Teller in 1949 (the mind boggles), paying out on a two-dollar bet to John

Wheeler that the tau and theta mesons were distinct particles—the anecdotal material is sparser and less evocative. His version of the intellectual history of this more recent period, also, is rather more cryptic: like the conventional synopsis of the plot of an opera, it succeeds in being accurately concise but fails to convey an intelligible story.

Perhaps that was inevitable, for the half-century of "post-modern physics" is as rich in intellectual achievement as any comparable era in science. The chronology at the end of this book is just as crowded with important discoveries in the 1960s and 1970s as it was in the 1920s and 1930s. Much of that achievement is in the technique of experimentation, which Pais—himself a theorist—does not describe in detail but which he sincerely admires and respects. Much is in imaginative insight—inventing the novel concept that resolves a paradox or surmounts an antithesis. But the goal of the enterprise, and the major part of what it has achieved, cannot be separated from the mathematical language in which it is formulated. Pais apologizes innocently to "the general reader who has come this far" (that is, to 1948) for not giving a compressed account of the actual mathematical techniques that were invented at that time to "renormalize" quantum electrodynamics. Some "general reader" that, who has already digested the Dirac matrices and will have to swallow the SU(3) group before the feast is over. The physical world is there, certainly, for all humanity to understand—but he or she will need to have a PhD in theoretical physics to comprehend many things—for example, "the Higgs mechanism for generating W and Z masses"—which are now central to the whole subject.

The history of "modern" physics has been told many times, although seldom with such insight and affection. Will the full history of "post-modern" physics read like more of the same? I doubt it. The whole style of work has changed. This is perfectly obvious on the experimental side. The apparatus that one of Ernest Rutherford's students could have set up in a few months, with the help of a laboratory technician and at a cost of £50, has been replaced by an immense technological structure, built and operated over a period of years by a multinational "collaboration" of hundreds of physicists at a cost of £500 million. The intellectual goals and achievements look just the same, but the human dimension has been "collectivized" beyond redemption.

Theoretical physics remains the realm of the individual, but in a much larger, denser, community. How many "atomic physicists" were there in the world in 1935? A hundred? How many "particle theorists" are there now? Several thousand? In the fierce competition for mere professional survival, Bragg's successor in Adelaide must publish from the first or be declared redundant. Yet I have been in Abdus Salam's office in Trieste and heard him talking on the telephone to Geneva, excitedly discussing the results of an experiment that might have fitted his theory, just as Bohr might have done with Pauli on the Berlin railway platform, 60 years ago. The more it changes....

Serendipity

Chance is such a big factor in research. The chronicles of discovery, even up to the Nobel Prize, apparently celebrate as much good fortune as determined enterprise. We say, quite correctly, that fortune favors the prepared mind. We note the fruitless perspiration that is eventually rewarded with inspiration. What more can be said? By definition, we do not know what it is that research is going to find: the unexpected is constitutional of our profession, and we do well to put our trust in fate.

And yet, with extraordinary confidence, with extravagant resources, we set up the most elaborate research projects. Apparatus is designed to the last refinement; experimental protocols are laid down in detail; whole institutions are established to undertake this or that specific investigation: ponder on three decades of fusion research for a wonder of rational intention.

This polarity of carefully conceived plans against wild chance is one of the gratifying tensions of the scientific life. Confounding the philosopher, we learn to accept the "irrationality" of serendipity, without falling into mental chaos. We acquire a Micawberish confidence that something will turn up to make a success of our plan. Looking back, we trace a path through critical moments of discovery or insight, and see how knowledge has evolved by exploiting such opportunities.

It can scarcely be supposed, however, that every opportunity offered by fate has actually been seized. The more intimate history of science, the more honest biography, records many such occasions. I met a most distinguished physicist the other day who had been near to three famous discoveries in his time—and there is nothing in that to his discredit as a scientist. Fate does not announce its gifts to us, like Premium Bond prizes, by recorded delivery. They come only as intimations, at inauspicious moments, within the routine commotion of life, and are easily discarded into the waste paper basket of oblivion.

If such a "serendipitous intimation"—an anomalous observation, an experimental "bug," a fanciful conjecture, an unanswered question, an inexplicable phenomenon, a bump on a curve, an odd marking on the photograph, etc—is to be exploited, it must catch our *attention*. It must "make us curious," triggering off further investigations, testing for reproducibility, exploring attendant circumstances, verifying tentative hypotheses or disposing of trivial explanations. A scientific discovery is not just an instantaneous unpremeditated event; it is a process whose initial promise remains unfulfilled until it is completed by deliberate action.

The question is: do we dare to take this action? Do we have the means for a novel investigation? Does it lie within the terms of our employment, or our research contract? Do we have time today to follow it up? Should we give it priority over an existing project? Might it be rather inconvenient to change our personal plans in this way? Can we be confident of our own capacity to undertake this research? Does it lie within the normal scope of our specialty, or is it "somebody else's" business? Could we persuade a grant agency to fund it? And so on. Thus we reflect upon it, and make our decision.

There is nothing irrational about such considerations. They call for sober calculation of capabilities and interests. They are constrained by the practicalities and realities of existing institutional arrangements, material facilities and personal objectives. Yes, our curiosity is aroused—but if the constraints are too tight and rigid, it may seem too risky to take up the challenge, and the opportunity will be lost. The research process cannot be designed to generate serendipity, but it can be kept open and flexible, ready to respond to the lightning, wherever it may strike.

Irreversibility

"No one will take me seriously," complains the scientific pioneer, exploring far ahead of the pack. We fully sympathize, but it is not easy to "take seriously" a surmise that seems wildly at variance with our comfortable notions of reality. "The Earth going round the Sun? Fiddlesticks!" "Men descended from Apes? Pshaw!" "Drifting continents? Whatever next?" How deplorable to scoff, and yet how difficult to pick out the one such idea in a thousand that is not, after all, as wrongheaded as it first seems.

To some degree we must be influenced by the reputation of the proponent. Galileo lost many friendly ears by seeming too clever by half, whereas a modest and sober scholar like Darwin had to be listened to carefully. The relevance of that reputation to the subject in question is also an important factor: many famous scientists have been tempted to blow the trumpet of their idiosyncratic prejudices far beyond their expert knowledge. But Ilya Prigogine won the 1977 Nobel Prize for Chemistry for precisely the sort of theoretical work on which his book *From Being to Becoming* is based. He is not only a Professor at both the Free University of Brussels and the University of Texas: he is also a man of wide cultural interests, very energetic and effective in public affairs related to science and education. His novel ideas concerning the time factor in fundamental physical theory must be taken very seriously indeed.

That is not to say that this is a simple or easy book to understand. Philosophical arguments, expressed verbally, are interwoven with the concepts and theorems of advanced theoretical physics, expressed in the appropriate mathematical form. This mathematical symbolism and terminology is used without affectation or undue elaboration: it is quite essential to the theme of the book. The genuine clarity of its style might not be apparent even to "the general reader with some background in physical chemistry and thermodynamics" to whom the book is somewhat optimistically addressed. The only approach to "the mystery of time" through the physical sciences is a steep and narrow staircase of formal education, winding interminably upward to a viewpoint overlooking the classical quantal and statistical formulation of theoretical mechanics in sophisticated mathematical terms. The central problem is that "physical" time is completely reversible. It is quite unlike "psychological," "cultural" or even "biological" time, in that it has no arrow pointing inescapably from past to future. If we are to believe Newton's Laws of Motion—and the whole immense apparatus of dynamical theory constructed upon them— "time" is simply a co-ordinate which might equally go one way or the other. That which goes up might equally well be coming down; those things that are being done could as well be being undone. Forward and backward look alike: it just depends on which way you put the film into the projector—not on something intrinsic in the dynamics you portray. Indeed, as Prigogine puts it, the Newtonian universe is static—"a universe of *being* without *becoming*"—frozen into a film whose successive frames could be run back and forth, at any speed, by the eternal, omniscient god of Christian theology.

Fortunately, life is not like that. Our deepest experience is of the reality of irreversibility: "The rose that once has blown forever dies," Running a film backwards produces farcical fantasy, not an acceptable, alternative tale. How can we accept a mathematical theory that seems to contradict this salient fact both of human existence and of the universe within which it has evolved?

The conventional answer to this challenge to classical mechanics was given, about a hundred years ago, by Ludwig Boltzmann, who drew attention to the practical irreversibility of any process generating disorder. For example, when a new pack is opened, the cards are in standard sequence, number by number and suit by suit: shuffle it a few times, and that order is lost. However long you go on shuffling the pack, up to zillions of times, the same sequence will never return; the transition from order to disorder could never be reversed. Yet each particular shuffle step, like each interaction between the particles of a gas or liquid, could perfectly well be run backward without looking absurd. In other words, the apparent asymmetry between past and future in the ordinary macroscopic world of clocks, chrysanthemums, cousins and chronologies is an illusion, and does not necessarily apply amongst atoms, electrons and similar "microscopic" entities.

For almost all practical purposes, this answer has served quite adequately. In particular, Boltzmann showed the formal connection between this vague philosophical notion of growing disorder and the precise thermodynamic principle of ever-increasing entropy which had already been developed to explain an infinite diversity of natural and artificial processes—the weather, steam engines, electric batteries, chemical reactions and so on. One of the most gratifying moments in studying theoretical physics is to learn the proof of "Boltzmann's H-Theorem," which shows how a statistical approach to mechanics, where one averages over the properties of, say, the vast numbers of atoms in a tiny bubble of gas, leads precisely to the familiar equations between such everyday quantities as pressure and temperature, energy and volume, heat that flows away and work that might be done.

Until recently, Ilya Prigogine's theoretical investigations lay within this well-established paradigm. In particular, he immensely extended our conception of the role of irreversible thermodynamic processes in the natural world. The traditional association of such processes is with boring uniformity, decay and death. We had been warned of the inevitable decline in the temperature of the Sun, of the tendency in all things towards more and more disorder and randomness, culminating in the "heat death" of a universe as empty and cold and dull as the canteen of a redundant factory on a Sunday evening in winter.

But Prigogine showed up the constructive role of such processes as the conduction of heat and electricity, or a succession of chemical transformations within a single fluid medium. Usually, these processes go ahead smoothly and uneventfully, but when they are driven hard, they often generate remarkable spatial patterns, such as the regular "streets" of cumulus clouds that develop by convection on a sunny afternoon; or they may keep stopping and starting, with uncanny regularity, marking out time like a slowly beating heart. In other words, the forms of living beings are not static equilibrium patterns like the rows of atoms in a crystal, but are ordered dynamically and maintained in a steady state by the tremendous irreversible flux of energy from the Sun, passing through every cell in our bodies, to be lost in breath, and warmth, and bodily wastes. In the end, the heat death must surely come—but until then the eating and drinking and kissing need not stop.

This is a profound and inspiring insight whose fundamental truth is slowly diffusing into all branches of physics, chemistry and biology. But now Professor Prigogine wants to go further. The conjectural aspect of his book is the suggestion that the observed asymmetry of time, the familiar irreversibility of all natural phenomena, is more than a "statistical illusion," and derives from some specific mathematical feature of the primary physical laws. At the microscopic level, also, there should surely be as much of "becoming" as there is of timeless "being."

The conventional wisdom is certainly vulnerable at one point: Boltzmann's "proof" of the "H-Theorem" is not quite sound. It is easy to see, for example, that, once in a zillion hands, the pack of cards can be shuffled back into its original order. There is a finite probability that the transition from order to disorder will have been reversed. In practice, this possibility can be discounted, but it cannot be ruled out in principle. This "scandal" at the very heart of theoretical physics has been the theme of much careful research, but never fully resolved.

To exploit this weakness, Prigogine draws upon some quite new work on some quite old problems. The question is: does a dynamical system, such as a planet moving round the Sun or an atom bouncing around in a gas, go through all possible orbits in a more or less random manner, or does it tend, after a while, to repeat some previous path? This turns out to be a subtle question, whose answer depends extraordinarily sensitively on the exact set-up. Some systems are "integrable," and thus effectively predictable. Others are "ergodic," and hence would satisfy the conditions for the H-Theorem fairly well. Sometimes these two types of behavior are mixed, so that some trajectories are cyclic whilst others are apparently random, depending upon the initial conditions. Think of a squash ball, endlessly bouncing around in a perfectly cubical court: if it were aimed absolutely squarely at one wall, then it would bounce back and forth along a single line—otherwise it would go all over the place, eventually passing through almost every point in space. Or remember the Mikado's punishment for billiard sharpers—"on a cloth untrue, with a twisted cue, and elliptical billiard balls."

This fascinating field of mathematical physics is only now being systematically explored. It certainly has very important consequences for statistical mechanics at the most fundamental level. But I am not fully convinced by the final steps in Prigogine's argument that this opens the way to incorporating irreversibility into the basic equations of motion—into every collision between atoms, say, or into the formal description of an unstable elementary particle. He sketches a possible theoretical scheme, first making the standard transition from classical to quantum language and then defining an operator that could be identified physically with "entropy" on the microscopic level. He and his collaborators seem to have made some progress in the mathematical representation of such a scheme, but I simply could not tell, without reference to the detailed literature on this subject, whether it will bear the weight of interpretation that he puts upon it. I am not even sure that I see the necessity for any such development: the essential unpredictability of every quantum process may already have built into it all the irreversibility of time we could ever want.

Prigogine here moves from familiar hard ground out onto the thin ice of conjecture.

For the moment, I do not think he is asking us to accept uncritically the whole line of thought which he presents so reasonably and unpretentiously. But we should certainly make the effort to clarify and strengthen it at various points, extending the mathematical analysis and using his conceptual scheme wherever it helps our understanding of these very deep questions. For his part, it will be essential to go beyond a purely formal theory, which only transforms the conventional equations into more relations whose observable consequences are exactly the same as before. We shall be expecting from him some predictions of data or phenomena whose experimental verification might confirm—or disconfirm—his novel and imaginative hypothesis.

A Billion Years a Week

Acomputer is a *tool*, working the intentions of its designer or user. It is no more malevolent than the village clock whose chimes wake us in the night, or the car whose failed brakes run us down. We invest it with personality because it is an instrument of the mind, rather than of the hand. It extends and mimics the very function that has always seemed to distinguish us biologically from other organisms— the capacity to reason. At times, it almost seems as if, inside the black box, there is one of us. Computers are humanoid, too, in their versatility. Almost any computer can be instructed to do almost any one of the enormous variety of different things that computers in general can do. There has been nothing to equal it since the abolition of slavery.

From the willing slave to the impassive robot is a small step for the imagination, but a big step for the art. Philosophers have insisted for centuries on the sovereignty of reason among the faculties of the mind. In 1936, Alan Turing proved that all formal reasoning could—at least in principle—be represented in terms of the operations of an extremely elementary device akin to a digital computer. That was just a piece of highbrow logic, for such devices did not then exist: but the way to "artificial intelligence"—"AI"—was theoretically open. By 1950, when the first primitive electronic computers were already ticking and buzzing, Turing had turned his old argument around, and asked by what practical test an immensely elaborated logic-chopping machine at the other end of a telex line could be distinguished from a human reasoner.

David Bolter in his book *Turing's Man: Western Culture in the Computer Age* is somewhat scornful of the AI movement. He sees it as yet another instance of the age-old project of "making a human being by other than the ordinary reproductive means," by which man seeks to "raise himself above the status that nature seems to have assigned him." The Greeks, for whom the essence of being was in the outward form, dreamt up the fable of the sculptor, Pygmalion. Cartesian man, obsessed with mechanism, would fashion for himself a clockwork automaton. Nowadays, it is the disembodied *mens sana* we would ape, torn from its rightful place *in corpore sano*. The ostensible goal of AI is clearly unattainable in a practical sense. Whether it even produces useful applications along the way is beside the point. In Bolter's opinion, the serious consequences of the AI movement are not technological: "By promising (or threatening) to replace man, the computer is giving us a new definition of man, as 'information processor' and of nature as 'information to be processed.' I call those who accept this view of man and nature Turing's men."

David Bolter is a Classical scholar although he has a degree in computer science. He writes engagingly, encompassing the sweep of history right back to the Greeks.

> Computer technology is a curious combination of ancient and Western European technical qualities. Developing through modern science and engineering, it nonetheless encourages its users to think in some ways in ancient technical terms. Turing's man has in fact inherited traits from both the ancient and Western European characters, and the very combination of these traits makes him different from either. Those of us who belong to the last generation of the Western European mentality, who still live by the rapidly fading ideals of the previous era, must reconcile ourselves to the fact that electronic man does not in all ways share our view of self and world.

If I read Bolter correctly, it is not only the humanist who is thus challenged. "The issue is not whether the computer can be made to think like a human, but whether humans can and will take on the qualities of digital computers." There is no reason not to replace "human" by "scientist" in this statement.

Most people believe that computers do already "think like scientists" and that scientists are able to "think like computers." Scientific pundits have a touching faith in the efficacy of early computer training as a basic education in scientific method. There is no doubt about the scientific *origins* of electronic computing. The hardware of all electronic technology came originally from pure science, guided into foreseeable applications. The systems and the software evolved to deal urgently with the very big numerical problems of science-based weaponry in the Second World War. The computer did not create the scientists' hunger for gargantuan calculations and continually fails to satisfy that growing appetite. The most advanced computers now are still those devoted to scientific problems.

The scientificity of a computer is popularly associated with the logical precision of the instructions that have to be given to it. It is that most infuriating of creatures, the perfectly literal servant, who fails to understand what one "really" had in mind when one instructed it to clear all the dishes off the table. This perfect formalism is as intrinsic to its workings as the precision of the gears of a mechanical clock, or the rigid drilling of an 18th-century army. Every computation has the detailed transparency and overall opacity of algebra. Algebra is effective because each of its manipulative steps is perfectly clear and absolutely compelling. The logical validity of every step along the way makes the result of a long and complex calculation equally compelling. But that result cannot be immediately "obvious," for otherwise we should not have had to rely on symbolic or electronic techniques to work through the argument. In so far as computing is strictly logical, and formal logic is mathematical, and mathematics is a science, then all computing is, indeed, "scientific."

There is also a vast amount of science in the marvelous combination of physics, chemistry and engineering that makes up the chips and circuitry. But this "scientificity" of the inner workings of the black box need be of no more concern to the user than the "scientificity" that the physiologist can discern in our own heart and lungs. These organs, too, obey a natural logic, perfectly consistent with their material construction, but our only concern is that they should perform their functions unattended.

At the other extreme, considered as a component in a larger mechanical, industrial, commercial or military system, a computer is no more "scientific" than the clerks and messengers who used to "process information," "collect and collate data" and "transmit decisions." If it is "friendly" enough to us we see and act through it, as if it were an instrument of our will. Tapping away at this word processor, where every letter or paragraph can be rubbed out and corrected, is more like writing with a pencil than using a typewriter. Give Turing's men another twenty years, and they will invent a speech-recognition device to revive the oral spontaneity of the Greek poet. Simply operating a computer is not particularly "scientific," but writing programs for it might be. This is a highly disciplined craft, in which there can be no breaking of rules, no transgressing of boundaries of time and memory space. If Euclid, algebra, and Latin grammar were good exercises for the intellect, then programming is the ideal training for the convergent scientific mind. Was that also characteristic of Greek pottery or temple architecture, as Bolter suggests? However, this is to see programming solely as technique, as an entirely self-contained glass bead game. Weave and embroider the electronic "messages" as you will, they convey nothing unless given meaning by the programmer. It is the nature of the inputs and outputs that we are concerned with, and especially the language in which they are communicated.

Bolter traces our attitudes in and towards language, and invokes the magic to be found in the diversity and uncertainty of reference of a natural language. As Plato appreciated, poetry is an oral medium, cooled by writing; Aristotle's logical analysis froze the meanings into forms. A modern computer language such as FORTRAN or COBOL seems to realize Leibniz's dream of an artificial language in which all ambiguity has been defined out, all relationships are as precise as they are in algebra itself. That project failed, because it could only be used between humans, about topics that humans took an interest in: now, with computers to commune with, just such languages have had to be devised. Is that how Turing's scientist must talk and think, totally impervious to poetry, puns or passion?

Formal computer languages certainly affect our attitude to natural language. Chomsky would have every sentence parsed through the computer's logical processor, and transformed into its deeper structure. Linguistics, the epitome of a humanistic science, is reduced to its scientistic skeleton. Bolter is right to rejoice at the failure of the project for artificially translating natural languages into one another, which exposed the nakedness of Chomsky's pretensions. In reality, that project has to be turned on its head: the programmer "translates the language of the outside world into computer language, and then the computed results back into a language that the world can understand and use." He or she is in the position of the scientific theorist, trying to bridge two domains that have to mirror one another in some important aspects. One is the domain of theory—a closed, formal, precisely charged domain of abstract mappings and models. The other is the open, undifferentiated continuum of the phenomenological life world, as we observe it and live in it. Like successful science, "successful programming is a movement between two modes of thought, modes so different that we cannot yet determine whether the most compelling human problems can ever be expressed in the formal language of computers."

Turing's scientist would generate a simulacrum of nature within the abstract domain of his or her computer and set it running to see how things turn out. The dynamism of computer action is fascinating—but is it true to life? It could not be. No mapping could represent reality in all its detail and extent, except reality itself. The life world is continuously variable, real, infinite and totally connected. A digital computer program lacks all these qualities. It is "discrete, conventional, finite, isolated."

Bolter's objections on this point apply equally to all abstract schemes—that is, to all scientific theories. There has had to be a selection of qualities and aspects to represent and model. Boundaries have to be drawn, approximations sharpened into exact formulae, and statistical associations ruled to be rigorous laws. In that respect, a computer model of an aspect of nature is not different in principle from any general hypothetical proposition, to be manipulated and tested for its implications. The scientist who sets up a computer simulation to explore the implications of a theory is acting within the same tradition as Newton, who had only his new mathematical calculus and long arithmetic to do the same. Turing's scientist may, of course, be tempted to overelaborate his model, but that has always been the weakness of the zealous theoretician. More significantly, he is bound to set up models that are easily computable, and to represent nature in terms of such models. The "information processing" aspect of a human being is now emphasized above the "mechanical" aspect, because our computers are more powerful now than our clocks. Terms of art from computer technology appear naturally in scientific discourse, naively as fashionable jargon, but more profoundly because some of these terms stand for novel themata out of which new scientific theories might be made. There is no reason why we should not explore the imaginative realm thus opened up.

Nevertheless, Bolter is right to say that all computer simulation is essentially superficial. This metaphor of lack of depth relates to hierarchies of pattern and meaning. Consider one of the successes of computational effort—weather prediction. Immense quantities of elementary data—temperature, pressure, wind velocity, measured at a network of points around the globe—are assembled and made to interact with one another according to their physical nature. The simulacrum lives its own life at the rate of, say, six hours per half-hour of computing (I am reminded of a bleary-eyed Fred Hoyle complaining of loss of sleep whilst running the life-history of a star on EDSAC II at the rate of a billion years per week). We soon get a remarkably good forecast of the weather map as it will be in a couple of days' time. But look at that map, and see that all that has happened is that a "depression" or a "ridge of high pressure" or a "front" has moved in from the Atlantic and across the country. Could the computer have told us just that? Perhaps—but only because humans, with their capacity to recognize such patterns as distinct entities, would have told it to do so. Computer thought knows only its own prescribed level of generality: it has no insight into larger schemes.

The ultimate challenge of AI is to construct a simulacrum of man. But even if such a golem were feasible, if would surely lack one of the primary characteristics of all biological organisms: individuality. "Memory," Bolter points out, "might be the key to artificial intelligence," for "we live in the world we remember." It is quite possible, as Bolter indicates, that Turing's man, in our present conception, is facing in exactly the wrong direction. The more computers come to think like scientists, the less scientists need to think like computers. The practicable project is to generate more and more powerful versions of "synthetic intelligence," where the computer is used as a tool, directed and manipulated by a human intellect, strengthening those capacities for encyclopedic knowledge, instant recall, unchallengeable logic and tireless assiduity which such instruments possess. The generation we are training in our schools and amusement arcades have learnt to use these tools as extensions of their brains, just as they ride their bicycles as extensions of their muscles and nerves. They may turn out to be freer than any previous generation to exercise their individuality and imagination.

Processing Words and Thoughts

We were lunching recently with D. He is very old, very distinguished, and still very active in a walk of life where he must spend a lot of time drafting thoughtful and precise texts. The conversation turned to word processing.

D posed the question: why were so many of his friends and colleagues bubbling with enthusiasm for their own personal WPs? He could understand that they must save a great deal of office labor, but what was the point of a person like himself—a high-level "decision-maker"—sitting for hours squinting at a flickering screen tapping away unprofessionally at a keyboard, instead of allowing the thoughts to flow from his pen on to a nice clean pad of paper, to be typed up later by a dutiful secretary?

Our first shots at an explanation were deflected into the box labelled "labor-saving device." Immediate correction of typos; editing and redrafting; printing multiple minor variants: retrieving and merging texts—the facilitation of routine office work was obvious. But the idea that this might now be the best way of actually composing a literary text, a scientific report, a public speech, or a legal document, still eluded him.

After three years of timesharing on the household Beeb, we wanted to explain that it had become a tool of the mind as much as of the hands. We had each acquired a new faculty, as ineffable as the blind man's faculty for sensing the world through the tip of his cane, as if with his own hand.

An effective instrument quickly becomes an extension of the body. As I sit here, with my ultraportable Z88 on my lap, I am no more aware of the mechanics of entering words into it through the keyboard than I used to be of the movement of my pen over paper.

There is more to it than that. As I write, I look back over the previous text, checking for consistence and harmony, changing words, phrases and sentences, continually challenged into better ways of saying what I had only begun to realize I wanted to say. Did we get D to understand that our word processor now serves us as a personal text composer, beyond the power of any typewriter or pen?

On the way home, we talked ourselves down to a deeper level. There is a peculiar benefit from being able to sweep away junk at a stroke of a key. The text on the screen can be exactly as one now wants it to be, without crossings out or insertions or substitutions, without sentences scrawled sideways into margins or phrases ringed for transposition.

The benefit of an ever-clear text is not simply aesthetic. It speaks back to us transparently, with authority, as if from the printed page. "This is what you now think," it tells us, and we have to ask ourselves whether that is what we do think or mean to say. The very text that we have just composed becomes a partner in a dialogue, stimulating further turns up the creative helix.

The processed text becomes an extension of my working memory—the part that I use when I am trying to think out a new idea. I articulate propositions and revise them mentally until I get a notion of how they might be fitted together. But my memory for these components of a putative text is weak, and I cannot keep them all ready at hand. This is what my WP does for me, as I sit before it, tapping out a line or two, scrolling up into the previous page or down into my working notes, trying out a sentence or two for shape. "How do I know what I think until I write it down?" has become "How would I know how to think without being able to see clearly what I was thinking?"

The Eyes Have It

Two or more people cast their eyes on the same object, or are presented with the same scene; they quickly agree, "that is a such and such." Try showing the photograph below to almost any man, woman, or articulate child, and ask them what they see: the answer will be invariably the same: "It's a man riding a bicycle." One would be truly astonished if they solemnly affirmed that they saw a typewriter or an octopus or a potato. There is almost complete consensus in their reports of what they perceive.

It is true that an occasional observer, with the appropriate life experience, might add, "The man is Albert Einstein—whom I once knew," or "—whose face is familiar to me from photographs," and might be able to say much more about the background buildings, etc. On the other hand, an Eskimo or an Aborigine who had never before seen a bicycle might not be able to give a satisfactory answer. For the moment, however, let us accept this everyday practical skill in its normal form, without qualifications, and without attempting to analyze the neurophysiological or psychological mechanism that underlies it. The point to grasp is that intersubjective pattern recognition is a fundamental element in this creation of all scientific knowledge. This is obvious from a few examples.

At its most elementary and primitive level, science includes descriptions and classifications of natural objects (plants, minerals, stars, etc.) based upon visual inspection. (Or by using our ears, nose or taste. These other senses are not sufficiently different from vision that they need be separately discussed.) The practical problem then is how to convey these observations to other natural historians or astronomers. What is the best form of the message to be communicated or stored in the archive? An example of such a message is as follows:

> Deciduous shrub, glabrous or nearly so, with weak, trailing subglaucous, often purple-tinted stems, either decumbent and forming low bushes 50–100 cm high, or climbing over other shrubs, rarely more erect and reaching 2 m. Prickles hooked, all ± equal. L'flets 2–3 pairs, 1–3.5 cm, ovate or ovate-elliptic, simply, rarely doubly serrate, glabrous on both sides or pubescent on the veins (rarely all over) beneath, rather thin; petiole usually with some stalked glands; stipules narrow, auricles straight. Flowers 1–6, white 3–5 cm diam.; predicels 2–4 cm with stalked glands, rarely smooth; buds short…etc.etc.

What is this strange plant? In the drawing below, we at once recognize a species of rose—the familiar field rose, *Rosa arvensis*. It does indeed have the characteristics listed above; in the drawing, however, we perceive a pattern which the botanist learns to distinguish like the face of a friend.

Thus, although the verbal description draws attention to important features which might help us to place this plant in the correct genus, whilst distinguishing it from other species of rose, it is essentially incomplete without the picture. The technical words that are used in this description refer to other remembered visual patterns. How would one define the adjective "serrate," except to say that it was "like a saw?"

Try to imagine a purely verbal account of such an object, without actually drawing a picture or seeing an actual specimen. We need not

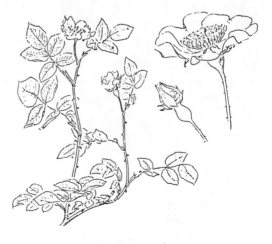

Rosa arvensis

be led astray by the numbers that occur in such descriptions. These merely summarize, in consensible language, the outcome of other sensory experiences, such as counting and measuring, and will seldom be amenable to the logical transformations of any mathematical theory. This example was drawn from *taxonomy*, where there is a long tradition of visual recognition and pictorial representation. The plant on page 62 from the sixteenth-century Chinese pharmacopoeia is immediately identifiable by a trained Western botanist as *Artemisia alba*. What could demonstrate more aptly the phenomenon of perceptual consensibility across the frontiers of almost unconnected human cultures?

It is true that this branch of natural history is neglected and scorned by many modern biologists, who prefer to concentrate on the more molecular aspects of the subject. But their research is inevitably parasitic on a whole body of taxonomic knowledge, accumulated by skillful visual observation and stored as specimens or pictures to be inspected, compared, or recognized from memory. Observation of natural phenomena, the accumulation of factual data, systematic comparison of specimens and of schemes of classification are essential activities in science, preliminary to, yet never entirely superseded by, theoretical unification at a higher level of abstraction. The ideology of physicalism blinds many philosophers of science to this elementary fact about what scientists actually do—and why. Indeed, the extraordinary fact that *consensual* schemes of biological classification can be arrived at by taxonomic research was the enigma to which Charles Darwin devoted his life and triumphantly resolved.

The recognition of the necessity for precise, critical, intersubjectively validated, visual observation revolutionized *anatomy* and *physiology* in the sixteenth century. In the Middle Ages, the medical sciences were dominated by theoretical systems dating from Antiquity. Occasional pictures, such as the accompanying medieval anatomical drawing (page 60), have no direct reference to nature; they are merely diagrams to illustrate the verbal descriptions of the human body inherited from Galen. The transcendental genius of Leonardo da Vinci is nowhere better exemplified than in his anatomical drawings such as the one of the human body reproduced on page 60.

Leonardo clearly perceived the inadequacy of verbal description: "And ye who wish to represent by words the form of man and all the aspects of his membrification, get away from that idea. For the more minutely you describe, the more you will confuse the mind of the reader, and the more you will prevent him from a knowledge of the thing described. And so it is necessary to draw and describe."

Leonardo was probably the finest scientific observer that ever lived; unfortunately his anatomical work was not published until centuries later.

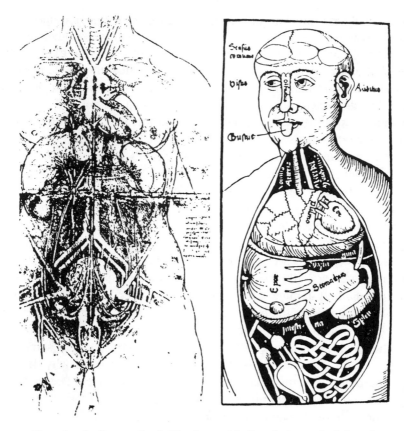

Drawing by Leonardo da Vinci   Medieval anatomical drawing.

His research program, which was eventually carried out by Andreas Vesalius, depends absolutely on the consensibility of visual perception. In assessing the "truth" of such work, anatomists should, in principle, take little account of theories; they simply carry out a dissection, and compare the drawings with what they can see with their own eyes. In practice, it is difficult for them not to see what they have learnt to see, under the influence of the accepted paradigm of the subject. But within that limitation of vision, the messages that anatomists communicate to one another, and store as "objective knowledge" in scientific archives, are drawings and photographs, to which the accompanying text is merely a commentary.

It would be a great mistake to suppose that the dominant role of visual observation in biology has now been overthrown by more "analytical" methods. The skilled eye and hand-lens of the old-fashioned "naturalist" were supplemented in the late seventeenth century by the optical microscope, which has only recently been overtaken in magnifying power by the electron microscope. But modern cellular biologists are as much concerned to discover and explain the spatial organization, the "shape," the "form," the "structure" of the "organelles" they can observe within the cell as were the Renaissance anatomists describing the organs they could observe by dissecting a cadaver.

A modern scientific paper on *microbiology, physiology,* or *pathology* is as heavily loaded with electron micrographs as any medieval herbal with its woodcuts of botanical species. The caption of the accompanying electron micrograph (page 62) does not read, "Look at this pretty picture!" The picture offers characteristic shapes which are highly significant and are to be interpreted scientifically. Thus, an immense amount of scientific information is gathered by the direct visual recognition of similarities or analogies between pictorial patterns, mediated solely by the enormous magnifying power of these instruments.

We may also reflect upon the fact that a great deal of experimentation in physiology and *biochemistry* is directed towards the elucidation of "invisible" spatial structures. In its way, the determination of the "structure" of a complex molecule by x-ray diffraction is a splendid example of the mathematical transforming power inherent in good, clean physics. But the goal of all the instrumentation, electronics, computation and scientific skill that go into such an investigation is to produce a three-dimensional model, or a series of pictures, of the arrangement of the atoms in space (like the drawing of an RNA yeast molecule on page 62). This is the piece of scientific knowledge that we shall eventually transmit to our colleagues or put into the archive—not simply for "its own sake," but because we anticipate that visual inspection of the model will suggest entirely new properties of the molecule that would not be immediately detected by mechanical measurement or analytical transformation of the corresponding data.

What is the goal of modern *chemistry,* but the creation of the perfect "molecule analyzer"? Feed a small sample of an unknown material into this hypothetical machine: five minutes later, out pops complete information about the *structure* of its constituents. Ideally, this includes a photograph of the arrangement of the atoms in each molecule—little spheres neatly labelled C, N, O, S, etc., or colored black for carbon, white for oxygen, yellow for sulfur, green for chlorine, etc., joined by neat springy "bonds," as in one of those delightful sets of molecular models that no teacher can resist, illustrating a little too concretely the basic principles of *stereochemistry,* whose imagining revolutionized chemical theory a hundred years ago.

Is it necessary to point out that *geology* and *astronomy* are almost completely dependent on visual observation for their primary data? As in biol-

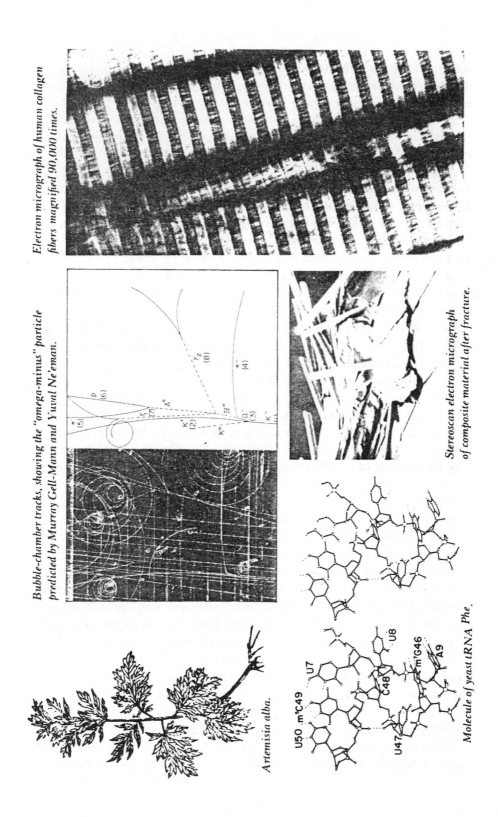

Electron micrograph of human collagen fibers magnified 90,000 times.

Bubble-chamber tracks, showing the "omega-minus" particle predicted by Murray Gell-Mann and Yuval Ne'eman.

Stereoscan electron micrograph of composite material after fracture.

Artemisia alba.

Molecule of yeast tRNA Phe.

ogy and chemistry, this does not preclude evidence obtained nonvisually by complex apparatus and interpreted analytically by the application of theory. But it is hard to imagine these sciences originating and flourishing in the country of the blind.

As an intellectual purist the physicist imagines that his science is free of the taint of observational subjectivity. In principle, the "data" of physics are numbers representing pointer readings or computer printout from mechanical or electrical apparatus, taken without significant human intervention. But consider the photograph on page 62—not, as it seems at first sight, camping gear bundled untidily in somebody's attic, but a "stereoscan" electron micrograph of the surface of a composite material after fracture. Under careful inspection, this picture reveals the way in which the reinforcing fibers are pulled from the matrix before they snap, and thus plays an important part in the physical interpretation of a complex but calculable phenomenon. Should such evidence from pattern recognition be denied to the research physicist on philosophical grounds? Or should we maintain that the quantitative study of the properties of materials is not really "physics" after all?

Visual perception plays a vital part in the most aristocratic branch of experimental science—the physics of elementary particles. One of the main instruments in this field is the bubble chamber, where high-energy charged particles make visible tracks that are easily photographed and inspected for unusual "events" (as in the bubble-chamber tracks shown in the illustration on page 62). Nowadays, to be sure, the eye of the observer is aided by optical and electronic devices, hooked up to a computer. Where the parameters of a large number of similar events are to be determined, these procedures may be almost completely automated. But significant scientific progress in high-energy physics did not have to wait until this approximation to instrumental "objectivity" had been achieved. Until quite recently, the fundamental link in the observational chain was a "scanner"—a person without deep scientific education who was instructed to search bubble-chamber photographs or nuclear emulsions for particle tracks of a particular type. It is significant that although individual members of a team of scanners might acquire a good deal of practical skill, there was no need for an elaborate "calibration" of their eyes and brains. This perfectly exemplifies the high degree of visual consensuality that scientists take for granted as they collect information about the "external world."

As a final illustration of the scientific application of the maxim that "seeing is believing," consider the photograph of a small Martian crater (page 64). Peculiar markings were observed on the surface of Mars, in the neighborhood of a great volcanic cone. It was suggested that these might be due to wind-blown deposits of sand. A model "volcano" was set up in a

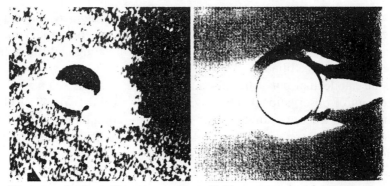

Photograph of small crater on Mars (left) compared with distribution of sand around model.

wind tunnel, and sand deposits allowed to build up around it under the influence of dimensionally scaled current of air. The equivalence of the "natural" and "artificial" patterns is quite sufficient proof that this explanation must be substantially correct. In other words, this matching of visual patterns has validated a physical theory without recourse to formal mathematical analysis.

Suppose, indeed, that an attempt were made to verify this hypothesis by actual calculation. The standard aerodynamical equations of motion of the atmosphere would be solved arithmetically in a large computer, with boundary conditions adapted to the supposed shape of the crater and the strength of the wind. From the theory of the transport of small solid particles in a current of air, the applied mathematician would then compute the distribution of sand deposits in the neighboring region. But how would he or she decide, or demonstrate that the results of this computation were in agreement with experiment? Ten to one, the mathematician would instruct the computer to print out a "picture" of sand deposits, which he or she would compare visually with the space-probe photograph! Here, once more, the pictures themselves constitute the basic "information" that scientists convey to their colleagues.

In this effort to demonstrate the ubiquity of pattern recognition in science, I have deliberately chosen cases where visual perception plays different methodological roles. But whether we are concerned with biological taxonomy, or with the use of "graphics" in the "interactive mode" of an advanced computer technique, we come back to the same fundamental point—that the bodily senses are the only link between a human mind and the world he or she inhabits. Visual perception, by its intersubjective consensibility, is an essential element in the creation and validation of scientific knowledge, and pattern matching provides a standard which is never completely superseded by more "objective" devices such as mechanical instrumentation.

Science in at Least Three Dimensions

Models of Science in Science Policy

S cience policy, like all policy, is no more than expediency, or a series of short term conflicts between crisis demands and inarticulate traditions, unless illuminated by a general conception of the nature of science and the society in which it exists. In the intellectual jargon of our day it is customary to delineate such a conception as a "model" or "system," thus giving it a concreteness which it seldom deserves. Nevertheless this practice does suggest that we should define our theory completely and realistically enough to convince ourselves and others that all the essential working parts are present. An aeronautical engineer who talked only of aerodynamic theory, and who never showed us even the merest sketch of a design of an aircraft, including wings, engines, control mechanisms, landing gear, etc., would not be a very convincing guide in technology policy—and the same applies to many theorists of science who seem only able to see the subject from one intellectual aspect.

What we may have learnt in the last few decades is that the scientific activity is not simply the application of a certain line of philosophy. Nor is it simply a highly evolved and sophisticated manifestation of the characteristic psychological quirk of curiosity. And the fact that it is clearly a social activity, involving a distinct community of professional participants, does not mean that it is simply a club, or a tribe, or yet another human group. In other words, it is impossible to make a proper model of science solely out of "philosophical" or "psychological" or "sociological" elements. Each of these aspects of science is of the greatest importance. Sound logic, personal creativity and social cooperation are indispensable features of the growth of scientific knowledge. What is more, they are not entirely independent factors but are as closely interrelated, interconnected, and mutually adapted, as the wings, engines and tail planes must be of any aircraft that can safely fly. It is not possible to make a satisfactory model of science in less than three dimensions—the *cognitive* dimension of knowledge, the *psychological* dimension of the individual scientist and the *social* dimension of the scientific community.

This is not a very novel thought. It is certainly implicit in a great deal of what has been discovered, or for the first time clearly stated, by many different philosophers, sociologists, and historians of science, in recent years. It can also be detected as an underlying consensual theme, even amongst those who argue most vehemently about particular details in the general field of science studies. But the model is seldom presented as a whole, either as an academic framework for students, as a well established foundation for further research, or as the "system" for which policy is to be made in the larger, political, economic, and social sphere.

It may be that science theorists have been reluctant to set this model up, or to rely on it for insight, because they have not fully appreciated the simple principle on which it actually works. This principle can be stated quite easily: *the goal of science is a consensus of rational opinion over the widest possible field.*[1] Such a simple principle naturally calls for comment and qualification, in all sorts of ways, so that it is less of a formal definition than a "characterization" of science. Nevertheless, by the three-dimensional form and its dynamic force it has the power to interact with, integrate, and drive the model along. From it may be deduced many of the familiar features of science, such as the so-called "scientific method" that harnesses individual curiosity to the creation of a body of public knowledge, the norms that govern the relationship of scientists within their community, the practice of peer review and other elements of the communication system of science, and so on.

For a satisfactory understanding of science policy in the modern era—that is, the external relations of what we have come to call the "R&D system" with society at large—it seems necessary to extend the model into a fourth "dimension" of economic and material resources. In this augmented model the distinction between science and technology, between industrial research and industrialized research, is blurred into insignificance. There seems no doubt that this is the way we should now be thinking, talking, and teaching, about science in its social context, in the effort to understand such great contemporary problems as the application of science to genuine human needs, the place of the scientific expert in the social polity, the validity of the "scientific world picture," in the philosophies and ideologies of our times, and the social, economical, and political, conditions under which science can continue to thrive and contribute to human affairs.

The Epistemological Challenge

But these are large issues beyond the scope of this paper.[2] And it turns out, when we look at them closely, that they drive us back into the more elementary three-dimensional model which might have been sufficient to describe the "academic" science of a previous era. Even this is a complex subject with many diverse aspects. What I propose to do, therefore, is to develop a particular theme in some depth, to illustrate the richness and fruitfulness of the multidimensional approach. Even this is only a sketch of an investigation with many fascinating ramifications.[3]

This theme is simply the answer to the fundamental epistemological challenge: "Is science to be believed?" With the decline of various philosophical fashions which, for one reason or another, argued that this question must necessarily be answered positively, or which took it for granted that this was, indeed, the case, this is no longer an issue that can be just left to the philosophers. In fact, when you come to think about it, it is a fundamental question about which every conflict between the supporters and opponents of science, necessarily rages. Asked more subtly—"To what extent are we justified in relying upon this or that item of scientific knowledge"—it is of very practical significance in every application of science in human affairs. Those of us whose main experience of science has been in the "natural" sciences of physics, chemistry and biology, are seldom called upon to explore these questions in any depth, in a general sort of way. But they can certainly not be avoided in the social behavioral sciences, which now make considerable claims to technical expertise and often assert an understanding of their subject matter that goes far beyond mere practical experience.

Science as a Social System

As with more mechanical and material devices, such as nuclear reactors or jet aircraft, it is much easier to explain how this model works, with a few very rough and schematic diagrams, than it is in the most perfectly limpid prose. The primitive features of the model are shown in Figure 1.

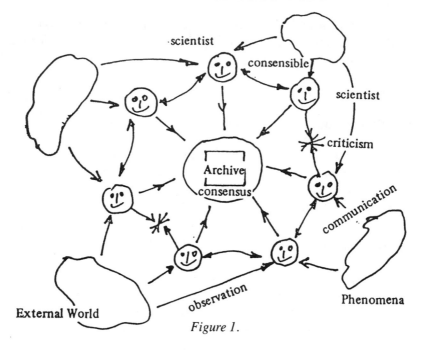

Figure 1.

In this picture one can see the individual scientists observing the "external world" and communicating their observations to one another. When these observations are not in disagreement they are transmitted to the scientific "archive" where they are stored as established scientific knowledge. If, on the other hand, the research communications of independent scientists, or groups of scientists, do not agree about the nature of the phenomena that they report, then there is criticism, debate, repetition of experiments, etc., etc., until a consensus is finally arrived at. In this diagram, of course, the scientific "archive" stands for the whole of the public literature of science, as it might be available in scientific libraries, whilst processes of communication and criticism would necessarily include all that goes on in and around scientific conferences and less formal meetings.

In exploring the epistemological theme it is convenient to take for granted the existence of the appropriate social institutions, such as learned societies, invisible colleges, research institutes, primary and secondary publications, and so on, within which the activities indicated in this diagram can, in fact, take place. From this point of view, the main functions of the behavioral norms of the scientific community and the customary practices of the scientific information system are simply to provide a stable, social framework, within which individual scientists have every opportunity and incentive to carry out their observations, communicate the

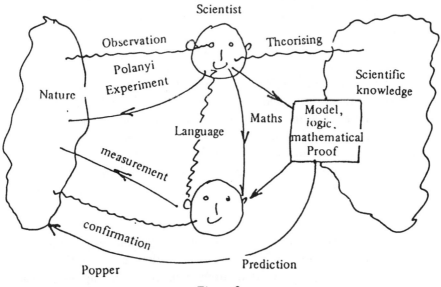

Figure 2.

results, and engage in mutual criticism, until there is overwhelming agreement on each significant point.

But, if this degree of consensus is to be achieved at all, there must be some restrictions on what is worth communicating as an item of "scientific" information. These items must, for example, be of such a public nature that they are "consensible"—i.e. that there is a reasonable prospect of arriving, in due course, at a consensual view. This is not a trivial limitation, since it excludes from science information that is essentially subjective, such as personal aesthetic values, ethical and moral judgements, spiritual and religious experiences, and the like. Notice, however, that the demarcation criteria between scientific and other knowledge is not the idealized abstraction of "objectivity" but the much more empirical characteristic of "inter-subjectivity" whose validity is not beyond practical test.

Experiment and Theory

The characteristic elements of the "scientific method," as traditionally discussed in the philosophy of science, are indicated in Figure 2. There, for example, we find not only the passive observation of nature, but also the actively contrived situations where phenomena are observed under reproducible experimental conditions. The condition on experimental results that they must be reproducible by independent scientists, is, of course, an elementary application of the consensibility criterion.

On the other side of the picture, mathematics plays an extremely important role as an unambiguous language of communication. By convention, mathematical terms and symbols are so precisely defined that there can be no doubt as to their meaning. A message from one scientist to another, reporting research results in this language, is thus perfectly clear and sharp; the question whether the substance of this message can be agreed upon, or not, is then much easier to determine than it would be if the various terms contained in it were vague and ill-defined. Thus, the rhetorical transactions of persuasion and criticism can be resolved much more successfully in a science such as physics, where the language of mathematics is given a central role, than in, say, psychology, where it is so very much more difficult to define the terms of debate.

Another important characteristic of the mathematical language is that it carries with it the consequences of mathematical research in the form of innumerable theorems which are agreed upon by all concerned. The possibility of achieving a consensus, incorporating the results of many different pieces of research, is thus greatly enhanced by using these theorems to transform the various messages into equivalent forms that can be tested for agreement. In other words, we have the possibility of incorporating a variety of particular observations in theories or models, from which these, and a great many other potential observations, can be deduced, as required.

This, of course, leads us into familiar areas of the philosophy of science where we may study, in some detail, the logical characteristics of mathematical models, the psychological conditions under which they arise, and the extent to which they are made plausible, or certain, by the way in which they agree with experiment of the extent to which their predictions are confirmed. A simple, well established, general theory is, of course, the ideal form of scientific knowledge, and the means by which we sometimes arrive surprisingly close to this ideal must always be emphasized. But notice that even the very best of such theories, such as the fundamental laws of mechanics and thermodynamics, are not to be given any higher metaphysical status than that they belong to a body of knowledge on which there is agreement by almost everybody who has ever studied it seriously.

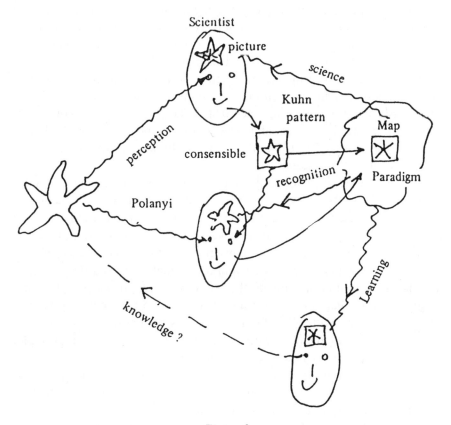

Figure 3.

Patterns, Pictures and Maps

It is important to recognize also (Figure 3) that not all the messages that scientists communicate to one another, or to the central archives, are expressed in words or mathematical symbols. It is a simple fact about science that a great deal of information is conveyed in the form of visual patterns—diagrams, photographs, maps, graphs, etc. In the biological sciences this is patently obvious, but it also applies, for example, to the structural diagrams used by chemists and to the bubble chamber photographs that are so carefully studied by physicists. Such communications are not, of course, so perfectly unambiguous as, for example, a thoroughly debugged computer program, but they are quite sufficiently consensible to play a major part in science. This is because, quite simply, we all have a highly developed facility for pattern recognition. We can agree almost as easily about the contents of a picture as we do about the meaning of words in a persuasive piece of logic, or even about the interpretation of symbols and operations in a piece of formal mathematics. If it were laid down, as a matter of principle, that scientific knowledge should exclude pictorial communications, then there would be not much left of our Scientific, medical, or engineering faculties, botany, zoology, geology, biochemistry, materials science, and large chunks of physics, would all have to transfer themselves hastily to the faculty of humanity or maintain themselves illegally by *samizdat*.

But, of course, the inter-subjective, consensual element that is recognized in a picture is always something of an idealization or abstraction. It is in this abstracted form that it is eventually stored in the archives, as an established item of knowledge. In a very general way, we may describe both mathematical models and consensual patterns as "maps" of the knowledge on which there is essential agreement. This is a very suggestive metaphor, in that it sums up the publicly shared, quasi-objective and schematic form that is characteristic of the best sort of scientific knowledge. By analogy with other familiar features of maps, we can also deduce quite a lot about the nature of science. It is obvious, for example, that there need be no fundamental contradiction between two different ways of representing a body of scientific fact (such as the quantum-chemical and valence-bond descriptions of molecules) any more than there is a contradiction between a map of an underground railway system that emphasizes the topology of the interconnections of the network and a street map of the same city that tries to show the details of distances and directions.

Public and Personal Knowledge

But the scientific knowledge that has been "mapped" in the archive is not merely a passive product of the research process. In order to play his part in the scientific community, each individual scientist must become fully aware of the current consensus of the subject to which he hopes to contribute. That is to say, when we learn a subject we internalize these maps as world pictures, or paradigms, against which our further perceptions of the world must, inevitably, be judged. It is the socially organized and supposedly well established character of such patterns that makes them especially compelling. The individual scientists of our model never really look at the world with childish naivety; they are bound to recognize in it just those models, or patterns, or maps, that they have been taught to look for.

It must be emphasized, however, that this restriction of vision is not absolute. The consensus of knowledge, in any field, is never complete. Paradigms are never fully articulated in every detail and there is, therefore, room for individuality in the way in which they are pictured by different scientists and in the processes by which they arrive imaginatively at new and more powerful scientific concepts. In the end, science is made up out of much more primitive "themata," whose roots lie much deeper in the psyche than the artificial constructs of contemporary theory.

This, in its turn, raises a number of extraordinarily interesting questions concerning the relationship between personal and public knowledge. To discuss these questions it is very convenient to use Popper's distinction between three different "worlds," (Figure 4). In what I would like to call the *material domain* there is, so to speak, all that actually *is*. In the *mental domain* of each one of us is our own perception, or cognition, or understanding, or knowledge, of that material domain. But, in so far as we are able to communicate with others, there is a whole body of shared public knowledge (again purporting to refer to the material domain) which might be called the *noetic domain*. This is the domain in which scientific knowledge is characteristically to be found—not only embodied in scientific books, papers, data, atlases, pharmacopoeia, museums, collections of taxonomic specimens, etc., etc., but also, more ephemerally, in the messages that are exchanged between scientists as research proceeds.

The Human Element

The fundamental issue for scientific epistemology is the relationship between the first and third of these domains. In our three dimensional model of science, we have not put in any direct connection between the material and noetic domains. All that science "knows" about the "external world" has come to it through the mental domains of the individual scientists— by perception inwards, and by the communication of observations etc., outwards to the public realm. This is an important point, because much philosophizing about science seems to take for granted the possibility of by-passing the human link. This, surely, is the objective of many attempts to show that certain scientific theories are logically necessary, or incontrovertible, or otherwise capable of validation by some means that does not depend on the uncertainties of human judgement. In other words, there is a happy belief that one day a super computer will be constructed which

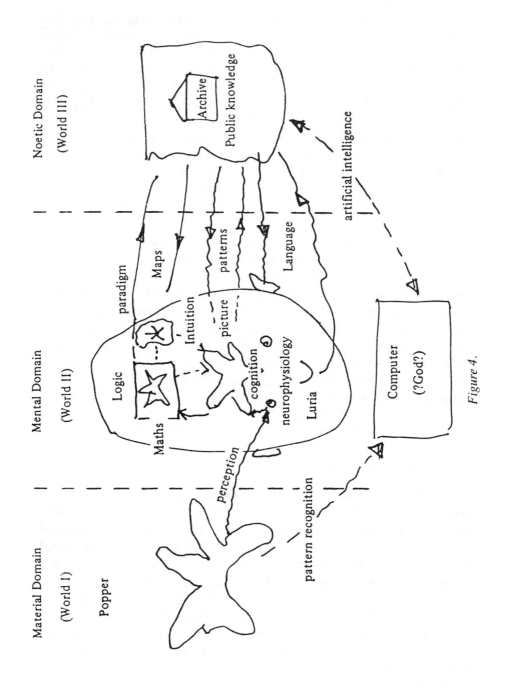

Figure 4.

will make observations of the natural world and turn them into scientific models, theories, or maps, without the intervention of human intelligence.

This possibility, in the very remote future, cannot be entirely discounted. If such a computer were constructed that would be a sort of final Judgement Day for science—the end of a worldly era full of error and sin, and the coming of a Golden Age when all would be revealed. But this appeal to a transcendental prospect has no relevance for our beliefs in science, as it stands to-day. We have seen, for example, that a great deal of science actually depends on the human capacity for pattern recognition, which has not been simulated to any significant degree by any mechanical system such as the digital computer. In other words, there is no computer algorithm—no logical formula—that can check what has been done every day, in a thousand laboratories and scientific libraries, such as looking down microscopes, scrutinising graphs, comparing one map with another, or otherwise using one's elementary human powers of perception and cognition. Every step in the scientific argument that depends on such processes is strictly non-logical; it cannot be validated by appeal to mathematics or symbolic logic, or even by the opaque and inexorable internal workings of a computer program that goes on for a week at a rate of one operation per nanosecond. It is only the extraordinary degree of inter-subjective consensus about the meaning of a picture that can be achieved amongst healthy, alert, human beings, that validates such steps.

This is not to say that logical analysis, and its mathematical equivalents, do not play an extremely important part in science. But this analysis can only be applied to the communications that flow between the mental domain of each individual and the noetic domain of science as a whole. These messages can be tested for possible contradictions or ambiguities, or transformed according to agreed rules, from one form to another. Such an analysis may suggest a single coherent mapping principle that represents the essence of a whole class of messages. Experience has proved the immense power of formal logical and mathematical languages for the analysis and representation of scientific knowledge, but this does not justify a more metaphysical belief that that is how the external world really is, or that non-mathematical knowledge is, in some way, non-scientific.

The thrust of the epistemological challenge to science is thus directed right into the human mind. As we have seen, this is not simply a polaroid camera, from which we can peel off every few seconds the human picture of the world. The rather private mental domain of each individual is in close and continual interaction with the public noetic domain of society. A great deal of what is "in our heads" comes from what we learn from others, read in books, or have been taught to see in maps and pictures. It is well known, for example, that the steps of a logical or mathematical argument may become so familiar that they are assimilated almost unconsciously into the operations of the mind, and manifest themselves in surprising forms as tacit knowledge or, intuition. It is here that we might search, seeking evidence from psychological experimentation, introspection, and neurophysiology, for the sources of the imaginative and critical creativity which drives science along.

The Credibility of Science

By now it must be quite clear that the original question "Is science to be believed?" can be given no absolute answer. The emphasis of the investigation then falls on the more discriminatory question "To what extent is science to be believed?" This calls for answers at three level—at the level of particular points of scientific information, at the level of well established general theories, and at the most general level of scientific knowledge itself, considered as the product of the activities of the modern research community.

At the lowest level, there can be no general answer. Some of our scientific knowledge, such as the freezing temperature of mercury, or the bacterial cause of tuberculosis, has been so thoroughly tested and checked by so many independent observers, for so many years, that it would be absurd, in practice, not to rely upon it. On other questions, such as the exact mechanism by which proteins are synthesized in accordance with the genetic code, or the reasons why certain types of metallic alloys are much more resistant to mechanical fatigue than others, there may be a great deal that is generally agreed, but there is also considerable uncertainty and controversy about a number of significant points.

In such cases it is no use looking in archives or asking for the scientific truth of the matter from some grand national academy. For the lay person, just as for the industrial magnate, or responsible politician, there is no alternative to seeking the opinions of various experts with various points of view, trying to get the hang of the various pieces of evidence on which these points of view depend, and arriving as best one can at some personal assessment of the situation. And, of course, when it comes to completely speculative questions such as what happened during the first few seconds of the Big Bang, or whether biological ageing is genetically programmed, the only rational policy is to listen to, and enjoy, the debates, purely as cultural events.

On the whole, we may expect the reliability and credibility of the answers to particular scientific questions to improve with time, since that is the task to which science is devoted. It is the job of every scientist, and of the scientific community, to chip away at evident contradictions and uncertainties, to find compelling reasons by which controversies may be resolved, and generally to improve the epistemological status of scientific knowledge in every detail.

But this cannot protect us from nasty shocks. Many pieces of scientific information which are thought to be entirely reliable have never been properly tested. Experimental results are not always carefully reproduced and theories are seldom improved until they are fully consistent with the data. It is scarcely surprising that some of these later turn out to be erroneous or fallacious. Every practicing scientist or technologist soon learns from bitter experience not to trust entirely to a fact just because it has been published in a scientific paper. The best trained research workers and the most sophisticated instruments can fall into error. The process of critical reassessment and correction of the knowledge communicated to the archives is often much slower and less systematic than scientists pretend or than most non-scientists would like to believe. This is probably the major pitfall in the search for reliable scientific knowledge at the practical level.

Scientific Realities

It is much more difficult to guard against possible defects of scientific knowledge at a more general level. In many fields of science, the whole program of research is dominated by a consensual paradigm, which is taken to be unquestionable. By education and other forces of socialization (Figure 5) the archival map that has come to be accepted in the noetic domain is projected deeply into the mind of each individual scientist. This socio-psychological phenomenon is so powerful that it gives all the attributes of reality to what may be no more than a speculative theory or model. For every scientist in this field the concepts and constructs of the theory—we might be thinking of electrons, or genes, or mental aptitudes—seem to have the same existential status as everyday objects such as tables and chairs.

This is an extremely interesting phenomenon which is very familiar to anyone who has gone deep into any branch of advanced science. From a psychological point of view, it is evidently analogous to the processes by which the "reality" of everyday things become established in early childhood. According to Piaget this occurs through the interaction of sensorimotor experience—i.e. the child's perception and manipulation of every-

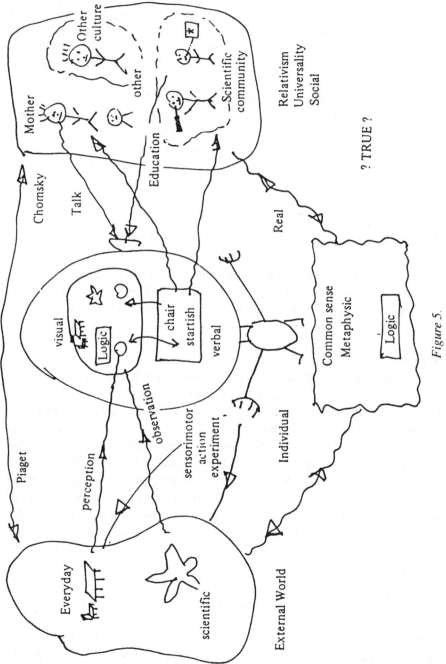

Figure 5.

day objects—with the introjection of the corresponding social categories as the child learns to speak. Similarly, for the student of science, there are parallel activities of experiment and observation, on the one hand, and theoretical education, on the other, which are mutually reinforcing. The student is given magnets and iron filings to play with just as a child might be given a ball or a stick; each quickly learns to see in them what others see—on the one hand, magnetic lines of force, on the other hand, objects that can be named and distinguished and separated according to the elementary logic of natural language.

But the sense of reality of everyday things, once established in infancy, does not change whilst we remain alive and sane. Scientific realities, however, have seldom been tested to the same extent and have sometimes proved illusory. In the history of every science there are unexpected discoveries and conceptual revolutions on every scale. These revolutions range from such grand climacterics as the transition from classical to quantum mechanics in the present century, down to, shall we say, the realization in the last twenty or thirty years, that a liquid is not just a very, very imperfect crystal but has a characteristic random structure all of its own.

It is necessary to realize, also, that paradigm fallacies of this kind do not automatically correct themselves as soon as anomalies are discovered. There is a striking degree of psychological resistance to paradigm change within the Invisible Colleges of an established discipline. This also is a familiar psycho-sociological phenomenon, from which even those who have committed themselves fully to the norms of the scientific community cannot completely escape. This is illustrated by the story of the theory of continental drift; it took half a century, from the time when it was first clearly stated, as an explanation of many familiar anomalies in the geological record, before this theory was finally accepted as the "reality" of the history of the continents and oceans.

Alternative Sciences?

If it is possible for such very large fallacies to become embedded in what we took for thoroughly reliable scientific knowledge, is it not conceivable that science, as a whole, in substance and in method, may prove to have been a grand illusion. This eventuality is certainly not excluded by the structure of our model. Although the scientific consensus is supposed to extend, in principle, to all rational beings, it is limited in practice to a small, self-appointing group of scientific professionals, most of whom have been carefully schooled in the principles of science "as we know it." It is an open question, for example, whether the scientific reality that might be constructed by persons brought up in quite a different cultural, linguistic, religious and metaphysical framework (for example, Mayan priests and Azande witch-doctors) would be essentially the same as the one that has been constructed within the cultural context of Western Christendom. The sociological and anthropological case against scientific relativism is difficult to counter in principle.

Nevertheless, if we look at the forms of argument and types of evidence that are regarded as acceptable in modern science, we see that these have characteristics that are universal for all human beings. The consensual foundations of the natural sciences are just those elementary principles of everyday reality—the logic of separable quasipermanent objects, located in space and time, capable of moving and acting on one another just in those ways that we discovered in childhood when we learnt to throw a ball or wield a stick.

In the face of the apparently non-intuitive properties assigned to hidden variables in such highbrow and all inclusive scientific theories as quantum mechanics, this seems a bold assertion. But the starting point of all branches of modern science has been directly empirical, and its arguments have never diverged very far from those of Aristotelian logic. The highly abstracted maps and theories that we have had to make to explain very peculiar and contrived phenomena may well be wrong. Nevertheless, it is hard to even imagine an alternative form of science—that is a well-organized consensual body of rational knowledge—that gave serious credence to what we now reject as entirely para-scientific phenomena, such as extra sensory perception, conversation with the souls of the dead, and so on. Our present scientific world picture is undoubtedly grossly inaccurate and incomplete. But it is conceived by scientists, not as an alternative to everyday common sense reality, but as an extension and refinement of the material reality that is shared by all who live in this world.

Other Features of the Model

This is obviously a very schematic account of the way in which the epistemological theme can be developed out of the three-dimensional model of science. But, perhaps, it illustrates some of the valuable features of this model. It is obvious, for example, that many contemporary insights and controversial issues of science theory can be accommodated easily within the model. Robert Merton's norms of the scientific community, Karl Popper's falsifiability criterion for scientific knowledge, Michael Polanyi's emphasis on tacit knowledge, Thomas Kuhn's concept of scientific paradigms and Gerald Holton's recognition of the underlying themata out of which most scientific theories are actually constructed—all these find their way naturally into the model, without grave distortion or contradiction. This is not just soft or trendy eclecticism; provided that we recognize that none of these concepts or principles should be treated as an absolute from which all truths must be seen to flow, then we quickly appreciate their value as powerful maxims and insights about science as it really is.

Another valuable feature of the model is the light that it throws on the credibility of the knowledge acquired in the social science faculties of academia. Whilst there is no strict line of demarcation in this model between the natural sciences and the social sciences, it is undoubtedly very much more difficult to arrive at a reliable consensual map about the behavior of human beings than it is about the behavior of magnets, or molecules, or microbes. The irreproducibility of historical social phenomena, the vague categories and fuzzy logic or discourse concerning human behavior, and the cultural diversity of perceptions of social relations, must severely limit any such program. This is not to say that we do not have a great deal of reliable knowledge concerning human behavior, but this knowledge cannot yet be organized and mapped as sharply and predictively as in the natural sciences. This is no news for most modern intellectuals; at least, there is nothing in our model to suggest the contrary.

Policy Perceptions and Implications

What is there in all this for science policy? As I said at the beginning, my main purpose is to illustrate the need for a multi-dimensional model of science, if science theory is to be put to any real use. The fundamental issue of all science is always the question "Can I believe it; is it really true?" The epistemological question thus strikes right to the heart of the matter; a theory of science that did not give a proper answer to this sort of question would be gravely misleading as a guide to science policy. On the other hand, there is a great deal to be gained, even for the politically sophisticated notables of the R&D system, from a more general perception of what they are really doing, and why it actually works.

If there is any general policy implication of this analysis, it is that scientific knowledge is not gained simply by the application of rational argument and blind instrumentation to the natural world. It is not so much the mechanical sophistication or personal brilliance of the scientists that guarantees the reliability of their results; the key factor is the peculiar social structure of science, which pits one mind against another, encourages and yet controls controversy, and thus keeps taut that linkage between the poles of imagination and criticism which is the sublime tension of the scientific process. If science fails and falters in the future, it could be because we have killed the goose that laid the golden eggs—that we have tried, in an excess of rationality, to streamline and reduce to preconceived order the anarchical and competitive social relations between independent minds which has been a constitutive principle of science in its historical development and is, I believe, absolutely essential for its effective working. The scientific notables know this very well, but have they made that clear to those higher political authorities on whose goodwill and understanding science now must depend?

References

1. J. M. Ziman, 1978, *Public Knowledge: The Social Dimension of Science*, Cambridge University Press.
2. J. M. Ziman, 1980, *Teaching and learning about science and society*, Cambridge University Press.
3. J. M. Ziman, 1978, *Reliable Knowledge: An Exploration of the Grounds for Belief in Science*. Cambridge University Press: detailed references to the literature relevant to the present paper will be found in this work.

SUBDIVIDING THE REALM OF KNOWLEDGE

Extreme specialization is the most daunting and incomprehensible characteristic of the scientific life. People cannot understand why a clever person should spend forty years concentrating on some perfectly tiny subject, such as the jumping muscles of fleas, or the cellular chemistry of titanium, or the algebraic polynomial invariants of a certain class of knots. It passes belief that a whole community, a whole enterprise, should be devoted to the study of a multitude of such patently trivial topics. Indeed, this fact is so familiar, so extraordinary, so easy to satirize, that it is seldom seriously discussed! Enough to know that a scientist must be a very blinkered personality, with no knowledge or interests outside a very narrow area.

This popular stereotype, alas, is not entirely a caricature. But it is not quite the same as the "lonely seeker" image, for it implies that each seeker is actually confined to a tiny patch, a goldminer's "claim" in the knowledge domain, and is surrounded on every side by other assiduous diggers. It registers each individual as the occupant of a clearly defined social niche, with property rights, responsibilities towards near neighbors and other communal considerations. Subject specialization, which seems so personal, is thus another connection between the individual and collective aspects of science.

From a collective point of view, specialization is an organizing principle for the division of labor. The abstract map of knowledge is meticulously divided and sub-divided into disciplines, sub-disciplines, fields and sub-fields, until there are perhaps no more than a few hundred scientists in the world working simultaneously on the same range of research problems. The major boundaries between these sub-divisions are strengthened by the cultural traditions and institutional practices embodied in journals, learned societies, university departments, academic titles and so on. Science is thus reinforced internally by a highly effective framework for fostering and coordinating the research expertise and productivity of an immense number of dedicated individualists. It is extraordinarily difficult to organize scientific research without reference to this potent and rigid social order.

Permanence and rigidity, however, are antithetic to the flexibility required to adapt to rapid cognitive and technical change. Scientific progress is not adequately captured by the stock metaphor of "pushing back the frontiers of knowledge." It is not just a matter of extending the established specialty framework into new areas. Better scientific understanding entails novel theoretical and instrumental connections, revised conceptual boundaries, intellectual mergers, and so on. Over a period of time, all the established maps have to be redrawn in detail, and all the established specialties have to be radically redefined. Apart from occasional grand revolutions, this is a continuous social process, governed by an "invisible hand" operating right across the knowledge domain, far above the level of individual or institutional influence.

Consider specialization now from the point of view of the individual. Scientists define themselves personally and socially through the answers they give, in various contexts, to the innocent question: "What is your specialty?" This is a topic with many ramifications that are relevant to the theme of this book. In effect, it is the fundamental classificatory principle in the microsociology of the research world. It is essential to any analysis of the working relationships between individual scientists, or between small research groups. For a great many purposes, your specialty, my friend, is who you are, and be careful not to forget it, or you will get lost!

Moreover, the interaction between the personal and communal aspects is uncomfortably dynamic. Scientists like to believe that their specialty niches are secure enough to sustain normal professional advancement over periods of several decades. But this belief is not consistent with rapid scientific change. The risk of cognitive redefinition or institutional redundancy is not just an abstract intellectual peril: it threatens a life-world position in which a whole career may have been invested. I have reported elsewhere on the way in which scientists talk about and react to such situations (Knowing everything about nothing: *Specialization and Change in Scientific Careers*, published in 1987 by the Cambridge University Press). This is a topic that deserves much more thorough study, since it is unusually sensitive to the tension between personal, cognitive and organizational factors.

Pushing Back Frontiers—or Redrawing Maps!

Specialization

The life work of a scientist is to add a few bricks to the edifice of knowledge, or to fit a few more pieces into the jigsaw. These clichés express a profound truth. Progress in science, like progress in industrial manufacture, is born out of the division of labor. Adam Smith's famous account of the many distinct operations that go into the making of a pin can be paralleled by any historical account of the many distinct investigations that go into the making of scientific discovery. A work that would be beyond the powers of any single person is easily accomplished by the co-ordinated efforts of many individuals, each performing a specialized function.

The traditional term for a scientific specialty is a discipline. But the division of labor in research is far more detailed than it is in academic instruction. In any mature scientific discipline there is a framework of established facts and theories within which specific research projects can be very precisely categorized. This is very obvious in the classification schemes used by secondary information services to retrieve scientific information from published papers. In physics, for example, this scheme is numerically ordered to four significant figures, with two further alphabetic indexes for the finest levels of discrimination (Figure 1). There is no generally agreed-upon nomenclature for the different levels of specialization in science. A paper such as I might myself have written when I was active in research could have been located unambiguously to within about one-tenth of a "subfield," which is itself about one-tenth of the "field" of "transport properties of condensed matter." But this field is only about

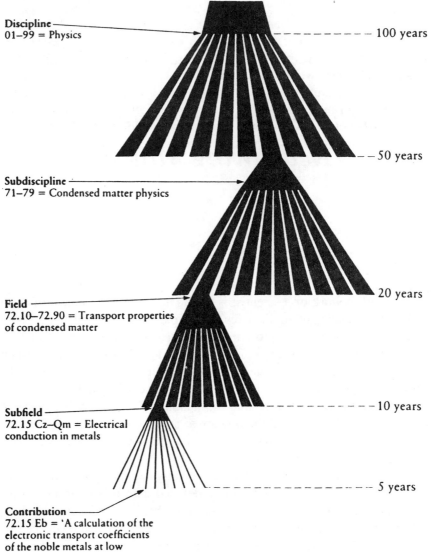

Discipline
01–99 = Physics

— — — — — — — 100 years

— — 50 years

Subdiscipline
71–79 = Condensed matter physics

20 years

Field
72.10–72.90 = Transport properties
of condensed matter

Subfield
72.15 Cz–Qm = Electrical
conduction in metals

— — — — — — — — 10 years

— — — — — — — — — — — 5 years

Contribution
72.15 Eb = 'A calculation of the
electronic transport coefficients
of the noble metals at low
temperatures'

Figure 1. What is your specialty?

one-tenth of one of the ten subdisciplines into which physics is conventionally divided. In other words, if you were to have asked me "What is your specialty?" I would probably have answered "Theory of electronic transport in metals," which would suggest that my research interests and activities were limited to less than 1 per cent of all that was going on in physics.

Physics happens to have a strong theoretical framework within which it is easy to define distinct fields and subfields of research. But research work in all the other basic sciences, and in many branches of applied science, is just as finely divided into specialties. Every discipline has its classification schemes, which reflect the various ways in which research workers and research groups define their particular research interests. From the lists of publications that our academic colleagues present when they seek promotion, it is evident that this differentiation into specialties goes down to the second or third significant figure. Studies of the citation links between published scientific papers also indicate that the characteristic extent of an active research specialty is of the order of 1 per cent of a conventional academic discipline. When we say that modern science is an extraordinarily specialized activity, we mean that research is going on in several thousand more or less distinct fields, each with its particular experts, its particular techniques and its particular topics for investigation.

Problems

That is not to say that every scientist becomes, and remains, a narrow specialist, working in the same field throughout his or her career. Some scientists do, indeed, confine their interests within a very narrow range, but others make significant contributions over a number of fields, or even move successfully from one discipline to another. But when you ask them what they are trying to do at any particular moment, they will tell you that they are trying to solve such and such *problems*. The scientist working in an established discipline does not face a virgin natural world awaiting exploration or exploitation; he or she is normally immersed in an intellectual, technical and social milieu in which certain "questions" seem to be demanding answers. Scientific progress is seldom achieved by boldly striking out into the unknown, or rashly speculating on the inexplicable: it arises primarily out of concentrated attacks on relatively well-posed questions that have arisen from previous research. If there is ever to be any hope of getting a reasonably convincing answer in less than a lifetime of effort, these questions must be of restricted scope. In other words, *a feasible research project can only be conceived in relation to a specific scientific problem within the bounds of a specific scientific specialty.*

One of the main tasks of the sociology of science is to understand how the labors of innumerable scientists, working away at innumerable scientific problems, are co-ordinated and integrated. One of the tasks of the philosophy of science should be to analyze the notion of a scientific problem. It cannot be just like a crossword puzzle, or mathematical exercise, since it is not known to have a unique solution. It arises out of the research process itself, and must therefore be both novel and open-ended. If there were any standard way of solving it, it would be a routine exercise, not a genuine scientific problem. And yet it must be formulated with sufficient clarity and precision that what would count as an acceptable solution can seem within reach.

The context within which a scientific problem is posed is thus just as significant as the question itself. This is what Thomas Kuhn really meant when he described "normal" science as "puzzle-solving" within an unquestioned "paradigm." In the terminology of machine intelligence, the context of a problem is its frame, defined as "a collection of questions to be asked about a hypothetical situation; it specifies issues to be raised, and methods to be used in dealing with them." Scientific specialties are just such frames for the formulation of potentially soluble scientific problems. Sometimes—as in the case of Darwin's problem of the origin of species—the frame is enormous, taking all natural history as its "problem area." In most cases, the frame of a problem is really very narrow indeed. At CERN, during the last four months, they have been spending vast sums to observe just half a dozen cases of a special type of particle collision out of the millions that are occurring in the apparatus. Of course there are very good reasons for thinking that it is important to answer the question "Can we find empirical evidence of the hypothetical W boson?," but this project can nevertheless be classified to three significant figures: it clearly belongs in category 13.80: "Phenomenology of hadron-induced high- and super high-energy interactions between elementary particles." All the skills of hundreds of physicists and engineers are concentrated into that narrow salient.

Discovery

It is one thing to formulate a good scientific problem; it is quite another matter to solve it. Only a small proportion of all the research projects that are started arrive at really significant results. Many fail to get any results at all. Others produce results that are later shown to be invalid. Of those that finally survive the successive stages of critical assessment and accreditation, the vast majority are of no great scientific interest.

Scientific progress is not made by the accumulation of routine research results: it is made by important scientific *discoveries*. This is another notion that calls for careful philosophical analysis. It certainly means more than the acquisition of information. Like the notion of a problem, a discovery cannot be defined without reference to its context. In common usage, the word carries with it an air of surprise. An investigation that arrives at exactly the result we expected can scarcely be referred to as a discovery: the information obtained must add to, or significantly change, what we already knew. In other words, it has to be considered in relation to a particular frame of existing knowledge.

Scientific discoveries, like scientific problems, have to be located in relation to scientific specialties. We automatically judge a discovery to be important or significant as soon as we see that it has implications for an extensive set of existing problems over a wide range of specialties. The other contributions to this meeting will undoubtedly show that this basic epistemological principle applies in almost all scholarly disciplines. It obviously underlies the practice of using citation counts as indicators of the quality of scientific papers and their authors. This is not, of course, a reliable index of the value of a particular contribution to knowledge, but it accords closely with the way in which all scientists think about the progress of their subjects.

But the specialty to which a particular discovery relates is not necessarily that from which it originated. As every scientist knows from experience, an investigation that was undertaken to solve problem *X* very often produces the answer to some rather different problem *Y*. Columbus set sail for China, and came upon America. Copernicus set out to improve the calendar, and discovered that the Earth goes round the Sun. It is not so much that many scientific discoveries are serendipitous: it is that they can have implications for quite a different set of problems than was originally foreseen.

Those philosophers who are scornful of what Kuhn called "normal" science have tended to see its sole outcome as the accumulation of minor discoveries that have little effect outside the fields or subfields in which they were made. They tend, therefore, to concentrate on those bold hypotheses or experiments whose initial problem frame extends over a whole discipline—Wegener's theory of continental drift, for example, or the contemporary experiments to test non-locality and/or causality in quantum physics. But an important discovery that can eventually revolutionize 100 specialties may arise out of the investigation of a very "unimportant" problem. Remember that quantum physics itself began in the narrow specialty of theoretical statistical mechanics, as a fudged solution to the problem of calculating the amount of light produced by a red-hot ob-

ject, and that all the sciences and technologies that use the nuclear transformations of matter arose from the peculiar phenomenon of radioactivity
observed in a few uncommon minerals.

Maps of Knowledge

The influence of a really important discovery is felt over substantial parts
of one or more scientific disciplines. Whole subdisciplines may be revolutionized by it, as seismology and volcanology have been, for example, by
the discovery of continental drift. We tend to think of a typical scientific
revolution as something that mainly affects knowledge about a particular
"subject," but the effects of a discovery are not necessarily concentrated
within a standard category of the classification scheme of a discipline.
The discovery of radioactive isotopes, for example, not only created the
new subdiscipline of nuclear physics: it has also had an enormous effect
on the experimental techniques of many quite distinct subfields of chemistry, biology and geology. If we were to mark these effects on the classification scheme of one of these disciplines, we should see them scattered
here and there, apparently at random, and not concentrated in any particular region of the scheme.

There is nothing mysterious about this. For excellent practical reasons,
information scientists construct one-dimensional schemes where each
specialty is labelled by an alphanumeric symbol, such as 72.15 Eb. These
symbols can be arranged hierarchically, or along a line, but this does not
correctly represent the neighborhood relations between the categories
they stand for. Thus, for example, in Figure 1, subfield 72.15 "Electrical
conduction in metals" is properly classed next to subfield 72.20 "Electrical conduction in semiconductors," but is quite a long way from field 74
"superconductivity" to which it is equally closely related. A branching hierarchy of this kind simply cannot do justice to the real affinities between
the specialties of a mature scientific discipline.

The next step, of course, is to try to represent these affinities on a two-
dimensional map, as in Figure 2. It is, indeed, possible to construct such
maps, where citation linkages form coherent clusters, corresponding to
recognized specialties. But although such a map may be good enough to
distinguish between specialties that would otherwise overlap, it does not
show all the connections that could be made between the problems that
are studied in these specialties. The topology of the network of affinities
between scientific specialties is really immensely complicated. A typical

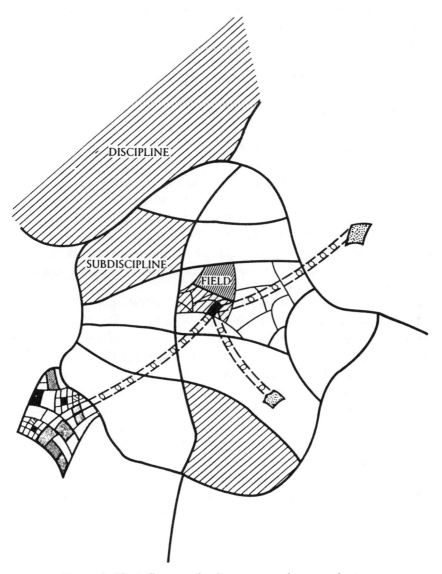

Figure 2. The influence of a discovery on the map of science.

scientific discipline is much more compact, much more tightly interconnected, than can easily be represented on a two-dimensional diagram. On
almost any subject of research, there is a dimension of theory and a dimension of fact, a dimension of methodology and a dimension of potential applicability. For the specialist there are further dimensions, or at least
perspectives, associated with different schools of thought, different ap-

proaches, and different research goals. This multidimensional complexity of the real "map" of scientific knowledge is, of course, embodied in computer techniques for searching the literature and retrieving relevant information. The computer is told to find any information that matches the items in a certain list of features—for example, noble metal, conductance, electronic, low temperature, theoretical—without regard to the order in which these features occur. In other words, it is conducting a search in a space of many dimensions.

These complications are especially obvious in a so-called interdisciplinary subject such as ecology. For a while, this relatively new domain of research will be subdivided according to the specialty structure of its component disciplines—biological ecology, mathematical ecology, etc. In a short time, however, it will be differentiated into its own characteristic specialties—by environmental habitats, for example—each of which must be multidisciplinary in its own right. The connections that develop between each of these specialties and the specialties of the component disciplines are of byzantine complexity.

For this reason, we should not be at all surprised when a discovery in one area of science turns out to have important implications in other areas that are supposedly quite distant from it. The surprise is due solely to our tendency to think of science as if it could be displayed on a simple map, with all the natural affinities clearly visible. In reality, there are hidden connections in higher dimensions—"tunnels," so to speak, beneath the map—along which influences flow naturally from specialty to specialty. The stereotype of the scientist as making a "breakthrough," and "pushing back the frontiers of knowledge," is misleading. These trite metaphors suggest that the influence of a major discovery flows outwards across this map, with a continuous expanding frontier. The actual process is often much more like the spread of an epidemic, where new foci of infection appear unexpectedly at points that are far away from previously affected regions.

Transpecialization

The influence of a discovery transcends the conventional boundaries of specialties and disciplines. A novel fact or concept can apparently leap across the classification scheme or map, to affect research in quite a different problem area. That is, it can change the formulation and solution of problems in that area, and thus become a new element in the paradigm or problem frame of that specialty.

This is obviously of immense importance in the progress of science. Historians and philosophers of science have often drawn attention to the power of analogy in the work of discovery. It is sometimes argued that all intellectual progress can be ascribed to the transfer of an existing model or concept from its conventional specialty to a new problem situation, into new combinations with other facts and concepts. But this is only one form of a general phenomenon of transpecialization which also includes the transfer of empirical facts, experimental techniques and other problem-solving resources from specialty to specialty, without regard for academic frontiers.

All scientists are taught that they must not think of their discipline, subdiscipline or field research as a closed domain, self-sufficient for its own development. But this precept is usually interpreted rather narrowly, and applied only to the immediate interspeciality neighborhoods. Theoretical physicists, for example, are taught that they should watch for new developments in applied mathematics, and sociologists of knowledge are now laboring to keep their minds open to new trends in epistemology. But this conventional policy does not take into account the influences that can travel from great intellectual distances and prove themselves fertile on new ground. Theoretical physics, for example, occasionally gains valuable insights from certain branches of very highbrow pure mathematics, and also from experience with very practical problems in engineering.

A fundamental lesson of the history of science is that every specialty should keep itself open for such latent influences from other distant specialties. This is, of course, a counsel of perfection, since it seems to call for just those acts of imagination that are labelled "creative." Nevertheless, it can be fostered by certain strategies of education and research management. There is some evidence, for example, that the influence of a discovery is transferred more by outward push than by inward pull. Individuals who have been "infected" by new ideas, techniques or facts carry them with them when they migrate into new problem areas. This would speak for a policy of encouraging intellectual mobility in the research system. It would be interesting to document cases of the other kind, where researchers resident in a particular specialty hear of some new discovery in a distant field, and realize its potential applicability to their own problem. Such events would presumably be increased by an emphasis on breadth and diversity of interests in training for research.

Modern science is moving from individual to collective modes of research. Institutes and project teams are created to study valuable natural objects, such as forests, or to develop artificial devices, such as nuclear reactors. The problems of mission-oriented research and technological development are unavoidably transpecialized, if not transdisciplinary, and can be tackled only by groups of people with diverse specialized skills and experience. The daily intellectual transactions within such a team undoubtedly encourage the transfer of new knowledge from specialty to specialty. This is not just a one-way process, from the basic sciences to their multidisciplinary applications. The problems that arise in engineering, medicine, agriculture, etc., set further tasks of fundamental comprehension for the basic natural sciences, and thus influence the paradigms of their specialties.

Redrawing the Scientific Map

Every scientific specialty is being transformed continually in internal discoveries or external influences. Sometimes this transformation is slow and uninteresting, as in the accumulation of data by routine investigations; at other times, an exciting discovery opens up many old problems for solution, and suggests many new ones. In the latter case, the subject may grow rapidly, both in the amount of research being done and in its intellectual scope.

But a scientific specialty cannot expand by a large factor without rupturing the classification scheme by which it has been defined. It is not enough to obtain a whole lot of new answers to old questions: if there is to be genuine progress, this information must be brought under intellectual control, and made the basis for new theories, new explanations, new concepts—and new questions. But the original definition of the specialty, and the way in which it is embedded in any larger classification scheme of the discipline, embodies current scientific opinion on just such matters. A "map" of a discipline or field is itself a paradigmatic theory of that field, and is subject to revision as new knowledge is acquired. In other words, the notion of scientific progress is inseparable from the generation of new theoretical structures, new paradigms and new definitions of disciplines, fields and other categories of specialized knowledge.

Any major transformation of the problem frames and problem areas within a particular field of research can be described as a revolution of thought within that field. We need not enter here into the great historical and philosophical debate concerning the way in which such a revolution occurs, and whether it represents a change to a state of knowledge that is incommensurable with what it was before. What we can assert is that the way in which knowledge is conventionally mapped within that field has been significantly altered, to the extent that the older specialty frames no longer guide the researcher on problems to be solved and the means that might be used to solve them. Whether this change has been almost imperceptibly slow, or frighteningly fast, the time has come to delineate new specialty boundaries, and to draw new "maps" of the subject.

When we are considering scientific progress in the large, we are bound to emphasize major paradigm changes that revolutionize whole disciplines. But the same process of growth and change occurs at every level of specialization, from the narrow little subfield to the major discipline. In my own former specialty of theoretical physics, for example, the subfield concerned with the electrical conductivity of liquid metals was revolutionized some 20 years ago by the discovery of a simple formula that at last made sense of all the old experimental data. I have not been following the subject recently, but I have no doubt that it has gone through several more changes of understanding and interest since that time. On a much broader scale, the modern subdiscipline called "the physics of condensed matter" came into existence about 40 years ago, with the application of quantum theory to the behavior of electrons in metals. But this was only one episode in the vast revolution of physics by the discovery of quantum phenomena at the beginning of this century.

As this example shows, the narrower the definition of a specialty, the shorter the time-scale on which it is likely to change. Citation analysis shows that the average "half-life" of a primary paper in physics is about five years. If it has not made its mark by then, and been incorporated into a larger movement of change, then it will probably fall into oblivion. My guess is that the lifetime of an active "subfield" of physics is about 10 years, and that "fields" of research grow and decline in periods of the order of 20 years. A typical subdiscipline, such as nuclear physics, probably retains its identity for 40 or 50 years, while the transformation period of physics as a discipline is something like a century.

These time constants are, of course, purely notional. Some specialties change much more rapidly than this spectral analysis would suggest, while others continue unchanged for long periods. The main point is that the map of every science—and of the sciences as a whole—is continually being redrawn at different rates on different scales. The occasional grand revolutions that are so visible to the historians and philosophers do not cover all the radical progress that is actually being made.

Adaptation to Change

It would be interesting to look at the historical development of several different disciplines, to see whether they all redraw their maps on similar scales at similar rates. One of the fundamental differences between a natural science and a branch of humanistic scholarship may be that the latter changes much more slowly, and retains its basic problem areas and frames for much longer periods. It may be, also, that the classification scheme of a discipline is not a significant indicator of its internal structure. Such a scheme is never more than a conventional device for putting books on library shelves or finding items of information in a subject index. At any given moment its categories are only roughly representative of patterns of specialization that are continually in flux. Every active research worker has his or her own perceptions of the shapes of the specialties and frames within which research has to be planned and discoveries made.

But from this point of view we see science as a vast collective undertaking, where growth and change occur on a larger scale than any individual can hope to influence. Whatever we attempt to achieve by our own efforts, the "moving finger" of progress will write, and move on, at its own rate, and we must harmonize our own lives and careers to it as best we can.

The structure of specialties, and their characteristic rates of change, is a subject with important implications for scientists as individuals and as members of scientific institutions. It is reasonable to assume that the discipline into which we were first educated and by whose methods we do our scientific work will outlive us in its main outlines, although we may have to move into new interdisciplinary areas as these develop. For most scientists, indeed, it is possible to maintain an active research role within a major subdiscipline, such as solid state physics, or seismology, or neurophysiology, or even econometrics, over a long research career. This can be our part of the edifice: the wall, or pillar, or buttress, or ceiling which our stones have helped to build.

But any scientist who persists unduly in a narrower field of research is at risk of being left behind as specialties change, and the active research front moves elsewhere. And there can be no security at all for a scientist attempting to specialize on a finer scale. At the subfield level, there are no really stable specialties at all. As influences flow in from neighboring or distant fields, all is in flux. Paradigms of theory, knowledge and technique change so quickly that they cannot be treated as stable frames for the definition of problems or the means for solving them. There is little hope for a scientist who cannot set up a research program within a somewhat broader range, and adapt intellectually and technically to the imperatives of progress and change within that range. This, surely, is an important message for graduate students, for their supervisors, and for all who manage, or are managed by, the research enterprise.

What Is Your Specialty?

W hat is your specialty?" How do your answer this question? It depends so much on the circumstances. Perhaps you are on a long flight, chatting with a not unattractive person strapped into the seat beside you. Will "Oh, I'm a sort of academic" be too off-putting? Or should you be a little more specific? "Actually, I'm a scientist" might sound slightly more interesting, suggesting a vocation rather than a mere profession. But even if the conversation became more serious and intellectual, it might be risky to admit to being a physicist. This happened to me once on a flight to Australia, and I had to deal with the electrical problems of his yacht for one person and cope with a scheme for perpetual motion from another.

At an academic cocktail party, of course, the presumed spouse of a presumed colleague would at least expect you to state your discipline, thus firmly placing you in your faculty and department. "I'm in physics," you respond, hoping that he or she is into something more interesting than chemistry or civil engineering. Inside the physics department itself, everybody is supposed to know who you are, so the question would seldom be asked directly; but the new secretary compiling the faculty list for the student brochure might have to be told that you should be put down as being in solid state physics, rather than in another subdiscipline such as nuclear physics or plasma.

Imagine yourself now in Washington, at your first meeting of the NSF review panel on materials. You will be introduced as primarily a specialist in the field of the electron theory of metals, but competent to deal with the basic semiconductor theory as well. Over lunch you will insist that this is a slight exaggeration. For the past ten years, you say, you have been working on the electrical properties of liquid metals, and you have just come back from Grenoble, where you have been an invited speaker at the

Third International Conference on this subject. In fact, it was a very useful meeting, as you were able to offer a job to a French post-doc who had just finished a thesis on the theory of the Hall effect in liquid transition metals at high temperature—a problem area, where the band wagon was just beginning to roll.

The answer to the simple question: "What is your specialty?" is not so simple after all. It is rather like being asked "Where do you live?," and having to decide whether the answer should be "England," or "about ten miles from Oxford," or "Oakley. It's a small village half way between Thame and Bicester," or "Little London Green," or "Number 27." When we were children, we were so excited by this discovery of our own individuality that we carved it on our school desks, adding, of course, that we were also on the Earth, in the Solar System, in the Galaxy, in the Universe. That is not necessary in Academia, which is united in believing that it is the only planet on which there are sentient beings.

But it is a large planet, of which one can know only a very small area. By convention, it is divided into a dozen or so "faculties"—Science, Medicine, Law, etc. Within each faculty there might be ten or twenty Departments, each devoted to a distinct discipline. Each discipline, in turn, could be differentiated into, say, ten sub-disciplines. But even a sub-discipline, such as solid state physics, covers an immense range of knowledge. My guess is that it would have to be subdivided further, by another factor of ten, to get down to the area which an experienced scholar would come to know well. Unfortunately, there is no standard word for a domain of knowledge of this size. I am going to refer to it here as a "field," whether or not domains of this extent really could be defined naturally as coherent "subjects" or "problem areas" in a particular academic context.

It is a cliché that modern academic life is highly specialized. The specialties into which knowledge is fragmented for research are said to be innumerable. We now see that this is not quite true. Multiply ten "faculties" by ten departments, by ten sub-disciplines, by ten "fields." It seems that a carefully chosen assembly of about ten thousand scholars would be competent to speak with authority on everything we think we know. Of course they would speak in raucous contradiction. With another couple of powers of ten one would get down to the ultimate level of differentiation—the number on the gate of that doctoral dissertation where the latest research problem might have been mastered. In that sense, each of us is the only real specialist on the particular question we are studying.

Academic specialization is nothing new. In every generation it is deplored, yet it grows continually narrower and more intensive. This is inevitable as long as the whole scholarly enterprise continues to expand in numbers and intellectual pretensions. More and more people are employed to discover more and more facts about more and more subjects. What can they do, poor lambs, but concentrate on gathering whatever crops will grow on the tiny patches of academic territory on which they happen to stand. After all, this must surely be the best that most of them could ever hope to put on show in the fiercely competitive marketplace of scholarly reputation.

Yes, we affect to scorn specialization—and yet we are utterly committed to it. Research is often a very personal undertaking, but in its outcomes it is a social product. In the production of knowledge, as in every other advanced industry, quality can only be achieved by the systematic division of labor. What we are trying to create is a closely woven garment, whose every thread is strong and supple but each thread is of a different material, spun to a different thickness, dyed a different color, crimped to a different texture. Such threads cannot be mass produced. A lifetime may be needed to master the craft of making any one of them. It is only by the coordinated efforts of tens of thousands of such expert craftsmen that the garment can be woven to its full strength and use and beauty.

The primary means for the division and subdivision of academic labor is specialization by "subjects." As we have seen, academics define themselves, and are defined by others, in terms of the bodies of knowledge they are supposed to be familiar with.

Can we make this notion more precise? In the literature of the sociology of science, there is much discussion of the way that specialties emerge, or can be discovered, in the normal processes of research. Clusters of cognitive commonality are constructed by elaborate computer analysis of the way in which scientific papers are linked by citing the same words or other papers. These scientometric techniques may, in the end, give us a useful procedure for defining a specialty formally, but I suspect that we would get into a circular argument if we tried to link this definition with the concept of a distinctive "subject." We would start saying something like "A subject is a set of facts and ideas all connected with..."—and then realize that this was just about the same as a co-citation cluster. In the end, we should probably come back to a set of terms that was not very different from a hand-made classification scheme constructed by a human indexer, based upon an intuitive grasp of the relevant cognitive entities.

Nevertheless, subject classification and specialization is used consistently and coherently in all academic work. Observations, experimental data, concepts and theories are classified for publication under subject headings. These headings appear in the titles of articles, journals and books. Academic institutions are divided and managed along the same lines. To be appointed to an academic post, one must become an authority on some subject, both by making research contributions to it and by being able to cite its literature. In some countries, such posts are firmly labelled by their subjects: "Lecturer in Quaternary Palaeobotany," "Reader in Sanskrit Poetry," "Professor of Quantum Chemistry," and so on. One will then have to teach ones subject to students, supervise research and attend conferences on it. One will be asked to referee papers and research grant applications on it, and represent it in the deliberations of a learned society. Younger scholars in the same specialty will respectfully request advice on their research and their careers, and there may be questions to answer about its industrial or military applications. In due course, one may even receive a public award for all that one has done for it, and go down to posterity as that great expert on the "electron theory of metals," or whatever it was.

What I am getting at is that in most aspects of academic life each scholar is cast in a single, consistent social role—the role of the specialist in a particular subject. The institutions, the norms, and the ethos of Academia are neatly dovetailed together to make a setting for a community of individuals performing these roles in a wide variety of subjects. Whether we like it or not, this is how the whole system works, and it is difficult to imagine it otherwise.

I would not go so far as to say that this type of social organization is an absolute functional necessity for the advancement of knowledge. But original scholarship can never be an easy, routine job. It demands extreme concentration of effort over long periods, and extremely refined, thoroughly practiced skills. To see a little further than past giants, one has to clamber laboriously up on to their shoulders. For most of us, life is too short to climb more than one such monument.

Subject specialization is the driving force of professionalism. It gets the best out of people by setting attainable standards. Individuals compete directly with their peers, and are assessed by them, on the basis of their actual performance within a well-defined arena of techniques and ideas. They can take pride in the essential virtue of the sound scholar—reliability. What they tell us may not be of enormous interest, but at least it is real. The threads they spin may not be brilliantly colored, but at least they are strong.

At last we have reached the central theme of this address. Specialization is the heart and soul of scholarship. The question is:—what does it do to the hearts and souls of scholars? We affect to deplore it: is this because it distorts what we know or because it warps the knowers? What are its psychic benefits and disbenefits to the individual? Many scholars do not live all their days in Academia. What happens when their career paths lead them off into outer space, and they discover other worlds?

I have to say at once that I am not going to talk about those few scholars who seem to have been specialists from birth. From the earliest days they can remember, they have cared only for prime numbers, or ants, or the writings of Jane Austen, or the search for the site of Troy. All they ask of life is freedom to indulge their monomania. They may appear warped or emotionally crippled but they do not seem to suffer for what they never seemed to want.

The majority of scholars do not come into the world as specialists, but are made so by their professional formation. Over a period of ten or a dozen years, they have been filtered through successive stages of a narrowing funnel of education. The first stage might have been a choice between the humanities and the sciences. At the next state it will have been necessary to give up chemistry, biology, and all the other scientific disciplines in order to specialize in physics. the decision to work in a particular sub-discipline such as solid state physics (rather than, say, nuclear physics) might be considered Stage III. This will be followed in due course by a fourth stage where a very narrow sub-field or research, such as the theory of the electronic transport properties of liquid transition metals, will seem to have expanded to occupy the whole mental horizon.

The actual phasing of these *rites de passage* varies from culture to culture. In Britain we continually deplore an educational tradition that encourages very early and rapid specialization. Many school children really do have to choose between "arts" and "science" at the tender age of 14, and celebrate the next stage at 18, when they are enrolled in a university course leading directly to "special honors degree" in a single discipline. Three years later, if they do well in their exams and decide to go to do a PhD, they will be pushed through a further stage of specialization in a year of perfunctory postgraduate courses, and then squeezed abruptly into the narrow space of a thesis topic. Yes, we do sincerely deplore the way we treat our young people, yet seem quite powerless to change our own peculiar customs.

Rumor has it that in other European countries they actually postpone the earlier stages of academic specialization by several years. As a result, their scholars and other professional people get a slightly broader school education than in Britain. It would seem, for example, that even the most magisterial of French historians would have had to show some competence with the differential calculus before entering the university, and a Dutch physicist at CERN could not be assumed to be quite as ignorant of

European history as his British colleague. They also take longer over their bachelor degrees and doctorates than we do, but I get the impression that although these are deeper and more thorough than ours, they are not substantially wider in scope.

As for the way that scholars are trained in the United States, it seems to be generally agreed that its school education, although much broader and more diversified than England's, is seldom very rigorous. I understand, also, that the bachelor's degrees are not necessarily specialized, although it is not clear to me what weight is attached to elective minor subjects outside the block of major courses needed to cover a department discipline. What I do know is that graduate programs last longer than ours, and are both less specialized and more thorough. In your system, the final stage of academic specialization seems not to come until the student is already in his or her middle twenties, and then builds on a broader and more solid base at the level of a sub-discipline.

Nevertheless, in spite of cultural and disciplinary diversions along the way, the career path of every scholar must eventually pass through the long, deep, narrow defile of a doctoral dissertation. This is the means by which subject specialization is instilled and institutionalized throughout Academia.

It is not just that the would-be scholar has come at last to the front line of research, and must now take personal responsibility for the minutest details of fact or inference. The whole structure of the PhD exercise is designed to reinforce a specialized approach to the advancement of knowledge. A well-defined scholarly problem has to be tackled and solved, starting from cold, in just a few years. That is only feasible if the problem is tightly framed. It must already be imminent within, and contained within, an existing sub-field, or sub-sub-field of learning. The required concepts, data and techniques must be within reach of an assiduous student, and capable of being taken on board in the short time available. It is a test of intellectual convergence. Divergence of mind, however valuable for the more experienced scholar, is merely distracting. Total concentration of effort within a bounded region is the order of the day, of the week, of the months, of the years.

But it is not just contact with a particular set of documents, a particular shelf in a library, or a particular piece of apparatus, that makes a subject specialist; it is contact with a particular set of *people*. An academic specialty is not just a body of knowledge: it is a social institution. At first, this contact may involve no more than regular participation in a graduate seminar, vying with other graduate students for the approval of the professor. But soon there will be learned society meetings to attend. Remember that first ten-minute paper in parallel session 17, in room 34, at 3:45 p.m., just before the tea break: the audience includes some of the faces behind the names that have already become so familiar by citation. Their questions are not unkind, and over tea they discuss one or two points that did not come out very clearly in the paper. At last, the defile of solitary research is beginning to open out into the Promised Land

It will take years to become fully established in that land. But from this point on, specialization is continually reinforced by socialization into the academic community. Each step up the academic ladder depends on enhancing one's reputation as a specialist on a particular subject. That reputation can only be won within a peer group—the "invisible college" of the specialty. The transactions that govern the internal sociology of science— the "recognition" that is exchanged for "contributions to knowledge," the critical assessment of such contributions, the respect accorded to authority, the protocols of citation, the competition for priority, etc.—all these transactions take place within the social space of each separate specialty, and seldom involve larger groups of people.

Let me add, by the way, that there does not seem to be any particular connection between versatility and scholarly ability. Some of the most eminent scholars have been extremely narrow in their interests, whilst others have been remarkably broad. In every generation, of course, there are scholars of such prodigious learning and diversity of intellectual concerns that they are hailed as polymaths, but these are not necessarily the most fertile minds in all the fields that they cultivate.

Even the most blinkered scholar must, of course, have some intellectual competence beyond his or her research specialty. To do any sort of job as an undergraduate teacher, for example, one has to know a bit more about a discipline than one might have learnt originally as a student. It requires some knowledge of an even wider domain to speak for a department or faculty in a more general institutional context, such as the university library committee or the student amenities committee. But this multi-disciplinary expertise is seldom taken into account in what scholars refer to as "their own work," in their curricula vitae, or in most of their activities on the national or international scene.

From the outside, an academic specialty may appear a somewhat arid domain of life and action. It is comforting, of course, to have ones own little plot on the academic map, in a named region that one can call home. What the outsider cannot appreciate is that this region offers psychological amenities that are not charted on any cognitive map. To extend the metaphor, it is the site of a *village*, clustered around a *marketplace* for scholarly goods. There one brings the produce of research, to be picked over and appraised before it is bartered for academic recognition. The village houses a small *community*, to which one belongs by right. The competitiveness and anomy of the vast world of Academia is relieved by the feeling of being amongst friends—of being a known person in a small human group bound together by common intellectual interests, common practical problems, common traditions of thought and understanding.

This communal spirit is reinforced periodically by meetings of a *club*—the occasional conferences where progress in the subject is reviewed, and the senior members of its "invisible college" gather in conclave. The cosmopolitanism of such corroborees is astonishing. One's most immediate colleague is not the person in the next office down the hall, or even in the next university down the road. See them here on the conference photo: that learned man from Tokyo, that lively woman from Palermo, that clever devil from Adelaide, that ponderous person from Santa Barbara, that bright lad from Leningrad, the shrewd old character from Bangalore. In endless conversations in the local bistro, reputations balloon or are torn to shreds. There are also jobs to be filled, and resources to be allocated. The network of colleagueship ripens into alliances and friendships, or—as in any real village—festers into bitter rivalries.

Every specialty also has its *castles*. Their stout walls dominate the surrounding landscape. They may be only little castles, but each one protects an estate where its owner is without rival—the leading world expert on the subject, or at least one of the three world experts on the subject, or at least the absolute expert on some particular aspect of the subject. Or, to change the metaphor, it is a *pond* where one can eventually grow into the biggest frog, with the most threatening or beguiling croak.

Scholars are, alas, very susceptible to the sin of pride. But vanity is not the only motive that attaches them so firmly to the tiny topics which—quite by accident—they have come to call their own. To the outsider it is always surprising how completely scholars commit themselves to seemingly uninteresting subjects, and how vehemently they defend their views on seemingly trivial issues. How could any sane person get so excited about, say, the classification of woodworms, or the political leanings of 18th century Shropshire parsons, or the interpretation of electron localization in amorphous semiconductors in terms of percolation theory? The fact is, however, that a specialty is the site of a *workshop*, where technical virtuosity is valued for its own sake. It imposes its own standards of excellence on those who work in it. What Gerald Holton has called the "Appollonian" scholar gets a peculiar satisfaction from this discipline, this exercise of exquisite craftsmanship in a microcosm. The acclaim of colleagues is merely confirmation that one has indeed achieved those impossibly high standards.

Another amenity of a specialty is that it provides a *quarry* for research projects. However deeply one goes into any scholarly question, one always finds more questions for further study. Scholars appreciate their autonomy, their freedom to undertake research on problems of their own choice. This autonomy is particularly gratifying if that choice is exercised in a field that seems narrow to the outsider, but which expands to a whole universe for the specialist.

The outsider's view of a scholarly specialty is thus quite misleading. For those inside it, it is rich and fruitful domain. We often forget that all scholarship is drudgery, and all scholarship is fascinating. Research on the structure of the universe turns out to be just as laborious and tedious as research on the structure of coal. Conversely, the minute problems that exercise the mind in the latter field can be just as "exciting" as the questions that hit the headlines when the origins of the cosmos seem to be at stake. Science, as Medawar remarked, is the art of the soluble. Every field has the fascination of a good crossword puzzle, which must, surely, yield to our attack if we go on long enough.

Sustaining the attack! Yes, persistence is one of the prime scholarly virtues. But it is also a psychological peril. One of the sad realities of scholarly life is the promising career that runs into the sands. An established scholar of good standing is seen to "go stale." The flow of publications dries up, communal activities are ignored, and specialized academic responsibilities such as the supervision of graduate students are rejected. More general psychological pathologies may also develop, such as withdrawal from the company of colleagues or gross neglect of teaching and administrative duties.

The aetiology of this syndrome must surely be very complex. But it is often associated with extreme specialization over a long period—with what Daryl Chubin has called *undue persistence* in a narrow field. Why should this be such a danger and can it be avoided? The primary reason is straightforward. The maps of scholarly activity are not graven on stone. A "subject" cannot be relied on as a permanent feature of the intellectual landscape. Just as it came into being through some discovery, some breakthrough, some problem posed for the first time, some area of research opened up by a new technique or concept, so it may fade away when it seems to have been adequately explored, or its problems seem to have been effectively solved or its practical applications seem to have been fully exploited. Detailed inspection of classified subject indexes reveals many deserted villages whose inhabitants have moved elsewhere, building new communities on virgin, uncultivated soil.

This is not just a consequence of spectacular breakthroughs that make hard-earned skills and expertise obsolete over night. Cognitive change is the vital force of scholarship, and is taking place all the time, on every scale. On the largest scale, the slow process of redrawing the boundaries of a whole discipline or subdiscipline seldom threatens the allegiances of a long personal career. For example, my former specialty of "condensed matter physics" is as much alive now as it was when I came into it, nearly 40 years ago. But the smaller the area defined as a specialty, the shorter its life as a recognizable entity. My guess is that a typical specialty of the size of what I have called a "field" has a lifetime of about twenty years. In that time it either expands so much that it has fallen apart into several new fields with many novel features, or it decays until it is no longer the site of an attractive and coherent research program.

What this means is that scholars of forty or so may have to face the fact that they have given themselves for life to a dying specialty. If they have not already yielded to the winds of change, by navigating confidently amongst the tides of new concepts and techniques, or by deliberately setting sail for new shores, then there is a grave risk that they will simply run aground. To return to the village metaphor:—the amenities on which they have relied will begin to decay. The market will no longer by buoyant with novelties. The club will decline into a bunch of old cronies. There will be little reputation worth defending within the castle walls. Worst of all, the soluble puzzles in the problem quarry will begin to seem trivial, whilst the difficult ones will loom as insoluble.

We often observe other people walking into this pitfall, but we are seldom aware of it under our own feet. The traditional amenities of scholarly life—academic tenure, basic research facilities, and autonomy in the choice of problems—conspire to screen us from obvious occasions for a major change of subject. Research projects intertwine, so that there always seems something to finish before we could contemplate making a clean break. There are professional responsibilities to graduate students, to colleagues, to learned societies, and to journals. Above all, there is our hard won reputation to think of. Would we not be giving up the credits accumulated in half a lifetime of research, and going into the world again as naked as any graduate student?

Nevertheless, in spite of these psychological barriers, it is imperative to make a radical move before the initial symptoms of undue persistence become irreversible. The same barriers stand in the way of any established scholar obliged for any reason to make an abrupt change of specialty. External circumstances, such as enforced change of employment, failure to get a research grant, or managerial decision are all too common nowadays in scholarly careers. It is hard to leave the village where we were born as a scholar, yet the pathways to other villages are not all closed. The stereotype of the *subject* specialist does not do justice to the actual research skills that an experienced scholar may have acquired over a period of twenty years or so. I recall an agricultural researcher that I once talked to. He explained that his specialty was the uptake of potassium by the roots of plants in clay soils, and he didn't think he could make a living at anything else. But later he mentioned that, although he had begun life as a botanist, he had, of course, had to learn a great deal of inorganic chemistry. Then, some fifteen years ago, there had been a revolution in experimental techniques, and he had had to teach himself electronics to keep in the field. More recently, some competitor had begun to build computer models of the systems they were studying, so, nothing for it but to take an elementary course of Basic at the local technical college and write his own programs to prove the other fellow wrong. Finally—almost as an afterthought—he remarked that this was not the sort of research that one could really do in the lab, so every now and then he had to put on his Wellington boots and trudge across the fields with the farmer, explaining exactly how to plant the crops he was studying.

As this example suggests, a narrow subject specialist can be a latent Admirable Crichton, the sort of person one needs beside one on a desert island, able to do many things quite passably well, and already competent to move in any one of a number of directions into new scholarly fields. It is well known that much research nowadays is essentially interdisciplinary, indicating that it stands at the junction of several different specialty regions. But the official maps of knowledge, organized hierarchically and classified decimally by librarians, do not show all the roads and paths that crisscross the landscape—or the tunnels into third and fourth dimensions which link apparently disparate fields of enquiry.

A radical change of specialty may thus turn out to be much less traumatic than it appeared in prospect. It is difficult from the outside to see the open spaces in an unfamiliar territory: it is also difficult to estimate ones capacity to make ones way into these spaces and occupy them. But a scholar of forty or so is not a graduate student, trying to learn an entirely new attitude to the acquisition of knowledge. He or she is a person who already "knows how to do research"—a complex of tacit skills, akin to "management" skills, which apply in every field of scholarship. The experienced scholar taking up a new problem knows, for example, that talking to the right experts is more useful than reading the right books. He or she also knows that an intellectual specialty cannot be fully mastered intellectually. Instead of trying to read it all up, the best thing to do is to roll up one's sleeves and get down to work on an actual problem in the laboratory or archives. It is essential to plunge in naively, just to get the feel of the real difficulties of the subject from the outside.

In fact, it is quite surprising how quickly a mature scholar can move into a new field of research and make some progress in it. Many that I have talked to about this reported that it was very hard work, but that it took only a year or so to be accepted as genuine inhabitants of the new village they had entered. They also reported that the experience had inspired and rejuvenated them. Of course, there are many who fail to meet this challenge. In some cases, they may have already gone too far down a mental cul de sac into staleness and anomy. In any case, elementary folk wisdom suggests that there are a certain number of people who are constitutionally unable to adapt to new circumstances. Some scholars manifest a peculiar rigidity of character which resists all forms of personal change, despite the support and encouragement they could be getting from those around them. Does the scholarly vocation attract people of this disposition, and simply confirm them in it, or does it, by its very nature, inhibit an attitude of openness to life and its vicissitudes? These are interesting questions for deeper study, but I would not dare to offer any answers to them.

What I do think, however, is that extreme academic specialization not only damages scholars emotionally: it also damages them intellectually. They become accustomed to plodding along in blinkers, looking neither to right nor to left. Everybody knows the importance of making connections between areas of knowledge that were previously thought to be unrelated. Nuclear physics is connected to historical research, through the techniques of carbon dating. Computer studies of artificial intelligence become significant for neurophysiology. The sequencing of DNA opens up new vistas in evolutionary taxonomy. These are notorious examples of a process that is continually occurring in detail, all over the map of knowledge. Yet the specialist in a small area who is expected to know only what is already in that area is not open to influences that might have come from adjacent areas, or even from much further away on that map.

A specialty is a parish: it cannot help being parochial. Sheer lack of interest in the activities of neighboring parishes is forgivable: prejudice against them is not. Unfortunately, a peculiarly narrow form of loyalty is almost institutionalized throughout the academic world. Pride in the achievements, and commitment to the research programs, of one's own specialty is transformed into scorn and denigration of other specialties. Humanists scorn scientists—and vice versa. Natural scientists scorn technological scientists—and vice versa. Physical scientists scorn biologists—and vice versa. Physicists scorn chemists—and vice versa. Theoretical physicists scorn experimental physicists—and vice versa. Theoretical solid state physicists scorn theoretical nuclear physicists—and vice versa. Theoretical solid state physicists working on electron theory scorn theoretical solid state physicists working on lattice dynamics—and vice versa. Electron theorists working on metals scorn those working on semiconductors—and vice versa. To the outsider, it is astonishing that such animosities can arise over such trivial details of concept, subject matter or technique. Perhaps this is only coffee table talk, which is not to be taken seriously. Nevertheless, it is symptomatic of the petty prejudices that close people's minds to unfamiliar ideas. Whether in a conversation amongst yokels in a village pub, or in a seminar for a selected group of scholars in an elite university, small-mindedness is the main obstacle to enlightenment.

Specialization benefits scholarship by dividing the labor of research into manageable stints. But it has its costs. Is the traditional role of the individual scholar as a subject specialist now being superseded by a more eclectic type of expertise? In the natural sciences and their associated technologies, sheer knowledge of the contents of a particular shelf of books and journals in the library may no longer be the primary stock-in-trade of the researcher. Problems drawn from the practical world of industry, agriculture, medicine, etc. do not define themselves neatly on the subject map. The ideas and techniques that may be needed to carry a project forward often come from many sources. The expert researcher has to have a nodding acquaintance with many different subject, and be willing to read further into them and apply them to the problems in hand.

Nowadays, a competent scholar in mid-career seldom works alone. He or she is likely to have become the leader of a project team, with members drawn from several disciplines. A wider view is needed, to integrate this diversity of skills into an effective research instrument. At higher levels of research management, the need to be a scholarly "generalist" becomes more and more evident. Only a small proportion of those who set out on scholarly careers become faculty deans, or college presidents, or directors of national laboratories, or research managers of industrial firms. But those who do rise to such worldly heights have to deal with issues that transcend the boundaries of any scholarly specialty, however broadly defined. In other words, they have to acquire a world view that brings together the contributions of hundreds of specialists whose individual achievements they could never hope to master in detail.

Modern scholarly institutions involve a great many other administrative tasks which call for general intellectual awareness rather than highly specialized knowledge of particular subjects. Every graduate dean will understand just what I mean by that. It is essential to know, for example, just where the work of every student and professor fits into an academic framework involving many departments and disciplines. And the indispensable technical services provided by libraries, computers, large-scale experimental facilities, etc. require quite different types of specialists with scholarly skills which cannot be located on the traditional maps of knowledge.

It is true that many scholars pass their whole lives happily enough as narrow research specialists without the distractions of administrative responsibility. Unfortunately, a few of these acquire such eminence that their opinions are sought on all sorts of subjects of public concern. It is then that the limitations of their understanding of scholarship itself become embarrassingly obvious. The wisest of them modestly admit their incompetence. But there are some who are always ready to volunteer firm views on matters of which they are profoundly ignorant. Taken for great minds by the world at large, they pontificate on politics, on education, or even on other branches of learning. Is this decline into punditry simply due to natural human vanity, or is it a peculiar product of their highly specialized training? A successful career in their own tiny field ought surely to have taught them the importance of genuine expertise: instead it seems to shut their eyes to the existence of equally genuine bodies of expertise in other fields.

Teaching specialties are, of course, wider than research specialties. A graduate program that tries to cover a substantial body of knowledge cannot do so by the aggregation of lectures and seminars separately covering each of its component fields. At least some of the teachers must have a comprehensive view, taking the whole of a sub-discipline or discipline as their "specialty." As one climbs back down the educational pyramid, the base of knowledge becomes broader and broader. At each level, every teacher needs to have some appreciation of all that the student is supposed to be taught at that level.

Yet the ideology of academic specialization denigrates the generalizing work of the university teacher, dropping research in order to put together a lecture course or a student textbook out of "secondary sources." This disdain is ill-advised. It is true that new ideas usually arise out of minute and specialized investigations for which extreme subject specialization is appropriate. But what we learn by crawling along the frontier of knowledge with a handlens is quite meaningless unless it is integrated and re-integrated into conceptual maps covering larger and larger areas. The work of integration is fundamental to scholarship, but is not always recognized as such, for it takes many forms—preparing seminal lectures, writing review articles and treatises, editing review journals, designing new teaching curricula, and managing new research initiatives. It should not be left to editorial committees or colloquia proceedings of mutually uncomprehending subject specialists, or to old fogeys who are pretty well out of touch, or to young fogeys sitting in airplanes on the way to the conferences where they are to give their prestigious review lectures.

Advanced scholarship is a social institution founded on detailed subject specialization. But that does not mean that each scholar must stay within a single narrow specialty all his or her working life. We should not be dismayed at the emergence of a new style of scholarly career, as specialized as ever in its earlier stages, but often broadening or diversifying in later years. The trouble is that a very high degree of *permanent* subject specialization is taken for granted throughout the scholarly world. Because this attitude is unquestioned, from the PhD onwards, it is embedded in the persona of every scholar, and inhibits positive manifestations of intellectual versatility. The traditional stereotypes of the academic life are no longer properly matched to the careers that scholars now have to follow.

People cannot be engineered to fit perfectly into social roles, especially when these roles are still evolving. Times have to change, and most people are remarkably skilful at adapting to them. In my own study of these problems, it was heartening to observe the sturdily affirmative response of scientists in their forties whose expectations of living out quiet, highly specialized research careers had been blighted by events beyond their control. They are, after all, much better off than unskilled workers thrown out of their jobs by the vagaries of international trade and high finance. In many cases, indeed, the enforced change may have saved them from the slow decline into undue persistence which is the fate of so many academic specialists.

We are bound to suppose that graduate education is the key factor in making scholars more or less adaptable in their later careers. But there is a lot of life to be lived beyond the PhD, and much experience to be gained before a scholar is fully formed. The traditional pattern was for a graduate student to continue his or her thesis topic into post-doctoral research and beyond. Sharper competition for academic posts is forcing many young scholars to diversify their research interests, and to take what is going in a series of temporary jobs in various fields until they finally settle down. This can be very disheartening and damaging if the changes of subject are too frequent, but a scholar who has had to learn many new things at this stage gains confidence in his or her ability to make further changes later on in life. Indeed, many industrial firms deliberately give their technical recruits a succession of diverse research tasks, to encourage them to think of themselves as versatile people able to move, if required, from field to field.

It is normal nowadays for scholars to spend a lot to time working with others on interdisciplinary projects. This can be a very broadening experience. At first, each member of the team thinks that all they have to do is to contribute their own valuable expertise, and that then all the different pieces of the jigsaw will jump into place. But as the project proceeds, each participant receives from the others a variety of perspectives on a concrete problem, and comes eventually to appreciate the real strengths and weaknesses of several different scholarly disciplines and research techniques.

On the other hand, professional advancement has become so competitive that there are fewer and fewer opportunities to talk at leisure with scholarly contemporaries in other disciplines. Collegiality is a primary function in scholarship. Contact with ideas from distant regions of academia is mediated by social contact with people who carry these ideas around inside them and bring them out in conversation. One of the depressing features of a very large academic institution is that each individual is surrounded by immediate colleagues of a rather similar persuasion—members of the same sub-discipline or academic department. I personally had the great advantage, in my thirties, of being a member of a Cambridge college with about 100 Fellows. Their interests ranged from English literature to molecular biology, from medieval history to plate tectonics. Through the college responsibilities and social activities that we shared, it was possible to get to know each other both as people and working scholars. We need more bridges like these, from island to island of our academic archipelagos.

Formal graduate education is thus only one of the experiences that can broaden or narrow the career path of a scholar. Unfortunately, most PhD programs are grinding mills designed to produce finely divided research specialists. There is no way of using them to reverse this process and turn out coarsely aggregated "generalists." In any case, this would take time, and graduate students already spend too long at school.

We need to think much more subtly about what might be done. There is no point in trying to broaden the curriculum by insisting that graduate students take courses on a wider variety of subjects if these courses continue to be taught as if to research specialists. For example, as a scientist with strong personal leanings towards the social sciences and humanities, I am not impressed by schemes for giving science students occasional doses out of those other academic bottles to cure them of scientism, or cultural illiteracy. Nor do I have much enthusiasm for additional courses on the history, philosophy or sociology of science if these are taught in isolation as metascientific academic disciplines.

If students are to become more generalized and integrated intellectually and professionally, then they need something that starts from their chosen specialty and generalizes and integrates it into large domains of knowledge and action. This can be done, for example, by first leading them gently into an adjacent multidisciplinary region and then taking them on a tour of some of the other ways of looking at the world. This is why I have long been an advocate of "science and society" courses for science students—and for non-science students also. They provide a natural medium for travel into a variety of fields of academia and of the real world, not necessarily to learn specific answers to specific questions but as a tour of the horizon, as an introduction to the bestiary of the intellectual forest, and as an occasion to rehearse some of the issues that they may have to face in earnest in later life.

We are only just beginning to realize that extreme scholarly specialization is worse than deplorable: it may, in the end, damage the health of those who indulge in it unduly. I started with the personal question, "what is your specialty?" and tried to show the personal implications of the answers it elicits. Perhaps the remedies we seek will spring to mind if we start asking scholars a slightly different question—"what do you do for a living?" as if they were not really all that different from people in other callings.

GUARDING THE FRONTIERS

Specialization is one of the names of the scientific game: generalization is the other. Since its adolescence in the 17th Century, science has claimed ownership of all valid knowledge of the natural world. Physicists, in particular, are taught that past research has already revealed general laws that apply without exception to "everything," and come to believe that future research will inevitably make progress towards a complete understanding—in principle—of all that could be understood in that spirit.

The ultimate goal of a unified science is as illusory as the Holy Grail, but it animates many noble minds, both to the labors of research and the delights of speculation. New scientific theories of the microcosm and the macrocosm seem to come closer than ever to traditional religious and philosophical themes. In recent years, some cocky theoretical physicists have even begun to talk once more about God, not so much as an all-powerful parent but as a slightly gaga senior partner from whom they would like to take over control of the firm.

Strangely enough, many physicists come back into the real world through metaphysics. They are expected to contribute their own brand of wisdom to public debates about the eternal verities, and to have something profound to say about the fundamental questions that puzzle ordinary people. Popular philosophies of nature are laced with concepts derived from the popular writings of eminent scientists. A few scientists—again, typically, theoretical physicists—even enlarge their imaginative conjectures into full-blown general theories about all and sundry, far beyond the writ of established scientific knowledge. A chance encounter with a would-be natural philosopher illustrates the influence of such writings in a cultural area where most scientists would personally fear to tread.

A comprehensive scientific world picture is also sought by more modest, more empirical means. The perennial mission of expanding and unifying the realm of scientific knowledge leads inevitably to the exploration of phenomena which have previously been attributed to "supernatural" forces—that is, to entities and influences outside the scope of current conceptions of the "natural" world. Although such marginal phenomena are usually insubstantial and irreproducible, and their proposed explanations contrived and unconvincing, they often engage the curiosity and investigative powers of very experienced scientists far outside their normal areas of expertise. Inside some scientists there is evidently a "parascientist" struggling to get free. In fact, it is quite difficult to define formal criteria for distinguishing scientifically between normal and "paranormal" phenomena. Like the normal science that it challenges, parascience claims that events in the everyday world are linked with operations in a more abstract world whose properties may be inferred by scientific observation and reasoning.

The scientific community closes its ranks firmly against such investigations. But as members of society at large, scientists are often expected to state their personal views on a variety of matters where it is generally thought that scientific evidence or reasoning ought to be decisive. A great many honest, sincere, intelligent people have complex but vague beliefs for which they would much like scientific backing. Their concerns are seldom satisfied by open-minded professions of agnosticism or shut-minded assertions of total disbelief.

The pejorative label "pseudoscience" is a cheap let-out; in practice, it usually raises far more questions than it purports to settle. Some of these questions relate to the abstract domain of scientific concepts, or to its connections with the life world domain of every day experience. But many of them penetrate to the foundation stone of science as a social institution— the credibility of the reports we receive from other people concerning experiences that we would ourselves presumably have if we were in their place. This cannot be determined by expert knowledge of a scientific discipline. Practical common sense, everyday psychological insight, and ordinary social competence are the instruments required to detect unconscious delusion or deliberate deception in our fellow beings.

Wilful deceit for personal gain is thus very bewildering and upsetting when it occurs in the world of basic research. Scientists, scientific institutions and scientific communities are baffled by such behavior, and do not know how they should deal with it. It is not just that the whole scientific enterprise relies enormously on the unquestioned personal trustworthiness of its participants. The scientific role is only scripted for life-world scenes where this assumption is valid. There are only the vaguest model texts for situations akin to those that are quite familiar—even to scientists—in ordinary social life, such as buying a used car, listening to a political speech, or sitting on a trial jury. In the eyes of the general public, the reactions of scientists to such situations are often embarrassingly inept.

It may have been this feeling of embarrassment, rather than self-serving defense of professional purity, that inhibited public discussion of scientific fraud until quite recently. I have to admit that when I broached it publicly in 1971,* I simply included it amongst "Some Pathologies of the Scientific Life," as if deceit were a much more unusual social phenomenon than it really is, at least in ordinary life. Recognition of this reality is one of the significant developments in opening up the interfaces between the scientific and everyday worlds.

---

* This address to the General Section of the British Association for the Advancement of Science was published in 1981 by the Cambridge University Press, along with a number of other "occasional pieces on the human aspects of science," under the title *Puzzles, Problems and Enigmas*.

A Natural Philosopher

I didn't get his name, so let's call him Tom. I picked him up one very wintry day where the A34 goes south off the Oxford Ring Road. He was making for Plymouth to visit his girlfriend, so I could take him part of the way. He was in his twenties, tidy and personable, open and unassuming. He turned out to be a remarkable person.

He told me that he was reading "natural philosophy" in the Bodleian. That was intriguing. I had never borne that venerable title myself, but in high and far-off times it meant theoretical physics—my own former vocation. Was he a student at the university? No, he just read books in the library, and was taking time away from other employment to perfect his own ideas. I disclosed my professional interest, and we settled down to a lovely conversation.

Tom's account of his Theory of the Universe and Everything was lucid and unpretentious. It began, if I remember rightly, in the domain of logic, with successive bifurcations, and picked up atoms and energy along the way. But its depth and richness were evidently not to be mastered at a single hearing. He answered queries precisely and patiently. I doubt if I've ever had a tutorial student of such refined intellectual sensibility.

For Tom was—is—an autodidact. He seems to have left school at 16, knocked around in the world of Alternatives, dropped in and out of A level studies, become a computer wizard, and so on. He obviously knows a lot about natural philosophy but it all comes from semipopular books. The building blocks of his theory are quarried from the writings of authors such as David Bohm, Ilya Prigogine and Rupert Sheldrake. His mentors have been distant voices, translated into the unchallengeable monologue and unresolvable ambiguities of letterpress.

Not having read much of this literature, I didn't feel competent to argue in detail against it. Why should I? Bohm and Prigogine, at least, are old colleagues, and scientists of the first rank. Anything they have to say about the nature of things is based on a thorough grasp of contemporary physical theory. I would respect their views, even when they go further out on limbs of speculation than I would myself venture.

But I did suggest to Tom that such books are inevitably misleading just because they do not lay bare their hidden foundations. Their whole point is to interpret conventional theoretical physics in novel ways, each seeking a point of vantage from which the whole scene can be taken in at a glance rather than explored step by step on the ground. In spite of their positive, logical manner, they are poetry, not prose. They might provide the inspiration for a completely new theory, but not the bricks and mortar with which it would have to be built.

I went on to talk about the pioneering days of quantum theory: about Planck, the epitome of a conservative German professor, puzzling over some little local difficulties with atoms and light; about Heisenberg getting results with home-made formulae that he didn't know were matrices; about Einstein, the godfather of modern physics, who could never accept quantum uncertainty. I tried to explain how it was that the physics always preceded the philosophizing.

I had to turn off at Bristol, so we didn't have time to argue this through. When I told this story at home, I was naturally accused of professorial pedantry, so perhaps I should say no more. But I remember once, as a cocksure postdoc, asking a colloquium speaker if there was any simple way of showing the necessity of the unexpected result which he had just spent an hour laboriously proving to us. He was Irish, and alcoholic: he leant forward over the lecture bench and beamed:"Do you mean, can you see it without doing any work? No!"

Out of the Parlor, into the Laboratory

Séances, ectoplasm, table-rapping, telekinesis, clairvoyance, psychic photography—we know about them as we know about antimacassars, tandem bicycles, smoking jackets and guineas. It is not exactly that they are obsolete; they are just decidedly old fashioned, and not to be taken very seriously.

The great days of the Society for Psychical Research (SPR) were before World War I, when it was not out of keeping for a reputable scientist, even the President of the Royal Society himself, to participate in these mysterious and undignified goings-on. Nowadays, respectable parascience (if that is not a contradiction in terms) deals out interminable hands of carefully randomized Zener cards, to be perceived extrasensorily by voluntary subjects whose guesses are recorded electronically and statistically analyzed for an outside-chance score. In fact the paranormal has moved with the times, into a graceless era.

This is the transition chronicled by Brian Inglis in his book, *Science and Parascience: A History of the Paranormal, 1914–1939*, from the pseudopods and phantom voices manifesting themselves around a physical medium, to the purely mental phenomena of telepathy and precognition. The history of the SPR itself, between the wars, is a chronicle of the decline of the traditional style. It was rent with internal conflicts, between the "high and drys," who took an austerely skeptical view of most reports of psychic phenomena, and those with a much more confident attitude to the possibilities of a breakthrough. The topics of controversy were, as always, the sincerity and integrity of mediums and their investigators, compounded by the ploys of would-be publicists and professional magicians.

The course of events in the United States, and in the rest of Europe was very similar, although not so politely institutionalized as in Britain. By the mid-1930s, psychical research had moved from the dimly-lit parlor to the university laboratory, with J. B. Rhine at Duke setting the style. Mr. Inglis does not pretend to cover in detail the period since World War II, although those of us who have watched from our dugouts the intellectual buffoonery and subsequent red faces generated by the Uri Geller skirmish would like to know what he makes of this reversion to the psycho-physical paradigm in manifestations of the paranormal.

Mention of a contemporary episode is not irrelevant to a review of an historical study because it is so extraordinarily difficult to assess the significance of past episodes in this odd but fascinating activity. I think I know what I believe about Uri Geller and the people involved with him because I know some of them personally, have read the works of others, and was able to follow the story as it unfolded on television and in the pages of *Nature* and *New Scientist*. This would give me some basis on which to judge whether somebody else describing the same events was grossly misinformed on such relevant matters as the character and competence of individual witnesses, the current climate of scientific opinion, the temptations of publicity and the professional costs of nonconformity. These contextual considerations are very difficult to weight up in relation to events that took place half a century or more ago, in a cultural climate that has changed, among people, however eminent, whose personalities are only known to us in silhouette.

Surely one should not have to take such a very localized view of events the significance of which must be universal if they did occur as reported? The essence of a scientific discovery is that it transcends its immediate context: however bizarre and contrived the circumstances, what is observed is only interesting scientifically if it can be interpreted as a particular case of a more general class of facts or phenomena. Alas, the study of the paranormal has not yet crossed this threshold into the archives of scientific knowledge. The movement of psychical research out of the parlor into the laboratory is obviously motivated by this fundamental principle of the philosophy of science, but is still without evident success. If we are to take any notice at all of the types of events mainly reported in this book, we have to accept that they have not proved reproducible, have not been integrated into a coherent explanatory scheme, and have too often occurred under conditions where deception or self-deception cannot be convincingly eliminated. In other words, there is no way as yet of thinking about them, except in the cultural, psychological and material milieux in which they are said to have occurred.

Mr. Inglis manages to paint in the background pretty well. He seems to have read widely in English and French, and quotes aptly from many sources. Drawing detail from the published accounts, he sets the scene vividly, sketches in the characters of the participants, and records succinctly the successive volleys of argument and counterargument about the credibility of what is supposed to have happened. His drama is played by an impressive cast of participants and critics, and he feels bound to mention a miscellany of cases from the spoof photographs of fairies that took in Conan Doyle to the premonitory dreams of J. W. Dunne. It is obviously going to be a valuable secondary source for scholars venturing into this little corner of cultural history, but it can also be commended for its fluent, sympathetic and thoughtful style. It is a surprisingly agreeable book to read, in spite of the fact that it describes phenomena that Thomas Mann called "preposterous," and Maeterlinck "vain and puerile," and that many of the people involved must have been thoroughly nasty or silly. Or perhaps, fundamentally, parascience is a sweet and innocent pursuit by comparison with the real science of designing cruise missiles or racing with other rats for a Nobel Prize.

But what you may ask, does he make of it all? Does he have an attitude towards the events that he chronicles? Of course he does—and he make no bones about it. Mr. Inglis admits, candidly, that his intention was to present the evidence about, not for, the paranormal but "found it difficult not to slip into the role of counsel for the defence." He gives the benefit of every doubt to the believers, and is uniformly scornful of the motives and capabilities of the skeptics. He regards as "unfortunate" the old SPR rule that all the results obtained with mediums later caught cheating should be set aside, and remarks at the "absurdity" of the assumption that when a conjuror shows that he can duplicate seance-room phenomena with the help of tricks, the phenomena cannot be genuine. His partisanship is so patent that unless one is also a strong believer one would be inclined to dismiss his account of events out of hand.

That could be a mistake. Would the account presented by a self-proclaimed skeptic by any more reliable or instructive? There is as much temptation to speculate wildly on how the effects were produced fraudulently by "natural" means as there is to conjecture on how they occurred supernaturally. Nobody will ever know, now, what really happened when that ghostly hand made its appearance in the gloom and rang a handbell far out of reach of the medium, or just what information might have been available to the clairvoyant who seemed to know so much about the intimate lives of total strangers. One might just as well start from the viewpoint of a would-be believer in the supernatural, whose motive is to bring such episodes to our attention, rather than that of the skeptic who would induce us to dismiss them as impertinences before our curiosity can be engaged. The parties are emotionally separated by a high wall, topped with iron spikes and broken glass, not a rail fence on which one can sit relaxed, scattering pearls of wisdom to either side. Mr. Inglis is sometimes ingenuous, but he is not deceived as to the controversial character of the evidences for paranormal phenomena and gives plenty of information on the general nature of the objections and where one could find them set out in full.

So the implicit challenge is to read it all up in the original sources, and make up my own mind. Must I really? As a scientist, I am evidently supposed to have a complete and coherent world picture, in which all the little local difficulties are carefully tagged as "anomalies," to be investigated with all the resource and sagacity at my command. No such picture has yet been produced. Scientific knowledge is shot through and through with all sorts of factual anomalies, and logical contradictions, which scientists struggle to resolve. There are some general laws of physics, of course, which are thought to be universal, but they have only been tested thoroughly in very special and contrived circumstances. The policy favored by both practitioners and philosophers of science is to proceed rather cautiously into unexplored regions of reality; and not to extrapolate too confidently from what has so far been well surveyed.

Some of the theoretical claims of some parascientists, such as precognition, or communication with the dead, are so much at variance with all scientific and everyday experience, so logically inconsistent with the firmly established structural conceptions of our being, that I do not find them useful as starting points for further thought or action. But other conjectured phenomena, such as telepathy, do not seem totally inconceivable to me in principle, even if I cannot see any means by which they could actually occur. Perhaps if I could devise some subtle experiment to test such phenomena a little further I would be willing to join this particular expedition of discovery, even if I had not the faintest notion where it might lead. But without such inspiration I beg leave to return to my dugout, grateful for the sincere endeavors of Mr. Inglis and others—believers and skeptics—who may eventually—who can be sure?—enlighten us further on the nature of the universe and of humanity.

Devastating the Dissemblers

D oes anyone around here remember Uri Geller? About five years ago his antics were all the intellectual rage. He appeared on television in Britain and America, demonstrating an uncanny ability to bend keys without apparently touching them. The august scientific journal *Nature* published a solemn account of experiments testing his powers of extrasensory perception. Several well-established academic physicists were convinced that these powers were "psychic," and defended this view in a televised debate against the American "magician" James Randi. Having been criticized for publicly losing my cool against all that rubbish in that debate, I get a certain amount of quiet satisfaction from the fact that Mr. Geller's conjuring tricks have been pretty thoroughly exposed, and that the most competent of his scientific advocates has publicly withdrawn his support. So although I would myself begrudge spending much time and effort on unmasking such patent frauds, I am thoroughly with Martin Gardner in treating them with the derision they deserve.

Mr. Gardner is known to millions for his inventive, amusing and learned mathematical column in *Scientific American*, and for many other articles and books about science for the general public. The present work is a sequel to his *Fads and Fallacies in the Name of Science*, which ought to have put a stop to most parascientific absurdities when it appeared some twenty-five years ago. But as he himself ruefully admits, irrationalism flourishes as never before, and seems never to be defeated by rational argument. Each of these reviews of ridiculous books, which originally appeared, over the years, in various journals, is as downright and factually devastating as one could imagine, often arousing splendidly irate responses which he also reprints and comments upon with sarcastic relish. And yet, in almost every case, his 1980 postscript reports that each pseudo-scientific cult still goes on, or has been replaced by one that is just as bizarre.

His hard-headed, commonsense style, based upon H. L. Mencken's sage advice "one horse-laugh is worth ten thousand syllogisms" is probably the best tactic. In most cases, indeed, the facts themselves—gross inconsistencies of argument, glaring loopholes in the test procedures, shufflings and evasions—are quite persuasive enough without further mockery. Gardner has a short way with dissemblers, but does his homework thoroughly on such details as the previous careers of self-styled miracle workers (Geller was a professional conjurer), their religious and political beliefs (a number of scientific investigators of parapsychology are adherents of Scientology) or the precise circumstances of the supposedly paranormal events (the conditions of J. B. Rhine's most famous case of "extrasensory perception" did not guard against a trivial form of cheating). As he points out, much of the careful research that controverted these claims by simply getting negative results is glossed over or ignored for its apparent dullness. Although, by its very nature, as an unedited collection of book reviews, this book is somewhat repetitious, it is immensely valuable as a source of reliable information on all sorts of queer fish and queerer notions.

It would be easy to fill the rest of this review with comical or tragic examples of human folly or fraud drawn from this source—but that would be a little stale at second, or third-hand. One should, if possible, read Gardner's reviews in their original setting, savoring the accuracy of his aim and looking forward to the next issue of the journal, where their living target is sure to make a bigger fool of himself by trying to reply. This is a prime blood sport, in which the satisfaction of the spectators is entirely justifiable.

Nevertheless, though every shot goes home, it fails to kill. There is something very puzzling about this whole business of pseudo-science. Although I entirely approve of Gardner's objectives and methods in almost every detail, I am surprised to find that I do not perfectly sympathize with his attitude in general, nor precisely accept his opinion on what is really at issue. It is all so confused and extravagant, on a more distant and wilder shore of the mind than I am normally accustomed to visit.

It is *impossible*, for example, that metal objects can be bent or broken without the application of mechanical force, or that an isolated person can correctly determine the suit of a card being drawn at random in another room, or that living fairies should present themselves to be photographed by two young girls, or that flying saucers loaded with little green men keep arriving on Earth from the planets of distant stars, or that an uneducated Brazilian peasant could diagnose and cure the diseases of thousands of genuine sufferers. Whether such events are utterly banal like spoon-bending, or utterly fantastic, like the landing of immense flying saucers, they cannot be reconciled with any of the world maps by which we usually navigate through life.

It is *impossible*, moreover, that there are "psi forces" that transmit information instantaneously from mind to mind over large distances, or that the subtle paradoxes of quantum mechanics are relevant to such matters, or that there can be contact with the minds of the dead (this particular doctrine is getting a bit out of date) or that the movements of the planets have may discernible effect on human history. However crude or subtle they may be, these explanations of the inexplicable are also totally incredible.

It is *impossible*, on the other hand, that intelligent, well-educated people, apparently in full possession of their senses, occupying posts of responsibility in education or science, could be utterly mistaken, or completely fooled, or temporarily blinded to such an extent that their reports of such events are entirely without foundation. Indeed, whether the observer is a trendily imaginative professor of theoretical physics, like John Taylor, or a tough-minded academic bureaucrat like Anatoly Alexandrov, the Rector of Leningrad University and now the President of the Soviet Academy of Sciences, it would be incredible if their testimony were completely false.

It is *impossible*, nevertheless, that any sane person contemplating such events should not think of the devices by means of which "magicians" contrive to "see through blindfolds" by peeking down their noses, or manipulate objects while distracting the attention of their audience by irrelevant actions, or affect compass needles or geiger counters with small magnets or radioactive sources hidden in their palms or their toe-caps, or "read thoughts" by listening for subtle cues in questioning, or pick up messages from hidden sources via their accomplices. It is incredible that there can be people "so supremely ignorant of methods of deception, yet so convinced of their personal ability to detect fraud" that "they will watch a conjurer vanish an elephant on a brightly lit stage, and readily admit they cannot explain how he did it. Next day they will watch an ex-magician move an empty pill bottle three inches and instantly declare that no conjuring techniques could possible have been used."

It is *impossible*, all the same, that a publishing house with a reputable scientific and technical list, a public or commercial corporation broadcasting scientific and educational programs on radio and television, or newspaper directed towards educated professional people, could also publish absolutely wild and incoherent pseudo-scientific claims, entirely at variance with every standard of scientific accuracy and validity, apparently without any effort to check these claims against expert opinion and regardless of the consequences for a reader or listener who might thus be persuaded not to take regular medical advice on a serious condition. No consideration of audience appeal or commercial profit could make such behavior ethically credible.

It is *impossible*, by the same tokens, that a hard-headed government agency, such as the United States Navy, could part with tens of thousands of dollars (or pounds, or rubles!) to support secret research on parapsychological and such like phenomena, to be carried out by persons who have already demonstrated their gullibility in open research on the same topics. Whether for grand strategic weaponry or for petty spying, any such application of pseudo-science is quite incredible.

It is *impossible*, in sum, in this day and age, that such irrational doctrines as those of astrology, biblical fundamentalism, scientology, transcendental meditation, faith healing, etc. could be taken seriously by thousands—even millions—of people, including sublimely talented scientists and frighteningly responsible statesmen. Whether based upon ancient religious revelation or modern scientific quackery, such doctrines are all quite incredible to an open rational intellect.

Like any sane, sensible, well-informed person, Martin Gardner yearns for a world in which all these incredibilities would be truly impossible, So, I guess, do I. But the real world is not like that—and as life-long disciples of the Red Queen we have sadly learnt to believe as many as six such impossibilities before breakfast, and then gone on fearlessly to lunch and dinner. That is to say, we accept the fact of the illogicality of human beliefs and behavior, however much we may deplore it and attempt to combat it. Such is the fate and duty of the responsible intellectual.

But that is not really the difficulty. It is the whole complex of interconnected "impossibilities" that makes the *problématique* of parascience so intractable. How can one think clearly about the relationship between actions and ideas all of which one personally finds incredible? How can one understand the behavior of people with whom one has no empathic comprehension on matters that they find compelling? There is a hermeneutic barrier that baffles the most enlightened philosophical, psychological or sociological analysis.

The only thing I can suggest is that one should start a little further inside science itself. There, after all, is where we feel at home, and have adequate personal experience of how it really works. Instead of seeking for elements of scientific rationality in the parascientific margins, we might recall that even the best of high science has its intellectual and personal pathologies which are only a little less fantastic and exaggerated than those of pseudo-science.

Gardner tries to make a clear distinction between the cases of the professional scientist obsessed with an eccentric theory and that of the complete ignoramus asserting nonsensical scientific doctrines. Is this distinction really tenable at the psychological level? The professional scientist who goes a bit dotty has two clear advantages over the layman: he has ready access to the formal media of scientific communication, and commands invaluable resources of professional terminology and established doctrine. With dignified channels through which to spread his opinions, and the means to obfuscate opposition by esoteric technical argument, such a person is almost impregnable. If he is sufficiently eminent he may even win some disciples, and cause a great deal of trouble by fighting a guerrilla war against "official" science.

It is not so easy, after all, to decide where unorthodoxy shades into irrationality, especially in a very abstruse or ill-explored field of research. I would be interested to see how Gardner would deal with Linus Pauling's theory of the therapeutic value of vitamin C, which might possibly prove quite as disastrous for some pathetic cancer patient as any other quack remedy. Or he might give his opinion on the story of "polywater," which cost millions in research grant money, before it was shown to be an experimental artifact, even though it was obviously completely "absurd" within the most elementary framework of thermodynamics. Such episodes, on a smaller or larger scale, are characteristic of the scientific life, and they often bring to the surface all those human weaknesses of self-deception, gullibility, doctrinal blindness, and even deliberate fraud, of which he is so scornful outside the scientific profession.

That is why I felt a little uneasy about his summary treatment of the current controversy about the linguistic capabilities of apes. It may be, indeed, that this is just another example of very subtle unconscious clues, as in the famous case of Clever Hans, the "talking horse." But I had the impression that this was a serious scientific debate which still remains open. There is another narrow line between the brave defence of good sense and the dogmatic assertion of accepted opinion. Mr. Gardner knows this as well as anyone, and he is right to scoff at the argument that, because Lavoisier did not believe that meteorites came out of the sky, we should all be careful not to dismiss any unorthodox opinion just in case it might later prove to be correct. I am all for expressing an honest view on any controversial matter, so as to be quite clear where I stand for or against—but only on the understanding that I take no special pride in being right and do not regard it as an eternal black mark against me if it eventually turns out that I was quite wrong. A touch of that sort of humility seems in order.

Fudging the Facts

A fraudulent scientist makes good journalistic copy, even if he is not as dangerous as a murdering doctor or a corrupt judge. The little list in "the scientific hall of shame" is now growing at about the same rate as the list of Nobel prizewinners, and attracting similar notice from the attentive public. Two or three times a year, some wretched researcher is shown to have faked his experiments, or fabricated his data, or simply plagiarized the published work of others. Occasionally somebody unmasks some old rogue, now mercifully deceased, who rose to scientific eminence on a pack of lies. It is all very distressing. It gives one the shivers to think that a fearless seeker after truth, like that admirable Dr. Jekyll, should really be a grasping, cowardly, lying Hyde, polluting the sacred archives of knowledge, and incidentally making a good living out of it at the taxpayers' expense.

William Broad and Nicholas Wade are two of the most experienced reporters on science policy and politics in the United States. The probably did not embark on their book *Betrayers of the Truth: Fraud and Deceit in the Halls of Science* to convey such an ingenuous message, but like most of us who write for gain they may have found that they lacked the moral fortitude not to play to the populist gallery. When their work was published in America last year, it certainly got a noisy reception; the hissing of official science was drowned in raucous cheering from the anti-science lobbies, delighted to have their suspicions confirmed and their resentments justified. The reactions of the popular media in this country give the same false impression that this is just a book of sensational revelations.

By lapsing occasionally into this tone, the authors have damaged what could have been an excellent study of a very interesting subject. Of course it is not quite so new as they seem to think: several well-known novelists, including William Cooper in *The Struggles of Albert Woods*, and C. P.

Snow in *The Search*, long ago exposed the seamy side of the scientific life in quasi-fictional form. Nor is this a work of original scholarship. Most of the real-life cases were reported at length in *Science*, the weekly publication of the American Association for the Advancement of Science, for which Wade has been and Broad still is a staff writer. But it was certainly a good idea to bring this material together into a readable book. The official ideology of the scientist as an "honest seeker after the truth" (etc. etc.) is unconvincing without some account of how to resist the temptations to these most grievous sins. The social psychology of science is a lop-sided discipline without some attention to its pathological aspects. Scientists and scholars are in need of a detailed description and interpretation of this deviant phenomenon.

The primary message is, of course, that scientists are all too human. That is such a familiar fact of life that it was scarcely worth emphasizing it by reference to the way in which even the truly great—Ptolemy, Galileo, Newton—were not above fudging their facts to fit their theories. The interesting cases are the contemporary ones where the authors are well informed on the essential facts and are not called on for historical insight into the mores of the distant past. And in all these cases it is not the deceivers who deserve study, but their victims. The deceiver, after all, knows the whole situation he is in, and acts simply out of rational self-interest: the deceived, by their gullibility and their reactions to its exposure, reveal significant factors about themselves as individuals and as social actors. There is a treatise on the sociology of science to be read from the story of Cyril Burt, who completely obfuscated the study of the inheritance of intelligence by his fabricated data on identical twins, or of Mark Spector, who bamboozled some of the greatest experts in cancer research with cunningly faked experiments.

The traditional ethos of science is unashamedly individualistic. It is not surprising that people are tempted to help themselves into a cozy appointment and even a bit of glory. But how is this to be achieved? Science, after all, is a hypercritical profession, where everything is supposed to get checked three times over. The pride of the academic world is that it is not bureaucratically held together by unwritten conventions and uncodified norms, where virtue is rewarded by public esteem and vices are discouraged by loss of personal credibility. In reality, it is a loosely structured oligarchy where responsibility and influence are irregularly distributed and ill-defined. This is just the sort of system where the confidence trickster can flourish for years without fear of investigation. The prime characteristic of most scientific deceptionists is their ability to manipulate people around them to cover up their quite evident deficiencies of perform-

ance. Elias Alsabti had already conned several Middle Eastern governments in various ways before moving from one US medical research center to the next, building up his publication list with a string of totally plagiarized papers on unimportant subjects in unimportant journals. As one
of his victims ruefully told a reporter: "We got taken in. But if other people think they wouldn't have, they're wrong." Even Burt was really playing the same game, using ruthlessly aggressive rhetoric against critics of
his theories in order to divert attention from their transparently bogus statistical foundation. Spector was an accomplished cheat and convicted
cheque forger, clever enough to exploit the wishful thinking of an elderly
patron who was proud of "the son I never had," and who claimed to have
corroborated his most cherished scientific hypothesis.

Broad and Wade offer plenty of evidence for the notorious tendency of
scientists towards self-deception, which often makes them absurdly gulli-
ble. They also emphasize the "attention to rank and pecking order which
makes members of the higher strata of science almost immune from investigation." Elitism encourages other psychological traits which protect,
and may even foster, deceit in the lower ranks. Confident in their opinions, and vain of their sagacity, top scientists are often peculiarly fearful
of admitting to a folly. Gratifying their egos with a bevy of students and
assistants, taking on too many research problems and juggling with too
many administrative responsibilities, they are tempted to skimp their duties as patrons and colleagues in research. Broad and Wade are too sweeping in their discussion of the allocation of credit between senior and junior
researchers working on the same project, which has always been an uncertain and contentious ethical issue, but they rightly point to the exploitation and corruption that can develop when scientific reputation is closely
coupled with managerial authority. In this context they might have referred to Mark Popovsky's strictures on Soviet science, where a strongly
hierarchical institutional structure is customary.

Psychologically speaking, it is thus quite inevitable that a certain
amount of outright fraud should occur in science: but what is its prevalence, and how much mischief does it cause? The authors list fifteen notorious cases in the past decade; these are an infinitesimal proportion of the
hundreds of thousands of scientists now at work, and are clearly of no significance in the enterprise as a whole. But Broad and Wade go on to assert
that this is merely the tip of a vast mountain of deceit, covered up by
clouds of administrative obfuscation and genteel denial. They also assert
that the increasing number of publicly admitted cases is evidence that scientists are being driven further into deceit by growing pressures of competition for grants and jobs.

Here the authors have given in to the journalistic temptation to sensational exaggeration. Their estimates of the prevalence of fraud are extrapolated far beyond reliable evidence, and are wildly inconsistent with the personal experience of most active scientists. Of course, there can be few scientists who live up to the ideal of utterly scrupulous, totally objective, painstakingly thorough and positive honesty projected by the traditional ideology, but that does not mean that they are oblivious to these moral imperatives and knowingly present false or fabricated results for personal advantage. It is wiser to count the minor sins such as failing to cite a relevant source, or presenting the data so as to favor one's theory, along with all the errors, omissions, follies, obscurities, misinterpretations and other unintentional imperfections that cloud the face of scientific truth.

Tolerance of such imperfections is not only humane; it also recognizes the sociological principle that the reliability of scientific knowledge does not derive from the personal qualities of scientists, but from their interpersonal relations. Broad and Wade correctly emphasize that science is ruled, not by a Hidden Hand, but by an "Invisible Boot," which eventually rubbishes and discreetly disposes of erroneous data, and disconfirmed theories. This boot does not always kick as hard and as accurately as is sometimes supposed. One of the most interesting general points to emerge from the study of scientific fraud is that experiments and observations are seldom deliberately replicated, as philosophers have usually supposed, so that false data are not immediately exposed. Even when a research claim is not confirmed in a later experiment by another scientists, there are usually innumerable better explanations to be considered before one would suspect deliberate fraud. Research is a taxing craft, where the uncertainty of one's own results is far more obvious than the dubious reputation of a colleague.

Nevertheless, as philosophers and sociologists of science are beginning to realize, the truth of science is not to be discovered at any single point in the map of knowledge, but belongs, if at all, to the map as a whole. It is the self-consistency of the overall network of facts and theories, bound together in all sorts of ways, that makes it strong and powerful. A gross fraud, such as Burt's, can distort this network seriously for a while, but eventually the contradictions with soundly crafted research become apparent; it is checked and given the boot. Broad and Wade spend too much of their breath attacking obsolete philosophical notions, apparently unaware of this new metascientific insight which would have made sense of most of their observations.

The absolute honesty of scientists may not be a matter for deep concern at present, but any suggestion that fraud is becoming much more preva-

lent has to be considered seriously. The direct evidence on this point is
very unreliable; it may be, simply, that emphatic demands for public accountability and social responsibility in the practice of research bring
more cases than before into the limelight. Science is now a much grander
enterprise, much nearer to the center of national life, and much more subject to public scrutiny. Perhaps it is also getting tougher and more competitive, and its traditional individualism has to adapt to collective forces
and structures.

If science is, indeed, getting less honest, can anything be done about it?
Nobody who understands how it works would suggest that it could be po-
liced by some formal institutional device going beyond the traditional
mechanisms of peer review, referee reports, and the open critical literature. But one should not discount the ethos of the individual researcher as
a major factor in the health of the enterprise as a whole. A great deal still
depends upon the superego fostered by years of training; the still small
voice insisting that one must not skimp one's work, one must get it right,
one must tell it as it is, however unprofitable to oneself in the short run.
People are not born like that, but there must be a base for the scientific
ethos within the person, as it is developed from infancy. The study of deceit in science should really expand into a study of the practical ethics of
society at large, since that is where scientists mostly live.

ENLARGING THE SCIENTIFIC DOMAIN

I t is time to return from the wilder shores of science to the main stream. We have uncovered many active social factors in the lives of scientists, showing some of the ways in which they relate to the knowledge domain, to their professional colleagues, and to the world at large. We have also learnt that the history of science is not just a chronicle of the independent advances made by a crowd of individuals: it is also an account of a collective enterprise, whose story is much more than the sum Of many biographies.

The central problem of "metascience"—that is, the study of science as a phenomenon in its own right—is how to interpret this story. How should we explain the immense dynamism of scientific and technological change? The history of ideas has a long and honorable tradition of describing successive stages in human understanding of the natural world, and connecting these with parallel intellectual and societal developments, but it has seldom analyzed, let alone answered, this key question. If we are to appreciate fully the interplay of personal and social factors in the lives of individual research scientists, then we must take note of the larger forces that govern the climate in which they live and work.

Most scientists would say that this climate is essentially local. They would relate what they were doing to developments in their research specialty, which they might then connect with more general changes in the knowledge base of their discipline, represented as a catalog of recent discoveries and novel theories. In other words, they would define scientific change in cognitive terms, and relate to factors operating almost entirely in the region of the scientific domain assigned to their particular scientific sub-community. But even what looks like a straightforward internalist account of a typical development in the history of science usually turns out to be extremely complex and many-sided. Significant philosophical, psychological and sociological aspects can be observed, even in the minute detail of events. Scientific progress cannot be represented in full by any one of the current theoretical models, whether this is cast in terms of paradigms, refutations, norms or themata. All these and probably other types of abstract entity are required to give a meaningful picture.

As I have emphasized elsewhere (e.g., in *An Introduction to Science Studies*, Cambridge University Press; 1984), the sociological dimension is vital, even in a strictly internalist analysis. When it comes to discussing external influences on the goals of research, the sociology of science itself has to be enlarged to include political and economic factors. Externalist accounts of scientific and technological change must cover a much larger segment of society, to include all the social machinery of industrial research and governmental policy, not to mention all the public and private institutions responding to environmental, welfare, defense and general consumer needs.

Enlarging the domain of science thus embeds it more deeply in the society that benefits from and nurtures it. So far from growing into an autonomous "estate" of the nation, science becomes increasingly dependent on the economy and subordinate to the polity. This incorporation of science into the whole fabric of our culture has a profound effect on the social role of the individual scientist. It is no longer convincing to insist that this role can be sharply differentiated from what the same person is expected to do as a citizen of a particular country, or even, benevolently, as a citizen of the world. Debates about the relationship between scientific and other values can no longer be treated as vaguely philosophical, theological, or otherwise ideological: they have to be answered urgently in practice, before or through personal action. One might say that the remainder of this book is devoted to some of the implications of this development.

Reconstructing the Reality of Scientific Growth

Towards the end of the nineteenth century, there was a great deal of interest in the radiation produced by, as one might say, a "red-hot" or "white-hot" body. The spectrum of the radiation was carefully measured by experimental physicists, and was found to have a quite characteristic form, depending, of course, on the temperature at which it was emitted. This spectrum ought to have been calculable from the laws of classical mechanics, or thermodynamics, and of electromagnetism, which were thought, by 1900, to be very well understood. But the mathematical physicists of the day could not seem to get the right answers until Max Planck proposed an entirely novel hypothesis. He showed that the observed spectrum could only be explained if there were a new law of nature by which the energy in the radiation could be emitted only in discrete "quanta." This was the starting point for a great transformation of physical theory, whose scientific, technological and philosophical consequences have still not been fully worked out.

The score or so of scientific papers in which these ideas were developed are amongst the most significant in the history of modern science. With scrupulous care and objectivity, Professor Thomas S. Kuhn, in his book, *Black-Body Theory and the Quantum Discontinuity, 1894–1912*, has analyzed them in detail, showing how Planck and others seem to have been thinking at each stage, in the light of public and private documents, subject to a variety of historical and contemporary influences. The book is as clearly written as it could be, but does not make easy reading. Radiation theory is an advanced topic in theoretical physics, where mathematical equations are more eloquent than words, and the whole argument may depend upon the legitimacy of replacing an integral by a sum, or of assuming that the phase angles of the coefficients in a Fourier series are perfectly random. The basic ideas of quantum theory can now be interpreted fairly simply in ordinary language, but their original faltering, stumbling, groping development can only be followed in the language in which they were first thought out. I cannot judge the accuracy of Professor Kuhn's English translations of the original German texts—whose key paragraphs he reproduces in the notes—but only someone completely fluent in theoretical physics could have made sense of this very subtle and technical subject.

Such a dry, specialized book, based on minute research, could be of immediate interest only for the professional historian of science or academic physicist. But it is an invaluable work, not only for its clarity and accuracy but because it illustrates, within a narrow compass, the reality of the growth of scientific knowledge. This reality is not so easy to uncover or depict. The dedicated work of the scholar is necessary, to get at the truth. Because quantum theory is so important, it is naturally referred to for case studies of various philosophical and sociological principles—and is as naturally susceptible to distortion and myth. There are, of course, a number of other books covering this period in the history of physics which do not tell it in quite the same way. But within the uncertainties of scholarly minutiae and interpretative conjectures, Kuhn has probably got as near to the heart of the matter, as anyone now really needs.

Let us assume that his account is entirely reliable: what light does this episode throw on our understanding of science itself? In recent years, the subject that used to be called "history and philosophy of science" has been enriched and enlarged into a general academic discipline. Within that discipline there have arisen various schools of thought, centered on various metascientific principles which are often considered to be in conflict. Is it possible to decide between these supposedly contradictory principles by appeal to the history of black-body theory? I would interpret the evidence presented in this book as follows:

In the first place, it is practically the archetype of a Kuhnian "scientific revolution." Professor Kuhn almost ostentatiously avoids that interpretation in the present book, but it scarcely needs to be said. Black-body radiation could not be explained within the paradigm of classical physics; in the effort to eliminate the anomaly, the paradigm itself was overthrown. This seems a reasonable description of the whole episode, which thus exemplifies the Kuhnian principle.

It has to be admitted, however, that Planck made this revolution without radical intent. His original derivation of the experimentally correct radiation spectrum in 1901 seemed quite orthodox; it was not until 1906 that Ehrenfest and Einstein independently proved that this derivation was not quite convincing, and that the energy transfer between the hot body and the radiation field must all be quantised, thus clearly contradicting the fundamental laws of classical physics. Nor is it correct to infer that Planck was desperately casting round for a way of escaping the anomalous consequences of applying the classical paradigm to this problem: it was only in 1905 that James Jeans, following up a comment on the subject by Lord Rayleigh, showed quite rigorously that the conventional continuum model allowed processes that would eventually swallow up the whole universe in one tremendous "ultra-violet catastrophe." Which of these papers was just "normal science"—puzzle-solving within the accepted paradigm— and which was revolutionary? Planck thought he was simply applying to the problem of radiation a method that had been successfully applied to molecules by Ludwig Boltzmann, 20 years before: was this method acceptable by this time as a normal technique of classical theoretical physics, or was it still highly questionable because of certain mathematical paradoxes that had not yet been resolved? Analyzed in this detail, the history of the quantum discontinuity seems less perfectly Kuhnian after all.

Our metascientific thoughts turn, therefore, to the obvious alternative—the philosophy of Karl Popper. We at once note that the classical theory of Rayleigh and Jeans was thoroughly falsified by the experiments of Lummer and Pringsheim, whilst Einstein's use of the quantum hypothesis in 1907 for a theory of the specific heats of solids was soon qualitatively confirmed. In his later version of radiation theory (1911) Planck gave a plausible, but certainly far from rigorous, argument that every oscillator should have a half quantum of "zero-point energy" in its ground state; this hypothesis, also, was capable of experimental confirmation or disconfirmation. Black-body theory is an excellent source for Popperian hypothetico-deductive case studies.

But this model again fails to fit all the facts described in this book. Planck's great paper itself is profoundly unsatisfactory from a Popperian point of view. So far from producing a falsifiable prediction, he worked backwards from the known experimental results to a fundamental mathematical theory of the entropy of radiation. As Einstein put it, in 1909:

> Delighted as every physicist must be that Planck in so fortunate a manner disregarded the need [to justify his choice of statistical assumptions], it would be out of place to forget that Planck's radiation law is incompatible with the theoretical foundations which provide his point of departure.

I would also have difficulty in explaining the methodological role of several admirable papers by Ehrenfest, Einstein and others, which simply rederived the standard classical formula of Rayleigh and Jeans by a variety of different techniques, when this formula was already known to be inconsistent with the experiments. And what are we to make of the great "H-theorem" paradox of Boltzmann, which purported to show that all molecular processes are in principle thermodynamically irreversible, which is certainly not strictly true as Boltzmann formulated it, but which could only be disconfirmed in an inconceivably unlikely off-chance? The network of physical hypotheses and mathematical theorems within which every theoretical physicist actually works is so subtly and delicately interconnected that the "logic of research" can seldom be disentangled from the novel concepts it is intended to capture. Every paper to which Professor Kuhn refers has an intelligible scientific rationale, but there are few that strictly satisfy Popper's demarcation criteria for valid science.

Following another metascientific theme, we easily discover a "research program" in the literature on black-body theory. The experiments begin in the 1880s, and arouse theoretical interest. Kirchhoff, Stefan and Wien in turn apply thermodynamical principles to the spectra observed at different temperatures and show that these satisfy various phenomenological laws. The problem of explaining these spectra in detail now seizes the attention of the scientific community. The classical approach of Rayleigh and Jeans is clarified by Lorentz, and shown to be unsatisfactory. Ehrenfest and Einstein then solve the problem by the quantisation hypothesis, which provides a basis for further researches on specific heats, the photoelectric effect, etc.

This is the rational order of the papers, as it might be reconstructed by a metascientific historian. But his reconstruction not only transfers the very influential paper of Lorentz several years backwards from 1908: it also has no place at all for Planck, who had arrived at the right answer, for not quite the right reasons, years before. Professor Kuhn demonstrates convincingly the underlying coherence and continuity of Planck's own research on radiation theory throughout the whole period covered by his book, but this is less of a "program" than a succession of movements from position to position using the available technical tactics to follow a vague strategy of, say, reconciling Boltzmann's statistical thermodynamics with Maxwell's theory of electromagnetic radiation.

It is interesting to note, moreover, that the "anomaly" of the black body was not regarded by the physicists of the day as a major problem but as one of those highly technical and specialized issues which Planck—a highly respected professor in his forties—seemed perhaps to have resolved, and which several clever young men such as Ehrenfest and Einstein were now getting excited about.

A striking feature of all this scientific work is its earnestness and sobriety. The leading participants were mostly German or German-speaking academics, very cautious and very thorough in their mathematical and verbal argumentation. Despite the intellectual novelty of the system of thought at which they finally arrived, they could never be accused of irresponsible conjectures in which "anything goes." Yet there is one example of just such an attitude in a paper by Johannes Stark in 1907, where, as Kuhn put it, "speculating widely and eclectically in an area he lacked the training or patience to enter," he anticipated Wien's computation of the wavelength of X-rays, and then arrived, independently of Einstein, at the correct formula for the photoelectric effect.

It is only a short step from philosophical metascience to the "sociology of knowledge." Black-body theory seems to be the purest product of academic science, with only the most tenuous links—back through thermodynamics to the steam engine; forward to electronics and nuclear power—with the technological aspect of things. It could perfectly exemplify the "internalist" model of science, where socially significant discoveries arise quite unexpectedly out of the search for fundamental understanding of natural phenomena. But one must not underrate those neo-Marxists whose remarkably sharp noses can smell out current social issues in the most abstract branches of science. The philosophically disconcerting hypothesis of discreetness and discontinuity in electrodynamic processes—of which Planck said eventually "one will just have to get used to it"—must surely be yet another case of the dialectical transformation of quantity into quality, as foreshadowed by Engels and others.

Professor Kuhn was not apparently seeking material for the more functionalist schools of the sociology of science. This was the high noon of academic science when the Mertonian "norms" were at their zenith. From their published papers and private correspondence, it does seem as if these German professors behaved with exemplary propriety towards each other in their scientific work. It is disturbing, nevertheless, to recall the unhappy fate of Boltzmann, who committed suicide under the stress of misunderstanding and criticism of his highly original scientific ideas. There is also some evidence that the norm of "universalism" was by no means fully respected; it was the conversion of Lorentz, a much loved "father figure" for the physicists of the day, that finally made the quantum hypothesis acceptable to the scientific community. Planck's own feelings about the reception of his ideas arc well known, and suggest that this whole episode could be as good a source as any for historical cases of unscientific attitudes by very reputable scientists.

Even for the thoroughly eclectic historian of science, following the events as they really happened, without any philosophical or sociological bias, the story told by Professor Kuhn is not without its problems. How, for example, should it be summarized in a few paragraphs, say, for inclusion in a larger work on the history of physics. It is quite clear now, from some of the issues to which I have just alluded, that the conventional account with which I opened this review is far too simple. Quantum theory was not just "discovered" by Max Planck; it crystallized in the consciousness of the physical sciences over a period of some years, as a result of the efforts and interest of a number of very talented mathematical and experimental physicists. In the spirit of "scientometrics" Professor Kuhn illustrates this process quantitatively, with graphs showing the exponential increase in the numbers of papers published within this new subspeciality from 1905 onwards. But the historian of ideas must go back to the original contributions, paper by paper. Can each of these be summarized, in a page or so? In the light of posterity, they can be related prospectively to the final outcome; does this do justice to the skill and originality with which they grappled with the unknown reality of nature, in the mist of confusion generated by other scientists attacking the same problem? Scientific research at this level of sophistication is not the routine application of a standard technique; in the published work of a Planck or an Einstein, every word, every sentence, every mathematical formula, is chosen deliberately after intense thought to persuade the reader to accept a scarcely comprehended insight. Can this concentrated essence of the human intellect be analyzed convincingly for its underlying motives and meaning?

The history of science is thus almost infinitely complicated on every scale. From the macrosociological to the biographical, from the biographical to the bibliographical sources, from each primary publication to its contents in detail, from the sentence or equation to the thought that prompted it—this is a cultural realm in which battles are often won, or lost, for the lack of a horseshoe nail! It is not even certain that all the relevant information has been disclosed in primary publications, or in the private letters preserved for posterity. Would the whole history of the subject have to be rewritten if we knew what ideas may have been transmitted or aroused in personal conversations between the principal actors?

By comparison with the well-ordered and very reliable knowledge that science itself obtains about the natural world, the historical, sociological and philosophical study of science yields disappointingly complex and uncertain results. Many scientists regard such study as essentially frivolous—until they come to reflect on their own lives and the work they have done. Then they begin to realize that they live and act in several dimensions of mind and art, with motives and repercussions beyond the walls of the laboratory or lecture room. Metascience is very far from being scientific, but it is highly relevant both to the pursuit of knowledge and to the wider world in which science exists.

The story of the origins of quantum theory tells us something about the way in which science is actually carried out. It is heartening to observe Planck working out his own strategy of research and persisting quietly against the received opinion of his day. It is instructive to read about Einstein's "extraordinary ability to discover and explore problematic interrelationships between what others took to be merely factual generalizations about natural phenomena." Every student of theoretical physics is likely to encounter situations in which he seems to be getting the right answers for reasons that he does not quite understand—or the wrong answers by what seem to be entirely cogent and unexceptional methods; here are cases that might lend wisdom to his response. When is an absolutely clear and precise proof called for—and when is it legitimate to hint obscurely at what seems to be an imponderable underlying difficulty? How should one judge between an unconventional idea and all the obvious objections that can be raised against it? Michael Polanyi has described the "tacit knowledge" and skill that guides scientific research. This expertise is as necessary in mathematical physics as in any other branch of science. It is to some extent innate, but must be refined by experience and the stimulus of skilful teachers and patrons. Such experience can be simulated by thoughtful study of such excellent examples as are to be found in Kuhn's book.

One must be skeptical of any claim that science is governed by strong metascientific laws or principles. For every historical exemplification of any such principle, we seem to be able to find an equally valid contrary instance. This does not mean that science is an entirely arbitrary human activity, but it is kept in order by a self-governing social structure and by a scarcely articulated metaphysic rather than by specific intellectual rules or formal conventions of thought and behavior.

Nevertheless, there is much to be learnt from philosophical, historical, sociological and psychological theories about science. Karl Popper and Thomas Kuhn, Robert Merton and Michael Polanyi have each looked at science from a particular point of view, and have each seen many significant features that were not apparent to routine historiography, biography or epistemology. It is not to anyone's discredit that the whole system could not, after all, be reduced to any single formula; rather, each should be commended for having grasped at least one important characteristic of this many-sided activity. As we have seen, black-body theory nicely exemplifies these various characteristics, which are, indeed, aspects of science as a whole.

Observe, moreover, that these characteristics are not at all mutually incompatible. It is a fair approximation to the truth to say that quantum theory arose by a Popperian "hypothetico-deductive" process within the context of a Kuhnian "paradigm," and that this was achieved by the exercise of "tacit knowledge" by scientists who conformed to the Mertonian "norms." As we have seen, this description is very schematic and imperfect, but the notion that these fruitful generalizations are logically, empirically, or ideologically contradictory is clearly disconfirmed by this positive historical instance.

What we need to get used to is that these generalizations, like the mechanistic positivism against which they were all reacting, are not absolutes. They should be understood, quite simply as maxims guiding our understanding and action where appropriate, but which are also, at times, ludicrously invalid. That they are valid even to this limited extent is a sign of the remarkable progress that has been made in the study of science in recent years.

This, to my mind, is the real metascientific message of Kuhn's book. It rings true, because this is the way that science really is: stretched eternally between the creative and critical talents of the human mind; blundering about in the dark, but eventually correcting itself by an appeal to the opinion of those committed to its progress; highly technical and sophisticated in detail, yet always seeking simplicity and clarity of understanding.

To complete the picture, it would be necessary to go yet deeper to another dimension of contradiction and dialectic. For Planck the essential problem was to apply the method of Boltzmann, devised for particles, to the case of a continuous radiation field. He and Einstein surpassed their contemporaries in that they understood that the problem could not be solved unless these opposed "themata"—as Gerald Holton has evocatively named them—could in some way be reconciled. Planck thought he had done this by a subtle analysis of the processes of absorption and emission of radiation; Einstein went further—far beyond the credibility of his contemporaries—to attribute particle-like properties to the radiation itself. Nowadays, of course, we are quite happy with electromagnetic photons and wave-like electrons. But the dialectic between the continuous and discrete representations of physical phenomena—begun by the Greeks, very much alive for Newton, enriched by the quantum mechanics of Schrödinger and Heisenberg—is no less of a challenge in an era of quark containment and super-symmetry than it was in those golden days when the whole world of microphysics was revealing itself to its humble conquerors.

What Shall We Look into Now?

N early half the essays in the book *The Advancement of Science and its Burdens* by Gerald Holton are about Einstein, and the way he searched for a unified *Weltbild*—a coherent image of all reality. His lifelong task as a scientist was to puzzle out the cosmic jigsaw. He succeeded in finding a link between the pieces labelled "electromagnetism" and "mechanics" and showed that the piece labelled "gravitation" belonged next to the one labelled "geometry," but he failed to fit them together with a single formula. The advance of physics since Einstein's heyday has not really solved that particular problem, even though two new forces have been uncovered and one of them is closely connected with electromagnetism.

Einstein is dead, but there is no awakening from the dream of a unified theory. It crept into the consciousness of the West in antiquity, was given flesh and blood by the Medieval Church, and was then redefined more starkly by Descartes and Newton. It became the leitmotif of many great scientists such as Faraday and Helmholtz—not to mention a host of lesser figures such as Hans Christian Oersted, whose work is celebrated in one of these essays. Unification is the Supreme Project of Science, metaphysical and religious in inspiration. Curiosity can only end with an understanding of the nature of all things.

The integrity of Einstein's scientific commitment was matched by his clarity concerning its means and ends. In his writings he is, as he said of God, "subtle, but not malicious." From them Holton teases out his presuppositions, his approach to theory and experience, his ways of conceiving new models, and his attitude to the realities thus represented. An eloquent chapter analyzes the accomplishments of the intellectual innovator as a shaper of the imagination of his era. Holton deals sharply with fashionable gurus who incorporate half-baked versions of relativity or quantum mechanics into their grand philosophical or theological schemes. Einstein's thought was elevated, but he was no mystagogue.

Nor, indeed, did he regard his work as revolutionary. Scientists are well aware that they build on the solid achievements of the past. A novel contribution is seldom more than an extension or a revision of what everybody thought they knew already. The strength of Einstein's bridge between electromagnetism and mechanics was that it linked structures that were solidly built and firmly based. Some modern philosophers of science—notably Thomas Kuhn—have since argued that such a development has to be considered a radical reformulation of the whole *Weltbild*. Einstein and his contemporaries always insisted that they could only see it as an evolutionary development within a continuous intellectual tradition.

In Holton's own view, firm threads of historical continuity are provided by what he has taught us to call "themata." In every era, scientists fall back on these "enduring elements...somewhat like the old melodies to which each generation writes its new words." Einstein, for example, was guided by long-established notions of formal symmetry and simplicity, of causality, completeness, continuity and cosmological scope. Another scientist—Niels Bohr, for example—might have favored somewhat different thematic principles, but there was only a limited list from which either could have chosen. In fact, the measure of the greatness of both Einstein and Bohr is that each added one or two new and powerful themata to the common stock.

Thematic analysis must surely be one of the ways to an evolutionary account of the advancement of science. But it is not an easy path to understanding. In this book, Holton does not clear up my uncertainty about the distinction between a themata and a metaphor. Some of his themata— "atom," for instance, and "field"—undoubtedly have metaphorical roots, but he insists that not all scientific metaphors are to be counted as themata, however much insight they may convey. Where and how does he draw the line? It isn't frivolous to point out that modern physics is using anthropomorphic metaphors, such as the "birth" of a "strange" particle, just at a time when psychology is studiously representing its findings in terms of strictly impersonal themata, such as "feedback" and "databanks." Is this not another manifestation of that striving for a unified representation of the world?

As an Einstein fan, I appreciate this sympathetic and sensitive approach to the philosophy of his work. In an interesting essay, Holton suggests that concern among scientists about the ethics of their profession has displaced their former interest in the deeper meaning of their discoveries. Perhaps they realize how difficult it has become to keep the philosophy of nature aimed at the moving targets of contemporary physical theory. It took us a generation to swallow the relativity of time: we can wait a bit longer, until we are sure we have understood what happened in the Big Bang, before we rejig our traditional notions of eternity. Nevertheless, something vital is missing from the particular philosophy of science which is assumed as a background for this book. It describes research as the activity of individuals, without regard for the relationships between them. Einstein was untypical, not only in his genius but also in his capacity to work fruitfully on his own. But even he, in his middle years in Berlin, behaved like most of his colleagues, exchanging information with them, trying out ideas on them and reacting to their criticism.

The continual interaction between individuals gives science its strength. We accept Einstein's solutions to certain problems of physics because they were convincing to other physicists who were puzzled by the same problems. The individual scientist is playing a social role in a collective enterprise devoted to the generation of a publicly agreed body of knowledge. Theoretical unification, for example, is more than a personal predilection: it is characteristic of science because it is an effective way of resolving the contradictions that can arise between schemes of thought developed independently by different groups within the scientific community.

The absence of a sociological dimension is a more serious defect in the second half of the book, where Holton deals with various aspects of the place of science in modern society. In these later essays his mind is still unconsciously focused on the part played by a small circle of almost unique individuals devoted to a particularly abstruse branch of knowledge. I get the feeling that he is looking down from on high at all the other people and ideas in science. In spite of his wide scholarship and humanity, he shares the tree-tops viewpoint of a certain class of academic physicists and cannot do justice to the thoughts and feelings that grow up from the grass roots of society.

For example, he carefully distinguished between "science," dominated by the imperative of "omniscience," and "technology," whose historical imperative he sees as "omnipotence." These two distinct circles, he says, expanded independently, until they began to overlap with one another, and with a third circle labelled "society." The historian of highbrow physics thus sets himself apart from whole disciplines of systematic observation and intellectual ferment which have never been separated from the practical arts out of which they were born. I am not here raising the controversial issue of the shaping of scientific world-pictures by socio-technical factors. I am simply pointing to the depth of understanding, width of coverage and historical length of such solid bodies of organized knowledge

as geology, say, or human biology. These are true sciences, more pragmatic and empirical than physics, perhaps, but just as subtle philosophically, and just as nobly motivated by curiosity.

In these essays, the great technologies themselves, such as engineering
and medicine, are treated with due respect for their power, but with little
love. The needs they provide for and the virtuosity they elicit are not quite
on the same plane as the desire and pursuit of the whole. We could have
been reminded that the equations of Einstein and his successors were generated out of the observation of exquisitely contrived events in subtly designed material apparatus, and that the most awe-inspiring products of
modern physics—the atom bombs, the lasers, the microprocessors—ex-
plicable and imaginable as they may be by reduction from tree-tops principles, were made workable and constructible by a synthesis of all the
deeply rooted crafts of engineering.

For this reason, it seems to me that Holton both overestimates the historical discontinuity between the "three previously separate fundamental
imperatives, those that animate progress, respectively, in science, technology, and society," and underestimates the degree to which this discontinuity has now been eliminated. At the academic level these circles may just
seem to be overlapping, but down below, in real life, they are interpene-
trating and fusing together. Look at any TV program on AIDS. The scientific, the technological and political (not to mention the ethical, the emotional and the aesthetic) are rationally interconnected in such a tight web
that it doesn't make sense to disentangle them.

This many-sided, multi-colored activity is what we now have to think
of when we talk of "science." What can we do with it, now that we have
got it? In the fifties and sixties, as new strains of technology began to
emerge out of the high sciences of quantum physics and molecular biology, it did seem for a while as if Bacon's dream of "affecting all things
possible" could at last be realized. To use one of Holton's own metaphors,
it was almost as if the Four Horsemen of the Apocalypse would soon be
routed by the four forces of physics. But that, too, was an illusion of top-
down vision. For those who look upwards from the bottom, even the bar-
est needs of life still seem far out of reach.

More sober now, we know that we can never succeed in effecting all
things possible. Yet we are not without hope in seeking the means to effect some that are desirable. What things are desirable? In what direction
should we try to steer the bulldozer, gouging a path over what mountain,
to what goal? At this point, where the real action ought to begin, Holton
retires into generalities. In one essay he advocates a combination of disci-
pline-oriented and problem-oriented research stimulated by the sense of

"an area of basic scientific ignorance that seems to lie at the heart of a social problem." Surely we have been hearing something like that for nearly fifty years, since J.D. Bernal wrote *The Social Function of Science*—indeed, for more than three hundred and fifty years, since Bacon's *New Atlantis*.

Applying science to social problems is easier said than done. As Holton rightly points out, if the Four Freedoms epitomize our social values, then there is no technological fix—probably not even a sociological one—for the loss of three of them. On the contrary, the advancement of science and its associated technologies is seen by many as the major threat to both freedom of speech and freedom from fear—if not to freedom of worship. In any case, somebody has to decide what problems ought to be tackled. That somebody has to have a very long purse. Science is much too expensive to be undertaken out of sheer benevolence. Its costs can only be met by bodies with very large financial resources. Industrial firms say, quite reasonably, that they have other fish to fry; in fact, British industry is not spending enough on R&D even to fry its own fish properly. It is now entirely accepted that science should be supported by the state and directed toward meeting the needs of the nation. But those needs do not define themselves: they are what the Government says they are. In the political circles where such matters are decided, commercial and military research gets a much higher priority than research related to "social problems," so that is the way the money flows.

The procedures differ from country to country, but the effects are much the same everywhere. In our political system, each government department can spend what it likes on research in direct support of its mission—defence, or agriculture, or whatever. The national agenda for science is not determined centrally: it is a side-product of the competition for resources in Whitehall. If the Cabinet is persuaded that we should spend a lot on buying guns, then lots of our science is about gun-making and gunnery: now that we have too much butter, dairy science goes down the drain.

The central organs of government—the Cabinet Office and the Treasury—exert a more general influence on science policy. The Thatcher Government became obsessed with the desire for a quick turn-around in our economic affairs. As a result, research that promised a definite commercial return within a few years was favored over research with vaguer, longer-term goals. Short shrift was given to areas of basic scientific ignorance at the heart of social ills whose cost cannot be measured in hard cash.

Current debate over science policy revolves around the allocation of public money to various sectors of the research system. Less military and more civil, says the House of Lords. Less to weak university departments and more to strong ones, says the University Grants Committee. More for basic science, says the Royal Society. British science is still thought to be pretty good, but could rapidly fall behind. More for all science, says everybody, except the Treasury, which says nothing and keeps its fist tight.

Holton, writing as an American, has deeper disquiets. He argues forcefully for urgent, sustained and generous action by the Federal Government to get American education out of the shambles into which it has fallen in recent years. This leads naturally into his Jefferson lecture, which gives the title to the book and which focuses on the gulf that has opened between scientists and citizens over the possibility or the desirability of technical progress. Democracy and technology must learn to respect and embrace one another if they are to live happily together for much longer. Pleas for better public understanding of the nature of science and technology are fashionable these days, but seldom so thoughtfully or eloquently expressed.

The trouble is that democracies are just as greedy about getting the golden eggs out of science as more autocratic forms of government. Science is expected, eventually, to deliver the goods. The academics have to explain why the country should go on paying people to do research simply to satisfy their curiosity. When the bluff of the old "what's the use of a new-born baby?" ploy was called, the notion of "strategic" research came into currency. It is not difficult, after all, to suggest possible long-term utilitarian implications for almost all research, however academic it may be in its immediate goals.

Once these hypothetical strategic aims have been formulated, however, the demand comes that they be achieved. In the name of political accountability and economic efficiency, all science must be directed to generate the material benefits and appropriable profits that seemed to have been promised. A tenuous practical argument for exploring "an area of basic scientific ignorance" is translated into a managerial objective. A large-scale research program is targeted on what happened at one moment to look like an "exploitable area." Vannevar Bush's Law comes into operation: "Under the pressure for immediate results, and unless deliberate policies are set up to guard against this, *applied science invariably drives out pure*."

The big political issue for science is not how to draw up a shopping-list of desirable research objectives, nor whether some sorts of science should be limited for ethical reasons: it is how to run basic science. Somebody has to decide what the questions are to be, and how they should be tackled. Everyone agrees that the latter can be left to the technical experts—typically, by peer review. It is the responsibility of a non-technical authority, political or commercial, to define very broadly the area to be investigated. But in basic science the researchers themselves usually have a much better idea of what questions are worth asking from a scientific point of view than outsiders who can only guess at the potential utility of the likely answers. Curiosity and skepticism are more effective than systematic planning.

The traditional institutions of academic science had a way of motivating individuals to compete and co-operate for the advancement of knowledge. That was how thousands of individuals far less talented than Einstein generated the great system of ideas to which he contributed so notably. Can this sort of loosely-linked structure still be maintained within a larger, more rigid framework of socially-defined and financed strategic goals? Our existing stock of knowledge and its applications is already so complex that we do not even know what it is that we are ignorant of. It is really going to be very difficult to devise new institutional arrangements for a professional activity that can have, as Holton puts it, "a just claim to more authority when, and only when, it is widely seen to honor both truth and the public interest."

Expansionists and Restrictionists

We all have a pretty good idea of what "science" is, don't we—at least within the restricted English-language meaning? But what are "values?" In *Between Science and Values*, Loren R. Graham is shrewd enough not to fall into the trap of trying to define a term that he interprets to the widest extent. His subject is the relationship between the natural sciences and the philosophical, political and ethical aspects of contemporary culture. In all conscience, that is a broad enough canvas, though there are interesting relationships between science and aesthetic values that would also be covered by this title. Or is not science itself an attractive manifestation of a certain type of "value," part utilitarian and part aesthetic? In Graham's title there is latent a paradox, if not a confusion, which surfaces from time to time in the text and is not fully resolved.

The whole subject would, indeed, fall apart if it were not bound together by a strong cable around the opposite poles of "expansionism" and "restrictionism." The archetypical "expansionist" is B. F. Skinner, who urges mankind to save itself by designing new values consistent with his psychological theories. The most articulate popular exponent of "restrictionism" may well have been Sir Arthur Eddington who "reassured the educated public" that "science (to him physics) need not take away our religious beliefs because it deals with only one of three worlds—the world definable in terms of clocks and yardsticks; the abstract world." This is a straightforward typology that can be applied fairly confidently to the view of the many scientists who have written about philosophy, religion and politics from the standpoint of their professional knowledge and experience.

What is interesting, of course, is that this polarity is not necessarily correlated with the particular discipline claiming external authority or try-

ing to defend itself within a limited territory. Relativity theory for example, is not intrinsically restrictionist or even "abstract." V. A. Fock, a Russian theoretical physicist of the same high scientific distinction as Eddington and an equally authoritative popularizer of this esoteric subject,
was at pains to emphasize the consistency of the "theory of gravitation"
with the materialism of orthodox Marxism. Graham analyses and convincingly interprets these contrasting philosophies against the very different personal and political backgrounds of their authors. In fact, from the
point of view of the sociology of knowledge, one might inquire how it
was possible for a socially isolated "bourgeois" Quaker and a highly effective communist intellectual to be in close agreement on the scientific content of their subject to the extent that their respective textbooks are equivalent
standard treatises on Einstein's theory of space, time and gravitation.

Einstein himself was essentially a restrictionist. His human liberalism
drew him into the agonized political arena between the world wars, but he
was careful not to assert scientific principles out of their context, for he
was "more aware than most of his colleagues of the positive functions
that this restrictionism fulfilled: it protected society from the eugenists
and the social Darwinists, while it also insulated science from the attacks
of critics who believed that the new physics undermined traditional standards of ethics and politics." But Niels Bohr, for equally wise reasons, was
"making an expansionist argument for a direct, non-metaphorical relevance of atomic physics to the problem of understanding life and free
will and he was, as did his father Christian Bohr (a Danish professor of
physiology), arguing that a mechanistic conceptualization was inadequate to physiology."

As for Werner Heisenberg, his initial enthusiasm for philosophizing out
of physics was tempered by the nihilistic obscurantism of official Nazism
into a sophisticated restrictionist defence of German cultural ideals, where
"the ideal solution to all problems, both scientific and political," was to
show that the contradiction between Goethe's and Newton's theories of
colors could be resolved by insisting that they must have been talking
about "two entirely different levels of reality." In other words, the differentiation of expansionists from restrictionists may not be so much a matter of *quot homines, tot sententiae* but a variant reading of *cuius regio,
eius religio.*

The tendency among biologists—at least among those who have written spaciously about human affairs—has been expansionist. This applies
as much to extreme philosophical reductionists such as Francis Crick and
Jacques Monod, flushed with the intellectual triumphs of molecular biology, as to the significantly more sociopolitical exponents of ethology and

sociobiology such as Konrad Lorenz and E. O. Wilson. But here again, the way in which the scientist tries to give a scientific rationale to extra-scientific "values," such as social conflict or moral altruism, depends more upon the value systems floating around in his cultural milieu than on what can be convincingly deduced from his special scientific domain. If one is not too fastidious about the reliability of his data and their theoretical interpretation, one can, it seems, put together a plausible "biological" case for almost any political system, from technocratic totalitarianism to anarchical naturism—it all depends what people really want out of life, or what they will put up with in a thoroughly unsatisfactory world.

This ambivalence of "science" with respect to social values is beautifully exemplified by the history of the eugenics movement in the 1920s. Evolutionary genetics seemed to assert, pretty firmly, that the "fitness" of the human race is bound to decline unless something is done to replace the "natural" selection that we now override artificially. Of course, as Alfred Ploetz saw at the time, the "value consequences of applying the science of human heredity depend more on *when* and *how* it is applied than on the question of what is inherent in the science itself." But as Graham shows, controversies over the very *principle* of eugenics flared up in both Weimar Germany and Soviet Russia, with liberals, Nazis, and various more or less orthodox breeds of Marxists on both sides of the fence in all sorts of intellectual and moral postures. Even though the political and social milieux have changed, the nature-nurture issue still has "enormous potential destructiveness." The paradox is that "Eugenics is now a word in disrepute but the use of genetic knowledge to benefit humankind is a far more viable possibility now than ever before."

At this point, the thrust of the argument shifts from the philosophical to the practical. Nobody, least of all a priest or a politician, need take any notice of the religious or political opinions of a few big-mouthed professors: it is when scientific techniques affect social actions that "values" are really at stake. The prime example is in biomedicine, where the revolutionary technical capabilities of amniocentesis and life-support machines seem to challenge most of our traditional ethics of birth, life and death. Graham discusses these issues very thoughtfully, but concludes that this challenge is not a manifestation of scientific "expansionism." What can be done with the use of science is to be classed as "technology," and is then subject to the primary ethical value systems of the society that puts this technology into practice. He admits that the new techniques may "play a role in shifts in our secondary or derivative values:...a person may modify his or her position towards abortion or terminal care" but suggests that this shift would be justified "by reference to more fundamental values that are still retained, such as responsibility and solicitude for one's children, relatives, or intimate friends." Other issues of public concern about science and technology, such as whether there should be limits put around scientific inquiry, are discussed in the same enlightened spirit. This discussion does not carry the analysis much deeper than the conventional contemporary wisdom, but it would be difficult to disagree with his summing up:

> In such areas as slippery slope technology and human subjects research it is obvious that the scientists and engineers directly involved in the work have no monopoly of wisdom about the ethical, psychological and societal impacts of their work. At the same time, we know that the assertion by lay groups of control over the determination of the inherent value of fundamental research could have disastrous effects....

Does my praise sound a little faint? The strength of Graham's book is in the case studies of individuals firmly placed in their historical circumstances. It is remarkably interesting to get a clear and just assessment of the cultural stance of the well-known scientific personalities who have expressed themselves at length on the world outside their own intellectual backyards. In fact there are several others who might have been given the same treatment—for example, Bertrand Russell, the most influential serious writer on science and philosophy of the inter-war years, or J. D. Bernal, who turned a whole generation towards the "social function of science." I would trust Graham's considered opinion on such controversial figures, for he keeps an open mind and heart, and is well aware that "seen in its social context, science is far from value-free," and that it is in the "relationship between science and society where the historically interesting value conflicts arise." In these respects his is an admirable work, in context, style and attitude.

Nevertheless, the earnest reader might come away from it a little disappointed and frustrated. Graham does not pull his punches against outright silliness: Henri Bergson and Teilhard de Chardin get the treatment they deserve, for obscurantist waffle and uncritical zeal. But the philosophical position on which he himself stands is not well defined. His notion of "values" is so extraordinarily diverse that it can only be described as whatever people think to be important, except what they call science. But then he does not characterize science as such, except by particular disciplines drawn from the standard academic list—the sort of subject that might be taken in the Natural Sciences Tripos of Cambridge University, augmented by selected topics from the Mechanical Sciences and Medicine. But there are many excellent scholars and teachers who claim that this traditional classification of academia is merely an administrative

boundary, which arbitrarily cuts across many well-founded interdisciplinary subjects, such as psychology, archaeology, anthropology, geography,
psychiatry, community medicine, architecture, ergonomics, and even such
"value-laden" disciplines as economics and history. This comes near to
saying that the title of the book begs the question of its theme. The distinction between "restrictionism" and "expansionism" might be no more
than semantic. Or it might refer to some party slogan in a local scholarly
controversy: "my psychiatric techniques are more scientific than yours
(and therefore must be right, etc.)." Without some attempt by the author to
state his own "demarcation criterion" for scientific knowledge and scientific work, his every word is up for the logical chop.

I am not suggesting that this deficiency could easily be filled. Sir Karl
Popper proposed such a criterion ("potential falsifiability") nearly fifty
years ago, but it has not proven a sharp enough knife to dismember the
body of human knowledge into "science" and the rest. There are reputable
meta-scientists who argue very persuasively that even mathematical physics is socially "relative," and hence essentially "value-laden." Strictly
speaking, this emphasis on the corrigibility of all science, and its ultimate
dependence on human judgment, is perfectly justified. But so also is the
opinion that some of our practical knowledge of human behavior, even in
its ethical and aesthetic aspects, is as reliable as any "scientific" result obtained with electronic instrumentation and computational statistical analysis. In other words, there is a deep and serious contemporary debate covering the whole field of the history, philosophy and sociology of science,
on how to think about the relationship between "science" and "values,"
which Graham scarcely touches.

For this reason, the cable with which he tries to bind this subject together is not as strong as it seems at first sight. By accepting the implicit
definition of science used by each of the writers whose work he describes,
he is bound to voice the "local," "folk" opinions of the scientists in that
particular discipline, whether or not those are consistent with one another,
or with opinions about the scope of science in other scientific fields. Is
there broad agreement between biologists and physicists as to the scope of
"science?" The reductionist program—the ultimate in scientific expansionism—is much more of a threat to neighboring disciplines, from chemistry to molecular biology, than to distant socio-political "values." Graham correctly seems many manifestations of scientism as the
appropriation of scientific knowledge or prestige in the realm of politics,
to bolster up or attack social values or prejudices: one should also see this
sort of argumentation as typical of what goes on at all levels, and on all
scales, within the scientific world itself. To put it in its simplest form: a

"scientific fact" is what we (and everybody else) are entirely agreed on; a "value" is what we all know to be uncertain, and therefore a matter of personal opinion or taste: it is just in the no-man's-land between them that the action is. A realistic policy of expansionism, with limited but attainable objectives ("The Art of the Soluble"), is the mystique of the scientific life: that is the social reality whose ideology is caricatured by extremists such as Monod and Skinner and which has had to be protected by apologists such as Heisenberg.

This takes us to a point from which one can survey the whole scene. Graham correctly draws attention, in various places, to the intellectual and ethical values associated with the pursuit of science—not merely what it does materially for society, but also for its "commitments to harmony, protection, truth-seeking, and elegance." But he does not comment on the supreme principle of the scientific ideology: the acquisition of knowledge is an absolute good. The familiar doctrine that pure science should be done "for its own sake" is entirely meaningless, except as a justification for a scientist intent on doing exactly what he likes—a harmless conceit, provided that somebody will pay him to pipe this particular tune. But it is often given the more totalitarian form of a license to undertake any research whatsoever, regardless of the consequences, as if this were an activity taking priority over all others. In the chapter on "limits of inquiry," Professor Graham himself implicitly repudiates this principle, showing how it impinges on a variety of other ethical principles that most of us hold dear. But he does not climb up this central peak in the territory he surveys—the conviction that many scientists hold that in their work they are living out the highest "value" of all human values, both in itself and for what it produces. That is the hubris that this book skirts around, and wisely combats in many forms, but somehow fails to investigate and explain.

THE WORLD
BEYOND SCIENCE

Scientific research is a very self-contained and self-regarding profession. Unlike law, medicine, or religious ministry, it does not involve daily contact with the ordinary lives of ordinary people. Even in their teaching roles, academic scientists often communicate with their students only as if with fellow explorers in the abstract domain of knowledge, ignoring the fact that they are also fellow inhabitants of the everyday world. The much-derided ivory tower is actually valued as a sheltered place where they can gather together and devote themselves, without distraction, to their infinitely fascinating and challenging vocation. They even take a certain pride in the monastic virtues that flourish in such closed communities.

For reasons that we have already discussed, this tribalism is no longer permissible. Scientists are having to take much more account of the demands and values of the world beyond science. But it is still remarkably difficult for them to relate their own self-image to what nonscientists actually think about them and their work. The difficulty of communicating the significance of this work to the general public is quite understandable. Their Faustian image is regrettable, but arises from concerns that they themselves often share. What they find incomprehensible—and perhaps invincible—is that the great majority of people have almost no interest in their achievements or motives. Scientists simply don't understand what it is about themselves that is seemingly misunderstood.

This deficiency of reciprocal empathy between scientists and other people is the missing term in many formulae for opening paths between the world of science and the "real" world. It cripples many earnest endeavors to improve public understanding of science, whether voluntarily through the media or compulsorily through formal education. This well-intentioned cause, which is now actively commended and supported by most scientific, industrial and political authorities, is really up against much more formidable obstacles than its advocates realize. But the deeper sociological analysis that the whole subject deserves would be a diversion from our central theme.

Scientists are also surprised to discover that their out of area operations incur grave social responsibilities. Tribal mottos such as "knowledge is a good thing in itself" or "truth is the ultimate virtue" are not acknowledged as supreme by the world at large, and come into conflict with other ethical norms, involving terms such as "love," "beauty" and "honor," that do not appear in the communal ethos of science. It apparently requires an effort of imaginative humility to realize that research is not in any way a privileged calling and that scientists are subject to exactly the same moral imperatives as other social beings.

The advocates of scientistic technocracy—Plato's philosopher kings armed with computers—have always glossed over this basic ethical principle, which has had to be rediscovered and defended in every new generation. Essentially the same issues were being debated, in somewhat different language, more than a century ago, when science was much less influential, but was generally believed to be more benevolent, than nowadays. At that time, it was less extravagant, and apparently less risky, to allow scientists to follow their own bent. Under these conditions, a favorable equilibrium between the costs and benefits of research could be obtained by a somewhat different balance between the rights and responsibilities of researchers than would now be desirable.

The relationship between scientific goals and social values is not, therefore, a historical invariant. By the same tokens, this relationship also varies from country to country. The sociologists of scientific knowledge would discover more about science if they shifted their relativist instruments away from its epistemological status and focussed on its cultural role. Scientific discoveries are judged by criteria that are as universal as we can possibly make them: research projects are assessed by criteria that depend significantly on the economic, political and social conditions under which the research is to be undertaken and its results applied. The place of science in a developing country such as India is not the same as it is in a wealthy country with an advanced industrial economy.* This ap-

---

* I have also discussed the growth and spread of science in developing countries in several of the papers reprinted in *Puzzles, Problems and Dilemmas* (Cambridge University Press: 1981).

plies, for example, to conditions such as the connections between the academic and industrial sectors of society. Consequently, it is unrealistic to pattern the rights and responsibilities of research scientists precisely on those that have evolved in Western Europe and North America, where these conditions are quite different.

Everywhere that science is practised, it is taught in schools and universities. But this teaching normally reinforces scientific elitism, technocracy, and the stereotype of the ivory tower. Those who successfully follow the traditional science curriculum are confirmed in their own commitment to science as an independent estate: those who reject or are defeated by this aspect of education become hostile or indifferent to a form of life that they feel they could never understand. A radical reform of this curriculum is essential, both for scientists to come to terms with the real world and for the citizens of the real world to come to terms with science. This, again, is a major topic, which I deal with at length in *Teaching and Learning about Science and Society* (Cambridge University Press: 1980).

Scientists—and Other People

Mice and Elephants

At MIT in December 1982, I was asked to speak to the theme "Science, Technology, and Everyman's Life." Having put a certain amount of effort into the study of the *scientist's* life, I felt a bit like the student who had prepared for a zoology examination by learning everything about the Mouse, and was then asked the question: "Write an essay about the Elephant." He was equal to the occasion. "The Elephant is *not* a Mouse," he wrote. "The Mouse is a small nocturnal rodent..." and so on. Well, here goes. "Everyperson is *not* a scientist. The scientist is a quiet, inoffensive person whose life is very different from that of other people..." and so on.

In fact, there is some virtue in starting like that. All science, including a great deal of what we have come to call "technology," begins amongst scientists, and is initially shaped by their personal experience and view of the world. It comes to other people in forms that may not be well matched to their experience, needs or comprehension. This mismatch is the source of many of the social dysfunctions of science, to which considerable attention is now being directed in all the fields of science studies to which Derek de Solla Price made such notable contributions.

The social environment of the traditional scientific life is restricted and specific. The typical academic scientist lives in a narrowly confined and remote domain of interests and activities. In the cognitive dimension, only formal concepts are recognized as valid; in the technical dimension, only the means for experiment and observation are required; in the educational dimension, graduate training for research is the main interest; the informal invisible college takes precedence over other associations in the social dimension; political questions are thought about mainly in relation to the policies of grant-awarding agencies; and in the personal dimension, of course, other research scientists are esteemed more highly than technologists or, say "mere administrators." The range of these interests in each dimension is evidently quite limited. On a notional six-dimensional "map" they would define a very small region—a distant "pole" of the life-world sphere—where the average scientist can feel at home. As one moves outwards from this region, in almost any direction, one finds oneself in less and less familiar territory, subject to forces and perils over which one has less and less control.

Within this domain, however, scientists live and work—usually with great enthusiasm and considerable personal gratification—at the activity of research. They have little concern for sophisticated philosophical or sociological theories of the nature of science: in their life world the process of "discovery" is highly personal and realistic. The disappointments and drudgeries of observing, experimenting or theorizing about some aspect of the natural world or about some specific practical problem are more than compensated for by a moment of inspiration in which some obscurity become clear, or some entirely unexpected phenomenon is disclosed. The results are written up according to the conventions of scientific publication, or are incorporated in some technical device or industrial product—and..."What problem shall we tackle next?" Of course this conception of science is flagrantly egocentric. It completely ignores the social environment in which this activity is fostered or permitted, and takes no account of the communal factors that shape the research process and its products. Nevertheless, this is the personal experience that scientists themselves have traditionally brought to the understanding of who they are and what they are doing.

So much for mice: now what can be said about elephants? The notion of "Everyperson" is, or course, categorically unspecific. For every "scientist" there are hundreds of different types of "non-scientist," in correspondingly numerous and diverse occupational niches—lawyers, sales clerks, deep-sea fishermen, TV newscasters, and so on. But all these very different people in all these very different callings share a common characteristic: they are almost completely unfamiliar with the scientific life. They occupy, so to speak, a whole continent, at the opposite "pole" of the six-dimensional life-world sphere from the domain of the scientist. In the cognitive dimension they acknowledge the primacy of imprecise "values"; in the dimension of technique, they are mainly interested in practical innovations; for them education is valued for breadth and generality; their activities in the social dimension of science are mainly through "interest groups," concerned about public political issues; in the personal dimension, they esteem the managers and administrative decision-makers in politics and industry whom they recognize as having far more influence than the technologists and researchers. In other words, most people live in a large domain that is a distant as could be from the home territory of the average scientist.

In innumerable other dimensions of the life world, academic scientists are not, of course, so very different from other people. They can be as chauvinistic, homeloving, class-conscious or taxable, say, as the best, or the worst, of mankind. Nor should we overlook the significant variations of habitat between academic science and, say, technological development in industry or the practice of clinical medicine. The point to be emphasized is, simply, that these specialized scientific life-world domains must seem alien and uncomfortable to most of the "other people" who are the eventual customers, or victims, of science and technology.

Diverse Conceptions of Science

What is "Everyperson's" conception of science as a social activity? To judge by newspaper editorials, company reports and political speeches, it is a mysterious black box, labelled "R&D," by means of which "problems" are provided with "solutions." Money and skilled personnel are poured into this box in surprisingly large quantities, and there go through various unaccountable processes in which there seems an awful lot of wastage of resources and effort. Sooner or later, some message comes out of the box, in the form of advice, or a blueprint for a new device, which is sometimes astonishingly effective in dealing with the issues in question, or may even provide the means to do quite novel things which had never been thought possible. Sometimes, unfortunately, the supposed problem is not solved at all, or the suggested solution leads to even more difficulties, but that is reckoned to be the price one must pay for "progress." In other words, the public conception of science is highly instrumental: it is focused upon the ends that might be achieved by science, and scarcely at all upon the means by which these or other ends might be reached.

This conception of science is obviously far from the whole truth. It may not be inappropriate for a high-level decision-maker—the chairman of General Motors, say, or the Deputy Minister of Trade and Industry—to think about research in that sort of way, within a larger context of corporate or national effort. On the average, over a long enough time, science can be relied on to produce some reasonably useful answers to some of the questions towards which it is directed, and to earn its keep with vast quantities of other information a small amount of which may later turn out to be extraordinarily valuable to somebody, somewhere, some time. For many of the issues of science policy, this naive instrumental conception of science may well be adequate.

But very few of us are in any position to influence policy on that scale, or to apply science very directly to the solution of our own personal problems. The political decision-maker's conception of science not only fails to take account of the personal and social processes by which the results of "R&D" are actually generated: it neglects important channels of interaction between science and society. The direct technological influence of science on all our lives is, of course, of immense importance. But there are less tangible influences of scientific knowledge and ways of thought upon the more symbolic aspects of our culture—our shared ethical values and our conflicting ideologies. And one of the characteristics of totalitarianism is the attempt to make science a willing instrument of the ruling group by blocking all the public channels through which requests for advice, unsolicited warning whistles, demands for responsibility, and plaintive cries for freedom flow back and forth between the scientific community and society at large. Considered simply as a major social institution, science is much more than a rather elaborate, expensive and uncertain means for carrying out the practical intentions of those who pay for it and claim to control it: it is connected in a variety of ways to people and institutions throughout society, in schools, churches, lawcourts, editorial offices, hospitals, council chambers, shops and barracks.

Indeed, when we take a look inside the "black box" of R&D, we begin to realize how complicated it really is. At its very heart, the discovery process of research is still ticking away, like the escapement mechanism of a clock, driven by the main spring of intellectual curiosity, wound up by practical need, and balancing back and forth under the essential tension between criticism and imagination. Imagination is a personal trait, but the critical forces are supplied mainly by the community of peers to whom the results of research are primarily addressed, and who reward convincing discoveries with recognition. In recent years, the dimension of *apparatus* has become more and more important in shaping and directing the research process: even highly academic science is now carried out in organized groups and teams, sharing facilities or collaborating in research projects. Science has always been much more of a social enterprise than it appeared to the individuals taking part in it; nowadays, we may say that it is being *collectivized* in a variety of aspects, despite the very strong personal motivations and gratifications of the process of discovery itself.

Further study of the R&D black box reveals complex mechanisms on a larger scale than is immediately apparent in the everyday life of the research laboratory. Collectivized science is normally carried out within the managerial framework of a more extensive organization, such as an industrial firm, a government department or a university. A research institute or laboratory is seldom an isolated, self-sufficient social entity: it has to be considered as an administrative unit, as a budget item, as a group of highly trained experts, as well as a source of potentially valuable knowledge. The primary mechanisms of research and discovery do not lie close to the surface of the black box: they are, so to speak, embedded in a matrix of managerial structures, policy considerations, media of communication and education, and technical capabilities, which insulates them from direct societal influences.

The image of science conjured up by these remarks is evidently much too complicated to be understood as a precisely articulated "model." But it does indicate the length of the list of topics that might need to be taken into account in any serious analysis of modern science in its social context, and reminds us of the various possible relationships and interactions that might exist between them. That is what makes it so difficult to teach properly about "science, technology and society:" one has to draw from many academic disciplines, including philosophy, psychology, sociology, history, politics, and economics, not to mention the sciences and technologies themselves. No wonder scientists, and "other people" find it difficult to relate to one another, from one pole to the other of the life world.

The Dense Interface Between Science and Society

Would such a long list of relevant topics, arranged in such a complex scheme, have been needed if we had been talking about this subject a century ago? Surely not. In many ways, the academic science of the past was much simpler than it is today. It was not so tightly bound to society at large, and was seemingly more isolated from salient social concerns. But that does not mean that it was uninfluential or unresponsive. Its societal connections were no less direct than they are today, but not nearly so thoroughly organized. All that heavy machinery of technological development, of research management, of science policy, and of science education had not yet been set up to drive science from the outside or to be driven by it. We can easily be misled about contemporary science if we think that it is somehow just the same as it has always been.

This applies, in particular, to the regions that separate the life-world domains of scientists and of other people. In the past, these domains may not have seemed any closer together than they do today, but the space between them was not, so to speak, so densely occupied. The biological investigations of Charles Darwin, the mathematical insights of James Clerk Maxwell and the inventions of Thomas Alva Edison were far removed from the ordinary comprehension of everyday concerns of their time. They must have seemed quite as mysterious to most people as the structure of DNA, the theory of black holes, and the manufacture of microchips do to most people today. Scientific knowledge was remote, but it was not wrapped up in layer upon layer of people, products and practices. It did not have to pass through such formidable and magisterial organizations as government funding agencies, multinational industrial corporations, university administrations, or the institutions of mass communication, before it could affect the life of Everyman or Everywoman. It is not so much that the modern scientist is screened off from society, and totally uninformed about its values, desires and needs: it is that these are viewed through a succession of semi-transparent screens, obscuring and refracting the transmission of sympathy and understanding.

This is a rather general point, which seems fairly obvious historically, and could perhaps be made even more obvious by reference to abstract social or philosophical theory. For the present, it serves as a convenient theoretical scheme by which to explain some of the difficulties that seem to arise in the relationship between scientists and other people. A few specific examples will show the value of this perspective in dealing with some familiar issues. In particular, it indicates that scientists' understanding of the public may be just as deficient as the lack of public understanding of science which is so often deplored.

Science in the Media

Consider first the notorious phenomenon of mutual misunderstanding and mistrust between working scientists and working journalists. Ordinary scientists are scornful of the representations of scientific knowledge presented on TV and radio, and in the Press: conversely, ordinary journalists are scornful of the efforts of scientists to explain their work to the general public. The good intentions of each part towards the other may be perfectly sincere, and yet there is often a pathetic mismatch in the transfer of information across this important boundary between science and society.

To understand this mismatch, we have only to look at the conditions under which scientists normally communicate their research results to one another. A primary scientific paper is a very peculiar, highly stylized communication. It is addressed to the very small group of people who are already knowledgeable on the subject in question. It must be written in the established terminology of that subject, and must be based as firmly as possible on pre-existing knowledge. It has to be phrased very cautiously and tentatively, so as to persuade the most learned and skeptical reader that the author has got it right. By contrast, a popular article or broadcast seeks the widest possible audience, and must therefore be expressed in the very simplest and most comprehensible terms, assuming the minimum of previous knowledge. Its main purpose is to arouse interest, or to entertain, or to convey boldly some more or less significant message about life and the world. The styles of expression that become habitual in these two distinct domains of human action and experience are so contradictory that the frontier between them can only be crossed by a deliberate act of imagination, which very few working scientists or journalists have the wit or facilities to undertake.

This stylistic incommensurability between scientific and popular writing has always existed. It arises from the logic of the situation, and cannot be wished away. But in the past the frontier regions and the paths across them were not so clearly established. A few leading scientists found that they had the gift of writing and talking about science in a way that other people could understand and appreciate. Through the medium of "popular" books and lecture they could address themselves directly to a relatively large clientele of moderately well-educated people who were eager for instruction and enlightenment. At the annual meetings of organizations such as the British and American Associations for the Advancement of Science, members of the public could actually sit in on technical debates between scientists on controversial issues, and thus get some feeling for the scientific culture in action, even if they did not fully understand what was really at stake. Not only did such events get a great deal of general publicity: they were also taken seriously as professional meeting places by the scientific community, and were not regarded, as they mostly are now, merely as occasions for carrying out the specialized function of showing off science to the people.

Nowadays, however, the popularization of science has been thoroughly institutionalized. It has proliferated into an elaborate apparatus of magazine, feature articles and broadcasts, produced and run by fulltime experts. Science journalism is an admirable and demanding profession, fulfilling a vital social role. The technical excellence of journals such as *New*

Scientist and of specialized science broadcasting units such as those of BBC TV and radio is an admirable feature of our era. But apart from an occasional star performer such as Jacob Bronowski or Carl Sagan, most science journalists and popularizers have had only limited personal experience of either the scientific life or of down-to-earth popular journalism. They may have had a scientific education and done a little research, but they seldom know what it is to go on for years collecting inexplicable data, or to be a member of a team developing a new technical device, or to have the conviction that one has made an important discovery. On the other hand, they may have worked for a while on a national newspaper, but only in the role of a "specialist" whose career is devoted to collecting material concerning this esoteric and not very entertaining aspect of life.

Modern science journalism does, indeed, act as bridge for the transfer of information about science from scientists to other people. But is does not really bring these two domains of life and meaning into closer contact. The information is not carried by personal messengers from the heart of one domain into the other: it is passed on from person to person—or even from team to team—and the messages are distorted and transformed along the way. A scientist may well agree that a popular account of his or her work is more or less accurate—except that it misses just those fine points of intention and achievement which made it exciting. On the other hand, the lay person may feel that the reported discovery makes some sort of sense—but may still remain quite uncertain as to its significance or relevance, in this world or the next. It is very rare indeed that one reads a book or watches a TV program that conveys an authentic image of the content, or manner, of scientific research. It is equally rare for such a book or program to throw scientific light on the deeper interests and concerns of ordinary people. Science journalism does its best to bridge the comprehension gap, but by its sheer professionalism and institutional specialization it may have the paradoxical effect of making the barrier between scientists and other people even denser and more impenetrable.

Science Education

The institutions of science education have a similar effect of connecting science more firmly to society and yet erecting a barrier between scientists and other people. There was a time when the teaching of basic science was a matter of considerable concern in the scientific community, some of whose most distinguished members might even be teachers in secondary schools or technical colleges. But science education is no longer very closely linked with the research community except as the final authority on examination standards. It has become a major profession in its own right, with its own training institutions and its own professional associations. There cannot be any doubt that this vocational specialization has proved absolutely invaluable in bringing an elementary knowledge of science to the mass of ordinary people and in producing the great cohorts of scientifically trained experts needed for our advanced technological civilization. But the science teacher, like the science journalist, has usually had only marginal contact with the scientific life, and cannot speak about it from first hand experience. Modern science education smooths the way to mastering the contents of scientific and technological knowledge, but says practically nothing about the way this knowledge was acquired in the first place, or how it relates to the lives of most ordinary people.

Although designed to instruct people in general about science, and to prepare a small proportion of students for professional research careers, the traditional science curriculum is completely silent about almost every social and psychological aspect of the research process. The cognitive contents of various scientific disciplines are instilled dogmatically, right up to—even into—the graduate school: then, suddenly, the student is expected to take on the entirely novel role of intellectual innovator, skeptic, colleague and competitor in the scientific community. Not only is there no preparation for this abrupt change of personal orientation: there is no attempt at earlier stages to show the meaning of this vocation to students who are not, perhaps, intending to become researchers but who are certainly interested in it as enlightened "other people."

Recent movements to reform the style and content of science education have been directed towards bridging this gap, but they succeed mainly in showing how wide it is. Some years ago, for example, an imaginative effort was made to convey the experience of "discovery" to school children, by putting them into a laboratory situation where they could re-enact the observations of, say, Robert Boyle, or Antoine Lavoisier, and then draw their own conclusions. But this effort has failed in two important respects, because this contrived situation lacked two important ingredients of a real discovery event. It has not proved satisfactory didactically, because of the impossibility of putting the pupil in the initial frame of mind where the "correct" interpretation will be generated without prompting. It also gives a false impression of the process, whose primary characteristic is that nobody knows what the answer is to the problem under investigation, or indeed whether it has any answer at all. This approach to the teaching of science can be modified in practice to perform its didactic function, but cannot avoid subtly misrepresenting the actual work of research.

Another movement, stemming from recent developments in the sociology of scientific knowledge, would have the student or school pupil learn that all science is uncertain and socially relative, to combat the positivism and scientism implicit in the way that most science is taught. This opinion concerning the status of scientific knowledge can be made plausible from a sophisticated epistemological point of view, but such teaching would certainly not bridge the gap between scientists and other people. Indeed those who have proposed it have even suggested that would-be scientists should continue to be taught in the traditional positivist style, so as not to destroy their confidence in their calling, whilst the "other people" (if they could be identified at this stage in the educational process) would benefit from a reduction in their confidence in scientific expertise. In other words, a wedge would be deliberately inserted into the gap—and hammered home. This proposal is not merely misconceived; it springs from the failure of many academic metascientists to come to an understanding of the way of life of the very group of people whom they are supposed to be studying, and also to appreciate the difficulty that most children have in learning to distinguish between a compelling argument (whether "scientific" or not), a dogmatic assertion, and a plausible speculation. Having taken up a site in the frontier region between science and society, with the avowed aim of providing a link between the two domains, this academic group is now establishing itself as a distinct discipline, claiming expertise in mediating across the boundary, but actually impeding understanding between the two sides.

Even the movement to make science education more "relevant" to everyday life tends to treat science as an instrument, without regard to its inner workings—and hence giving little indication of the uncertainties and imperfections of scientific knowledge in practice. For example, a course may be devoted to the social consequences—good or bad—of medical discoveries, without any discussion of the moral dilemmas of clinical research, or of the sheer impossibility of making sure that a new and powerful drug is absolutely safe. Such a course, taught dogmatically, can actually stand in the way of a closer understanding between medical researchers and other people concerning the nature and capabilities of their calling.

A factor that needs to be taken into account in science education is that mature scientific discipline constitutes a distinct domain of meaning which is quite foreign to the way of thought of ordinary life. This distinction is often emphasized by the academic exponents of the discipline, who urge teachers to train their pupils to "think mathematically" or to "think biologically" from the earliest stages in their education. But this rigorous attitude may be self-defeating, in that it may alienate children permanently from a subject that they cannot see the point of and they feel they will never understand. The characteristic revulsion that many intelligent and well-educated people express towards science (and even more towards scientists) is partly based on ignorance, but it can also be traced to excessive zeal by teachers who try to jump their pupils into the heart of a sophisticated scientific domain instead of leading them gently into it out of the land that they already know and love. Although science education is now an autonomous profession, and not closely linked to the research community, its standards and style are still heavily influenced by academic scientists whose own school days are long past, and who have never had the chastening experience of trying to teach their subject to ordinary children of ordinary ability.

Specialization

At the research frontier, scientific knowledge is so specialized and fragmented that it is very difficult to see the wood for the trees. The particularities of a subdiscipline or problem area loom large. Techniques and concepts that are "locally" successful are given great weight in the scheme of things, whilst unfamiliar scientific ideas are denigrated. It is not entirely in jest that physicists belittle chemistry, or that solid state physicists speak scornfully of high-energy physicists, or even that those who compute electron band structures in the *alkali* metals take a dim view of band structure calculation for *transition* metals! This scholarly parochialism is an almost inevitable consequence of the high degree of specialization in the research enterprise, where the rational division of labor is the secret of progress.

But what Everyperson wants of science is a world view, a coherent scheme of things, a map of the earth and heavens on the widest scale. It is all very well to say (as every genuine scholar is bound to insist) that sweeping generalizations are dangerous and misleading. So they are as a basis for further research. But people who have not the inclination or the opportunity to study science in detail may still reasonably demand a broad scientific philosophy of nature, to match the claims of religious, political, and aesthetic ideologies in society at large. If, as many popularisers of science insist, there is indeed a "scientific world view," then why is it not at least sketched out in schools and universities as a single, integrated topic? Why does science education still divide and subdivide into separate "disciplines," as if the atoms that combine into molecules in Tuesday morning's chemistry lesson were entirely different entities from those that bounce off one another in the physics lesson that same afternoon? Is it simply because scientists are so narrow in education and experience that they themselves have not learnt to see their own cozy specialties as elements in a much larger (and more beautiful) picture? This is not to underestimate the practical difficulties of setting up and teaching a more integrated science curriculum that would not be segmented along the joints of the specialized disciplinary traditions. The main point is, once more, that the institutions of science education, bridging the gap between the scientific and lay domains, have been slow in bringing these two domains into close contact on this fundamental matter.

The same general point is illustrated by the difficulties that research scientists get into when they are asked for expert advice on practical matters. To other people, the scientist who has spent half a lifetime researching into some obscure aspect of nature, unveiling many mysteries and disclosing many marvels, must be the ultimate authority on all questions concerning that particular realm of being. Even if they do not know the precise answer to every question that might arise, they certainly ought to be able to give a well-founded opinion on what sort of answer they would expect to get if the problem were properly investigated.

This expectation is often sorely disappointed. What the lay person seldom appreciates is that the research scientist is not paid to be either a learned scholar or a clinical consultant. Research is "the art of the soluble:" in any field, attention is focused primarily on puzzles or problems that can be sufficiently narrowly framed that they might yield interesting solutions, or open the way to further progress in new directions. The successful research worker needs to know only enough to identify worthwhile problems and think of the best means of tackling them: a great many other dull or apparently irrelevant facts are best ignored or forgotten. Scientists are aware of their duty to be well-informed about their subject in general, but it is so difficult to make any progress at all in real research that their attention is inevitably focused into an area that is much smaller than the knowledge that would be needed to answer many practical questions that have not been previously formulated and studied.

Consider, for example, the question whether the accumulation of aerosol propellants in the upper atmosphere would soon destroy the ozone layer that shields life from excess ultraviolet radiation from the sun. Although there were experts on the chemistry of fluorocarbon compounds, on the diffusion of gases in the atmosphere, on the chemistry of ozone, on the absorption of radiation by the various constituents of the atmosphere, etc., etc., this complex chain of reactions and interactions did not lie within the scientific grasp of any single individual, and seemed at first to lie beyond the bounds of atmospheric science. Of course, in the time since the issue was first raised a large research effort has clarified many separate steps in the argument, and a reasonably coherent answer can now be given. This is not to suggest that somebody should have anticipated this very difficult and highly speculative question long ago, and prepared themselves to give an answer to it. The point is that the logic of the research enterprise as a whole makes most scientists so highly specialized in their interests and expertise that they simply are not competent to give broadly based advice on most of the practical problems that arise in real life, and should not be expected by "other people" to do so. This may be just as significant for science in society as the much more familiar case of the scientist who fails to appreciate the political or economic aspects of such problems, and can only offer impractical solutions of a narrowly technical kind.

The difficulty of getting advice out of science is compounded by the "non-scientific" character of the questions that are usually posed. Research projects are addressed to questions that are framed in specific terms within the categorial scheme of the discipline—for example, "determine the rates of all reactions between fluorocarbons and atmospheric oxygen and nitrogen, in the presence of ultraviolet radiation, and hence calculate the concentration of ozone at various heights in the atmosphere." But what Everyperson wants is an estimate of the effects of this on the life expectation of ordinary people, or whether it would really benefit later generations if we all had to give up self-propelled shaving cream and deodorant sprays. The scientist's experience and intuition is that these questions are so lacking in precision, and involve so many factors of ignorance or uncertainty, that they cannot be honestly answered at all. A major political innovation of our times is the creation of institutions such as the US Office of Technology Assessment to give careful consideration to all such factors, and thus to mediate between the scientific and lay versions of expert technical advice. Such institutions, whether official or unofficial, are beginning to play a valuable role in public affairs, but it is not clear how an institutional bridge of this kind can succeed in making closer contact between scientists and other people along this important social dimension.

Disjoint Values

Even in the realm of material technology, where all scientific work is supposed to be subordinate to practical utility, there is still a big gap between the Scientist and Everyperson. What Everyperson values for what science can do, the Scientist values as the embodiment of concepts and techniques. No phrase of our era echoes more sinisterly than Robert Oppenheimer's famous reference to the decision to build a hydrogen bomb: "Whenever a solution looks technically sweet, there will always be scientists to work on it." A device or system or therapy that performs according to its design specifications is a successful confirmation of theoretical predictions, and of mastery of the art. The gratification that this can bring to the scientist/technologist is often far greater than it could ever give to any other person making use of the product or benefitting from its use.

In the end, no technology can succeed unless it is consistent with Everyperson's life—in sickness and in health, in good and also in evil. If this is not already part of the design philosophy when an invention is first conceived, then it must certainly be built in during development and manufacture. Nowadays, perhaps, there is a deliberate effort to mediate between the technical conceit of the engineer, producing a "clever" or "elegant" design, and the practical humility of the customer, seeking ease, comfort, safety and reliability of operation. Nevertheless scientists and engineers may be so carried away by the ingenuity of their creations that they may be tempted to push the values of other people in the same direction. The doctrine of technological determinism—that all things possible ought to and will be effected—is now out of fashion, perhaps because it aroused no significant response in Everyperson's heart. But it expresses the fundamental ethos of scientific technology, just as the doctrine of "science for its own sake" expresses the fundamental ethos of academic science. Here again, the movement for technology assessment can be seen as an attempt to bridge the gap between the domain of the scientist or engineer and the life-world domains of other people, whilst the more radical movement for "appropriate technology" calls attention to this gap, and seeks to close it up by subjecting the technological innovator to the immediate realities of everyday life.

Social Responsibility in Science

Finally, my general theme is perfectly exemplified by the issue of social responsibility in science. Everyperson sees that science has produced both benefits and disbenefits, and insists that in future it should act more "responsibly" by producing only benefits. Taking a purely instrumental view of science, this seems a perfectly reasonable requirement. Scientist, on the other hand, knows that research is a process of discovery, whose outcome is inherently unpredictable. To ask for "responsibility" in such circumstances is as illogical as asking for "profitability" in a game of roulette. If every step in research, technological development, and industrial innovation were planned according to cost-benefit criteria, it is doubtful whether anything new would evolve, in science or in society. Everyperson, valuing safety, is poles apart from Scientist, who values and exploits risk.

It must be an important objective of our civilization to build a bridge across this disjunction of values and attitudes. To some extent this is achieved by the large scale institutions of the R&D system, which permit the individual scientist considerable autonomy and "irresponsibility" whilst moderating and directing the research process as a whole towards desired ends. But this makes the very dubious assumption that large bureaucratic organizations such as industrial corporations or government departments are essentially more "socially responsible" than individuals—that is, that they are more competent at predicting the beneficial and damaging consequences of their actions, and more securely oriented towards the former than towards the latter. It also raises the old query about who shall guard these particular guardians. In other words, this is an inescapably political issue, which cannot be resolved by purely administrative devices. The institutions that mediate between scientists and other people now lie close to the power centers of society, and cannot be exempted from political conflicts and interventions.

Nevertheless, the plea for social responsibility in science is not altogether vain. The scientific life has developed a protective ethos, which purports to excuse scientists from many of the normal responsibilities of Everyperson's life. Technocratic elitism, the doctrine of "science for its own sake," the conception of research as the unfettered pursuit of entirely unpredictable discoveries, the idealization of the "scientific attitude" and other scientistic notions, combine into a powerful ideology which separates scientists from other people in the moral sphere. This ideology is foolish and dangerous, for although scientists are often very talented individuals, and although their collective contributions often have profound social consequences, they have no right to suppose that they should be privileged above other useful, creative, law-abiding citizens, such as lawyers, sales clerks, deep-sea fishermen or TV newscasters. In a pluralistic democracy it is important that their personal rights should be fully respected, and that they should be encouraged to express their opinions publicly on issues where they have special knowledge or special interest—but no more than Everyman or Everywoman, whoever they may be.

I have argued that "other people" should try to understand more about science, and be brought into closer contact with the peculiar way of life of scientists through education, the media, administrative arrangements, commerce, and politics. All that is perfectly true. But it is just as important that scientists should not be disdainful of other ways of life than their own, and come to appreciate other values, other crafts, other achievements and other needs than those of scientific and technological research. Elephants, one might say, need to know something about mice, of whom they are said to be quite inordinately frightened: equally, mice need to understand the ways of elephants, who have the power to trample them into the ground if ineptly provoked.

The Social Responsibility of Scientists: Basic Principles

Introduction

Why should scientists, of all people, have to be told to be socially responsible? By their upbringing, they ought to be highly responsible individuals, strongly oriented towards communal goals. Every scientist must go through a lengthy social process of advanced education. Every scientist must learn to take part in the intricate social activity of research. Every scientist seeks social recognition of his or her personal reputation from the scientific community. And yet, it must be admitted, these most refined representatives of our civilization seem at times to lack the most elementary sensibility of civilized people—consideration for the feelings and needs of their fellow citizens.

The conventional way out of this paradox is to say that scientists are not properly educated. It is said that they are not taught enough about the importance of value judgements in human affairs. This may well be true, but it applies equally to almost everybody in the modern world. In our democratic, pluralistic societies, we give very little formal public instruction in the basic principles of social responsibility—that is, in the basic principles of ethics, religion and politics. It is hard to believe that the average accountant, or magistrate, or factory manager, or town clerk, has had significantly more exposure to religious education, or church sermons, or party political broadcasts, or coffee house debates on contemporary issues of conscience, than the average lecturer, or medical research associate, or aeronautical engineer. The ethical codes and moral imperatives that bear upon every public issue are so convoluted, diverse and contradictory that we avoid them in academic teaching, and leave them to be acquired by personal experience. Some humanistic disciplines of higher education, such as philosophy and history, do offer specialized training in the analysis of theologies and ideologies, but at a far more advanced and abstract level than is needed for the real moral dilemmas of everyday life.

For this reason, the orthodox wisdom that scientists ought to be specially trained to be socially responsible is not so easy to put into practice. Scientists are often implicated in issues of great public concern, and may be called on to look deeply into their own hearts in deciding how to act upon them. It is important that they should be aware of this professional hazard, that they should be well informed of the wider social context of their work, and should have had some opportunity to rehearse in advance some of the characteristic dilemmas which they may have to face.[1] But that does not mean that we can provide them with a ready made code of all the basic principles of social responsibility—compounded, presumably, of the Ten Commandments, the Sermon on the Mount, the Koran, the Communist Manifesto, the novels of Tolstoy and Dostoevsky, the plays of George Bernard Shaw and Berthold Brecht, and the writings of selected moral philosophers, from Plato to Bertrand Russell—to meet every eventuality. It is good that scientists should be as well educated as any responsible citizens on such matters, but that is no solution to our paradox.

The real obstacle to social responsibility in science is not that scientists are peculiarly ignorant or insensitive about ethical questions, but that they acquire an armory of precepts by means of which they defend themselves from the discomforts and dangers of social action. These precepts are not all consistent with one another, but they tangle into a web of rationalizations which scientists pick up in the course of their education, and use automatically to shield their consciences from attack. Of course we all learn to excuse ourselves as best we can when we have behaved antisocially or irresponsibly—although such excuses do not cut much ice with us when offered by other people. The peculiarity of science is that the principles which are used to excuse social irresponsibility have been elevated into a more or less coherent ideology.[2] This ideology is not well founded, but by setting science itself above all other human values it has a powerful influence within the psyche of every scientist and in society as a whole. If we are to understand the real significance of the demand for social responsibility in science, we must analyze these principles and uncover their contradictions and limitations.

Science "For its Own Sake"

The basic principle of this ideology is that the pursuit of knowledge is the most worthy of all human activities. Simply to acquire knowledge is an end in itself. A scientist has a bounden duty to explore the universe to its utmost limits, to leave no stone unturned, to follow up every curious circumstance, and so on, regardless of any other consideration.

This doctrine is usually expressed in the form: research should be undertaken "for its own sake." That is to say, science is disconnected from all other human activities or concerns, and has significance only in and for itself. Since there is no implication that science could somehow be bad for itself, this amounts to a total commendation of all research, without reservation. The pursuit of scientific knowledge is thus absolutely justified, as if by a universal Law of Nature or a Commandment of God.

This is clearly a metaphysical doctrine which begs innumerable questions of interpretation and validity. But it is made plausible by a germ of psychological truth. A scientist who is deeply involved in the pursuit of a particular bit of knowledge can become entirely obsessed with this inquiry, as if nothing else in the world existed. In scientific research, as in other highly skilled professions, such as master chess or legal advocacy, excellence of performance calls for total concentration of effort and will for the task in hand. The excellence thus achieved is to be encouraged and admired, if only for its aesthetic value. Considered simply as a life-long game, research is personally beguiling and satisfying for those who can do it well. No wonder many scientists want to be left alone to do their own thing, to follow their own devices, to solve the problems that they set themselves, without interference by outsiders.

There is, however, a world of difference between a professional mystique and an ethical code. Good scientists may lean heavily on a mystique of personal devotion to research to strengthen their resolve and maintain high standards of performance, just as good soldiers lean heavily on the mystique of obedience to orders and personal devotion to duty. But such a mystique is essentially a "myth," which ignores the social significance of whatever is done in its name. Taken to its extreme, it is as individualistic as Nietzsche's cult of the superman, and just as antisocial.

All Science is Good Science

In practice, the circular doctrine of science "for its own sake" is challenged whenever a method of inquiry offends against conventional ethical norms. This was obvious, in the most notorious extreme, when Nazi doctors undertook medical "research" on the inmates of concentration camps, quite regardless of the additional sufferings their "experiments" inflicted. At a less sinister level, this is the thread running through all controversies about experiments on human subjects, leading to considerable elaboration of the protocols for obtaining "informed consent," whenever there might be risk to life and health. The same principle applies to experiments using living animals,[3] at least in any country where the welfare of animals is held to be of genuine moral concern. In all such cases, cruel, dangerous, or otherwise distasteful research techniques cannot be justified simply on the grounds that the acquisition of scientific knowledge is of absolute value, regardless of the circumstances. The blanket principle that "all science is good science" is seen to be threadbare and unconvincing as soon as it is challenged on a specific issue of this kind.

The argument really hinges about the traditional ends/means axis of moral philosophy. According to this principle, one may justify the use of a relatively dangerous or obnoxious *means*, if this seems the only way to reach highly beneficial *ends*. Unless all our actions are hedged in by absolute imperatives, such as the strict Jain prohibition against killing animals, we are permitted to strike an ethical bargain, in which the potential human benefits of research outweigh the perceived human costs. But this is not a grand contract, licensing all research whatsoever, on the grounds that science has, on the whole, proved of positive value to humanity. The balance of advantage must be determined, and shown to be favorable, for each specific investigation that we propose to undertake. That is to say, we are morally obliged to give conscious attention to the objectives of our research, not simply in terms of its "scientific merit" but also as a potential contribution to whatever is highly valued in our society, such as individual good health.

Of course any such calculation is exceedingly vague and uncertain. It is at the mercy of the unpredictability of the outcome of every experiment, and of the diversity and incommensurability of our most cherished social values. In many fields of science, the best that can be done is to establish a conventional code embodying a notional ethical balance between our slightly dubious means and our somewhat distant and hypothetical ends. It is quite clear, however, from our experience with such codes that research cannot be undertaken just "for its own sake," since it often requires to be justified on more specific grounds. This basic ideological principle is thus manifestly false.

Scientific Inquiry Can Know No Limits

An "innocuous" scientific investigation is not necessarily beyond question. Even if the technique of a research is entirely harmless and inoffensive, it may still be judged morally objectionable on account of its objectives. Although we are always at the mercy of accidental discoveries which might do great harm, we are wise not to deliberately court disaster by directing research into areas where such discoveries are most likely to be made.

Many scientists dispute this opinion. It is held that scientific inquiry can know no limits, and that there is no question that a scientist may not legitimately ask and try to answer. In other words, every scientist should be free to set the goals of his or her research, without external constraint.

This doctrine arises from bitter historical experience. From its beginnings in the 16th century, science has had to establish itself against other bodies of organized knowledge, such as revealed religion, which have claimed monopoly rights over certain areas of fact and opinion. Freedom of inquiry, which had to be established as of right against arbitrary intellectual authority, is now one of the most cherished freedoms of the open, pluralistic society. There is good, practical, social sense in the principle that nothing is altogether too sacred to be beyond factual observation and sincere criticism. As the epitome of critical rationalism, science has proved itself a peculiarly effective instrument for such investigations, and is therefore regarded as a peculiar danger by the protectors of obscurantism and privilege. Freedom of scientific inquiry is closely linked with freedom of opinion, freedom of speech, freedom in teaching and learning, and other basic human rights which must be constantly defended.[4]

Nevertheless, these rights are not absolute. "Freedom of speech" is not an open license for slander. The scientific norm of "organized skepticism" is not an open license for all research, regardless of the consequences. As we have seen, there is no way of justifying the means of inquiry without reference to its ends. If those are dangerous, or malicious, or otherwise socially undesirable, then any research directed towards them can properly be called into question.

A scientific investigation is not a purely private act of thought; it is a deliberate social action.[5] An experiment, for example, must usually be elaborately planned, within a rational framework of theory and technique, to pin down a particular item of knowledge. Even though this item of knowledge is not fully known in advance, the intention behind the effort to capture it, and the foreseeable consequences of doing so, are not beyond consideration. We are quite accustomed to judging the objectives of research projects in deciding their relative merits for financial support; the whole peer review process for funding academic science works on this principle.

Of course, a research grant proposal, addressed to a public funding agency, always claims highly desirable and socially beneficial objectives. But that does not exclude the possibility that somebody might secretly conceive a research project with highly undesirable and socially malevolent objectives. The wicked scientist, doing research towards an evil end, such as the mastery of the world, is not only a familiar stereotype of science fiction; he is an imaginative warning of the consequences of the doctrine that scientific inquiry can know no limits.

In such an extreme case, we would all know that it was our heroic duty to frustrate his plans. Normally, the balance of principle between freedom of research and its potentially antisocial consequences is much more delicate and subtle. Is it not possible, for example, that research directed towards achieving a dramatic increase in the human life span would prove so disastrously unsettling, if it were successful, that it should be most definitely discouraged or even forbidden? This is not the place to enter into the immensely difficult technical, political, social and ethical consideration that surrounds such questions.[6] But by recognizing that such questions are entirely legitimate, and should perhaps be asked more searchingly and more often, we show that we cannot accept the scientistic doctrine that scientific inquiry ought never to be deliberately limited.

Scientific Information Should Be Open For All

"Communalism" is the norm commanding that scientific knowledge should be a public resource, open to all. Scientific information should therefore be published in full, at the earliest possible moment. This norm is one of the main foundations of basic science as we have known it, and has its political counterpart in the general laws that protect freedom of publication. But it is not an absolute principle licensing the disclosure of information that could be gravely damaging to innocent individuals or to society at large. It never looked remotely plausible, for example, as a justification of the actions of "atomic spies" who deliberately betrayed military secrets to foreign countries. Normal legal and political responsibility cannot be waived in the magic name of science, and the genuine outrage that many scientists feel about the secrecy surrounding military and commercial research can never excuse acts of bad faith and treason. Information that happens to have been gained by scientific methods, or that claims some specially "scientific" status, is not uniquely privileged as to disclosure or dissemination.

Once more, the extreme case establishes the point in principle, whilst most practical cases are much less clear cut. The classical examples arise in medical research, where there can be a genuine ethical dilemma on a question such as whether to publish preliminary evidence suggesting that some particular treatment might prove effective against some dreadful disease such as cancer. The ultimate benefits to be derived from following the scientific norm of immediate publication may be outweighed by the distress caused by raising false hopes in a large number of very unhappy people.

Paradoxically, the conventions of the communication system of science are sometimes invoked against the publication of scientific information in a form intelligible to the layman. Scientists are often reluctant to explain briefly the essence of scientific knowledge on a vexed question, because such a statement may have to be expressed in language that lacks the formal precision of rigorous scientific argument. This prissiness has no epistemological justification; in the final analysis, no scientific statement is logically rigorous and unassailable. When information at stake has serious social relevance—as it might be, for example, in assessing the environmental effects of a chemical agent such as tetra-ethyl lead in petrol—there is a clear responsibility on the scientific expert to bring his or her specialized knowledge into the public arena in a form that is simple enough to be matched up with the less tangible and subjective costs, benefits and values that will also enter into a policy decision.

Science is True

The central pillar of scientism is that science tells "the truth, the whole truth, and nothing but the truth." Upheld by this doctrine, the scientist feels uniquely powerful and morally unimpeachable. Conflicting assertions of fact or interpretation can be brushed aside. The servant of almighty truth need carry no burden of personal responsibility. Once it has been "scientifically demonstrated," for example, that black people are intrinsically less intelligent than white people, who would there be to question the practices of apartheid that logically follow from this "truth?" Indeed, in the long run, if we pursue scientific truth far enough, we may succeed in reducing the notion of moral responsibility itself to a logical paradox in a scientific theory of games and social behavior.

This positivist doctrine is now thoroughly discredited. Although we have no reason to doubt the practical reliability of well-explored and well tested branches of the natural sciences, we are also aware of the vast extent of our scientific ignorance on most matters of real human significance. No philosopher now supports a "scientific method" that can carry all before it in every field of knowledge. We must always ask to see the credentials of whatever is offered to us as scientific truth, look at how it is generated and validated, and make up our own minds whether it is more convincing than what we might derive from practical experience, common sense, personal insight, or social tradition.

This comparison with other sources of relevant knowledge becomes more and more apt as we move from the natural to the social sciences.[8] The would-be hard-headed scientist tends to dismiss as nonsense or prejudice every consideration that has not been formulated and superficially tested by self-styled "scientific" techniques. But to talk about social responsibility at all, one must give adequate weight to ethical, religious, humanitarian or other precepts that cannot be derived from principles akin to the laws of physics. Just what precepts we ought to live by are matters for discussion and rational debate involving many other factors of thought, feeling and experience. It is morally irresponsible—indeed positively amoral—to refuse to enter into such debates just because they can never be decided by appeal to absolute scientific "truth."

Of course, a well-attested scientific fact, such as the prevention of dental caries by the fluoridation of public water supplies, is often central to a controversial social issue. The reputation of science and of scientists as the most reliable source of information within particular spheres is to be preserved at all costs. But this reputation for credibility and probity is not to be trusted in human affairs beyond the narrow limits where it has fully proved itself.

Science is Rational and Objective

Scientists present their observations and theories as precisely and logically as possible, to make them credible and convincing to other scientists. Science is a body of public knowledge that must continually face critical analysis and crucial tests. To give their discoveries the best chance of preliminary acceptance, scientists adopt a style of formal rationality, insisting that the conclusions they arrive at are logically compelling. This rhetorical device is assisted by an impersonal stance. Scientific papers are written "objectively" as if the author had no hand in the matter, but were simply reporting events and arguments in which he or she had no particular personal interest.

These conventional features of scientific communication have genuine communal functional value.[9] Science is validated by active consensus. What individual scientists discover or conjecture is subjected to collective criticism until everyone is persuaded that it must indeed be so. In this process, scientific knowledge must be purged of "subjective" elements that are not universally compelling, or which are only valid from a particular individual point of view. Scientific rationality and objectivity are the terms we apply to the consensuality and inter-subjectivity that this process achieves.

These characteristics of the creation and content of scientific knowledge are often elevated into supreme virtues. Science is held to be *perfectly* rational—i.e., logically irrefutable, from its observational premises to its theoretical conclusions, and *perfectly* objective—i.e., representing the point of view of an abstract intellect free from the defects and vices of any single human mind. Even if it does not go so far as to say that science tells the *whole* truth about life and the world, this doctrine obviously puts the scientists in a very privileged position in every practical argument. By its rationality and objectivity, his science appears to transcend foolish, fallible, corruptible, human interests and concerns. Within his own sphere therefore—say in the engineering design of an industrial production line, or the medical arrangements in support of childbirth—the scientific expert is easily persuaded that the "scientific" solution to every problem is much the best possible.

In reality, science grows by processes that are far more fallible, far more subjective, far less disinterested, far less logically watertight than this ideology would allow. Reliable as it may be in most of its main lines of argument, and in numerous details, science nevertheless maintains many errors of fact and interpretation. Much of its rationality is superficial—little better than special pleading for an interpretation that is far from proven by the evidence. Much of its objectivity is spurious—little better than a depersonalized abstract formulation of the prejudices and interests unconsciously shared by a particular group of scientists working in a particular field. Innumerable "crackpot" theories of health and disease have been rationalized by medical science on the slenderest evidence. Deplorable 19th century theories of racial superiority, strongly influenced by the doctrines of social Darwinism and highly advantageous to the politics of imperialism, were supposed to be scientifically rational and objective. Rather than proclaiming such doctrines, it is the responsibility of the scientist to show that they grossly misuse the authority of science to rationalize particular social positions.

Although scientific knowledge on a particular point may well be the most "rational" and "objective" available, these qualities must be proved specifically by reference to the evidence and the arguments, not just taken for granted. In any case, these may not be virtues that ought to claim automatic deference in all human affairs. The scientist who depends solely upon "scientific method" for his or her opinions tends to adopt an inhumane attitude which is not sufficiently responsive to historical circumstances, moral values, the diversity of human aspirations and other untidy realities that cannot be "rationalized" and "objectified" out of the way.

Science is Neutral

The basic objectivity of science frees it from close attachment to the interests of particular social groups. It can scarcely be said, for example, that the law of the conservation of energy has been devised so as to specially favor the institutions of capitalism and that socialists need an alternative theory of thermodynamics which will support their political views. Science, we insist, is "neutral;" it cannot be enlisted permanently on either side in a social conflict, although it may prove a very effective ally for any party with whose polity it is naturally consistent. The neutrality of science is a bit like that of the angels, on whose side it is always advantageous to be.

But the strength of this doctrine obviously depends on the degree to which science is genuinely detached from particular social interests. This is a controversial issue in the sociology of knowledge, where it is pointed out, for example, that the historical development of the physics of energy was in fact closely linked to the rise of capitalism and the Industrial Revolution. Although the laws of thermodynamics would remain as valid as ever under a socialist regime, the central role of these laws in 19th century physics was a significant factor in the process of industrialization, and was not an "objective" consequence of the purely internal development of the subject. Thus, the inventory of scientific knowledge available for social polemic in any particular epoch is not altogether "neutral" as between the various conflicting parties.

This applies particularly when research is undertaken deliberately to settle some controversial issue. It is extremely difficult to phrase the questions to be answered, and to choose the technique of inquiry, so as not to favor one or the other party. Although many components of "objective" science may be involved in the investigation, the circumstances in which the research is being undertaken, the source of sponsorship, the terms of reference given, the investigators and the form in which the results are reported, all have some influence on the outcome. It must be remembered that the relative objectivity of science does not derive from its sophisticated techniques, advanced theories, and rationally ordered argumentation: it is the produce of the social process of creative criticism within the scientific community, which minimizes subjectivity and the influence of particular interests. Thus, the scientist who naively believes that science is neutral in political and economic conflicts is very ill-prepared to bear the social responsibilities that these conflicts often demand.

The Scientific Attitude

Another major doctrine of scientism is that scientists, as a group, have special personal qualities. They have a "scientific attitude" which embodies such intellectual and moral virtues as logicality, open-mindedness, curiosity, detachment, skepticism, independence of authority, etc., etc. They should therefore be given a special place in society, with special responsibility for crucial social decisions.

This myth "personalizes" the qualities of rationality, objectivity, etc., of scientific knowledge in the abstract. If science has these properties to a very high degree, then it is assumed that those who produce science must be correspondingly gifted. But this is nonsense. Scientific knowledge is a *social* product, and may therefore turn out far more logical, or unprejudiced, or original than any of the individuals who have co-operated to generate it. The output of a motorcar factory is a far better vehicle than could possibly be designed and constructed by any single motor mechanic!

It is true that scientists must learn certain conventions and norms in order to co-operate and compete in the scientific manner. A professional research worker soon learns from experience that it is more effective to argue coolly than to indulge in personal polemics, and that it is advisable to anticipate all possible critical objections before publishing a new idea. Within the scientific community a delicate balance is often achieved between imagination and orthodoxy, between the wisdom of age and the enthusiasm of youth, between institutional authority and individual autonomy. It is possible that the way in which the "republic of science" conducts its own affairs might be followed to advantage by other social groups. But these are essentially professional mores, which do not necessarily fit scientists for active roles in society at large.

There are, of course, scientists, such as J. Robert Oppenheimer, who have given inspired leadership to their colleagues. Most research organizations are reasonably well managed, and science policy is as well conducted as most other government policies. But there is no foundation for the technocratic notion that scientists and other technical experts should be given special authority to determine general public policy. Scientists are often clever in peculiar ways, but otherwise they are usually quite ordinary people with quite ordinary talents. Their professional training and work experience does not fit them particularly well for high responsibility in public affairs. They have seldom had to persuade large numbers of people to support them in an uncertain enterprise; they are seldom skilled in the arts of bargaining and compromise; they have seldom had to take large decisions under severe pressures of time and ignorance; they are seldom even well informed on the larger circumstances of history, law, religion, economics, etc., within which political decisions have to be made.

The scientist who hides away in the peace and quiet of his laboratory, consoling himself with the belief that he is engaged in an honorable and beneficial profession, is not behaving quite irresponsibly. He may have judged, modestly and realistically, that effective social action calls for knowledge, insights, experience and fortitude that he does not have, and cannot afford the effort to acquire. It is not enough to be politically "enlightened," or to make an occasional gesture, like signing a manifesto or letter of protest. One must shed the arrogant notion that scientists are much wiser than politicians in the matters of peace and war, or poverty and wealth, and get down to the thankless business of understanding what is going on and trying to make it go right.

Science Has Nothing to Do with Politics

In its early days, science had to establish itself in a society riven with political and religious conflicts. In some countries, even now, it faces oppression from tyrannical governments. It is prudent, in such circumstances, to insist that scientific knowledge is strictly objective and neutral, and that the professional work of scientists—like the work of nurses and taxi drivers, say—is of no particular political significance. Research, we say, is just a specialized technical trade; scientists are just ordinary law-abiding citizens; learned societies and universities are just organizations for co-ordinating and advancing this innocent, useful activity.

The knowledge of what happened to science under totalitarian regimes fully justifies a general policy of "keeping politics out of science." The traditional norms of science cannot survive in the poisonous atmosphere of violent social conflict. A scientific institution, such as a university, that has been taken over by an anti-intellectual junta of colonels has lost its meaning and might as well be left for dead. It is obviously foolish to provoke such disasters by getting involved in politics without good cause.

But this sound practical maxim has been elevated into the doctrine that science has nothing to do with politics. This doctrine cannot, of course, apply to the response of individual scientists to the ordinary demands of citizenship. They must accept the responsibilities of life, like everybody else. It is morally offensive to suggest that they ought to have special social privileges because they are studying the eternal verities, or because they happen to belong to a professional community that transcends national frontiers. An elite of scientists who were not answerable to the laws of any nation would be just as socially irresponsible as a multinational corporation that paid no taxes and obeyed no safety regulations.

A real danger of pretending that "science" can be separated entirely from "politics" is that it inhibits the exercise of collective social responsibility by formal scientific organizations such as learned societies. Like the parallel doctrine that scientific knowledge is "objective" and "neutral," it can be used to conceal the intimate connection between research and policy, between thought and action.

This connection between science and politics is obvious in the area of "science policy." Modern science is not a self-contained social group like a minor religious sect; its resources come so completely from public funds, or from large private corporations, that it must be institutionally concerned about how these resources are supplied. In practice, even the most obstinately "non-political" scientific societies find that they must express a point of view on many controversial political issues relating to science, technology, the environment, education, etc. Fearful as they may be of the dangers of getting too heavily committed to one side or another in such controversies, the leaders of the scientific community know that their formal institutions cannot afford to appear totally unconcerned about matters in which they are so very closely involved.

Indeed, science is not simply one of the institutions of society, it is itself a social institution. The norms and conventions of scientific life are not independent of the norms and conventions of society at large. For example, freedom of communication, which is so essential within science, cannot be distinguished in practice or principle from general freedom of speech, publication, assembly, travel, teaching, etc. Modern science grew up during the period when these social and political rights were being formulated and safeguarded by law, and is designed to function within an open pluralistic society where these rights are respected in practice. It is hard to believe that science as we know and value it could long survive in a society with contrary norms—where, for example, scientists were not free to communicate their discoveries and to criticize each other's research in public and private.[10]

In other words, science is in politics up to its neck. For its very own existence, it must be ready to fight actively for the human rights of scientists—and hence, without distinction or privilege, for the human rights of all other citizens. The prudential acquiescence of the German scientific community to Hitler in the 1930s was a spiritual betrayal that availed it nothing; cowering behind the pretence of having "nothing to do with politics," it did not make even a gesture against a social philosophy that was bent on destroying it.[11] One must assume that scientists take some pride in their profession—otherwise they should get out and do something more worthy, such as nursing or taxi driving. A doctrine that inhibits scientists from acting collectively, through their own professional institutions, in defence of the long term interests of that profession, is thus peculiarly irresponsible and antisocial.

The Consequences of a Discovery Cannot be Foreseen

The outcome of research cannot be known precisely in advance; if it could be, then there would be no need to do the research. Scientific discoveries are often hit upon by accident; but if they could be arrived at intentionally, they would not be discoveries. The irreducible element of uncertainty in science is often called upon to free scientists from responsibility for the consequences of their research.

But the same principle applies to every action that we take in life. However carefully I plan, I cannot be sure that what I do will cause no harm. My decision to set out for work five minutes earlier than usual may bring death to the child who runs out in front of my car. Through the person whom I accidentally meet at a party and subsequently marry, I may become the parent of a terrorist assassin. All moral philosophy takes account of the unintended consequences of our actions, beyond a close horizon of rational foresight. It is only what we do knowingly, with our eyes open to the outcome, for which we must bear the responsibility.

The factor of uncertainty in science provides no general excuse for social irresponsibility. As already pointed out, research itself is a deliberate activity, undertaken with conscious purpose. Even in highly academic science, most research projects must claim definite objectives to gain material support. The scientist who drafts a research grant application in terms of such objectives cannot turn round later and deny any responsibility for the obvious consequences of achieving them.

In "knowledge-oriented" fundamental research, these objectives may be so indirectly related to human affairs that the responsibility for a scarcely imaginable outcome is minimal. Ernest Rutherford, for example, cannot justly be blamed because his discoveries in nuclear physics eventually led to Hiroshima. But the work of the great majority of scientists nowadays is "mission-oriented." Their research is designed to answer questions with specific ends. They are employed by organizations seeking military power or commercial profits, or public welfare, or other familiar practical goals.

Science nowadays is not just the pursuit of knowledge that might one day, just possibly, turn out to be useful. On the contrary, scientific research is largely undertaken as the most effective means of reaching chosen industrial, commercial, military, social or political goals which are beyond the scope of present-day understanding or technique. Science is seen as an instrument for "solving problems"—i.e., as a means of warding off the dark dangers of misfortune, and arriving successfully at a desired future state. No scientist employed voluntarily in such an enterprise can deny some personal responsibility for what comes of it, just as no member of a gang of thieves can get away with the plea that he didn't know what they were up to and that he only went along for the ride.

Scientists are Servants of Society

Since the great majority of scientists are employed by large organizations, they are strongly tempted to unload all the blame for what comes of their research on the policies of those organizations. In other words, they plead the irresponsibility of the subordinate for what is done in obedience to orders. Often enough this plea may not be unreasonable: "Should I have really risked my whole career just because I happened to think that our new cosmetic lotion was not justifying the advertising claims for it?" Sometimes, on the other hand, it may be grossly immoral not to "blow the whistle," regardless of personal consequences, on some socially dangerous development in a research program.[12] That is to say, obedience to orders is not an absolute shield against personal blame, but depends very much on the circumstances. And although many of the standard examples of the need for social responsibility in science revolve around this sort of question, they really belong to the much wider issue of personal responsibility in general. This is a traditional issue of moral philosophy and law. The responsibility of the research scientist in such cases is not different in principle from, say, the responsibility of the soldier for acts of terror carried out under orders from above, or the duty of a company accountant not to acquiesce in the presentation of fraudulent accounts.

My whole argument would, of course, be invalid or irrelevant in a society where everybody was in duty bound always to obey the orders of higher authority. In such a society, the only principle of social responsibility would be that this duty must be performed to the utmost of one's individual powers, regardless of the consequences. Although it is not difficult to find statements along these lines from various public figures in various countries, this servile doctrine is so contrary to the elementary realities of a complex industrial society that it cannot be taken seriously, except as a deliberate attack on the spirit of licensed dissent that animates the scientific enterprise and gives scientists their personality and individuality.

Scientists are Just Technical Workers

Organizations that employ scientists usually prefer them not to be concerned about the social consequences of their research. The job of the scientist is supposed to be simply to provide technical information and advice for the top policy makers who decide the commercial, military or political objectives of the enterprise. As the catch-phrase puts it: "scientists should be on tap, not on top." Having deflated the claim that scientists are peculiarly well fitted to take such decisions, we are in no position to insist that they ought to have more influence of this kind than they are in fact permitted, which varies considerably from organization to organization and from country to country.

But the supposed distinction between the limited "technical" input of the scientist and the more general policy-making inputs of other members of a large bureaucratic organization is not valid. It is simply a reflection of the myth that science itself is perfectly objective and neutral, from which it is deduced that scientists must be incapable of taking into consideration the human values and other subjective interests that enter into any policy decision in the real world. The principle that "scientists are just technical workers" is often used to excuse them from any responsibility for the consequences of such decisions—even though these may depend very heavily on their research results or professional advice—but it is really no more than the obverse of the doctrine that science transcends the realities of the everyday world and is quite divorced from normal human concerns. It is thus a very convenient doctrine for those who want to exploit the technical capabilities of scientists without raising questions about the morality of the enterprise in which they are engaged.

Science Cannot be Blamed for its Misapplication

We have now reached the heart of the matter. The fundamental question for the socially responsible scientist is whether the search for knowledge can be entirely separated from its later use. The ideology of scientism is shot through with fallacies and contradictions that can only be avoided by making a rigid distinction between the "scientist" who makes discoveries and the "technologist" who applies these discoveries in the form of useful devices and techniques.

A century ago, this distinction would not have been inept. In the heyday of academic science, there was little experience of deliberately organized "research and development" to solve specific problems, or to achieve pre-conceived practical goals. A simple "discovery" conception of science was appropriate.

But the world has been changed by science, and science itself has changed in the process. The "instrumental" conception of science, as a positive means of getting things done, is now a practical reality manifest in numerous mature social institutions, such as industrial research laboratories and government scientific agencies. And a glance at any of these institutions shows that they are no longer based upon a sharp division of labor between "scientists" and "technologists," between research to gain knowledge and the application of knowledge for practical use. It no longer makes sense to suppose that these two different conceptions of science are ministered to by distinct professions, carrying out quite distinct social roles. We can only understand the science/technology complex if we accept that the discovery of knowledge and its application are merely different phases in a single social activity, different aspects of a single, coherent social institution.

Is it really necessary to give chapter and verse for a social fact which stares us in the face every time we open a popular scientific journal or switch on a television program about "science?" Our most self-consciously academic institutions in the most exploratory scientific disciplines, such as cosmology or molecular biology, are caught up in projects with specific military or industrial objectives. Our hardest headed practical enterprises in engineering and medicine employ basic research scientists charged with the pursuit of knowledge almost for its own sake. In apparatus, personnel, theoretical foundations, education, professional societies, managerial structure, funding arrangements, and many other features, there is such a close convergence and interpenetration of science and technology that one can no longer distinguish them by their social roles and social responsibilities.

Science is no longer an elite vocation that can shrug off its responsibilities by reference to a traditional professional ideology. Within the great complex of the R&D system, individuals undertake innumerable different, often highly specialized jobs. Some have great freedom to follow interesting lines of research wherever they seem to lead, with very little external control. Others must co-ordinate their work very closely, in large teams directed towards prescribed goals. Just as there is this wide range of personal autonomy within the research professions, so there must be a wide range or degrees of personal responsibility for what is being attempted and what has been achieved by research. The social responsibilities of scientists and technologists are as varied in character and weight as those of other citizens in other walks of life—from those of the judge or senior civil servant, say, to those of the taxi driver or barrowboy.

Basic ethical principles tell us all plainly that we have a duty to be aware of the extent of our responsibilities, as they bear upon us in our individual lives, whether as citizens, as parents, or in the exercise of our calling, and to take them up to the utmost of our strength. Whether they like it or not, this duty now rests particularly heavily upon scientists. There is great public concern about many of the effects of science upon society and upon humanity as a whole. Out of this concern there now flows the demand that scientists must be more careful, must be more responsible, in what they do, for it could bring us all to disaster. Of all conceivable disasters that might overwhelm mankind, nuclear war is far and away the most likely and terrible. Unwittingly, perhaps, the scientists made such a disaster possible; inexorably perhaps, they are driven into activities that make such a disaster ever more threatening. But whether or not they should be individually blamed for what they have collectively done for the world, not one of them can now cast off a personal responsibility to think about these matters and to act to make this disaster a little less probable.

References

1. J.M. Ziman, *Teaching and Learning about Science & Society*, Cambridge University Press, 1980.
2. H. Verhoog, *Science and the Social Responsibility of Natural Scientists*, Doctoral Dissertation, Leiden, 1980. This work, which goes much more deeply into the connections between various philosophies of science and various ideologies of the social role of science, contains a very complete bibliography on the subject of this chapter.
3. D. Morley, Dumb Animals and Vocal Minorities, *The Sensitive Scientist*, SCM Press, London 1978.
4. Council for Science and Society, *Scholarly Freedom and Human Rights*, Barry Rose, London 1977.
5. J. M. Ziman, What are the Options: Social Determinants of Personal Research Plans, *Minerva* xix, 1–42, 1981.
6. G. Holton, R.S. Morrison, (eds), *Limits of Scientific Inquiry*, W. W. Norton, New York, 1979.
7. D. Morley, op.cit., 1978.
8. J.M. Ziman, *Reliable Knowledge*, Cambridge University Press, 1978.
9. J.M. Ziman, *Public Knowledge*, Cambridge University Press, 1968.
10. J.M. Ziman, Human Rights and the Policy of Science, *The Bulletin of the Atomic Scientists*, vol. 34, No.8, 1978
11. J. Haberer, *Politics and the Community of Science*, van Nostrand Reinhold, New York, 1969.
12. American Association for the Advancement of Science, *Scientific Freedom and Responsibility*, AAAS, Washington, DC, 1975.

Social Responsibility in Victorian Science

An Authentic Source

We tend to regard the contemporary problématique of science, technology and society as unique. Yet it has its historical antecedents and origins. Can we trace these back into the times of Alfred Nobel? Was there any consciousness, then, of science as an instrumental factor in society, as a means of achieving preconceived ends? What did the scientists themselves think they were doing, beyond advancing knowledge? What was important to them, outside their laboratories, lecture halls and book-lined studies? Questions such as these cannot be answered satisfactorily without reference to primary sources.

As a student of science in present-day society, with concerns mainly for the future, I have always regretted a personal lack of scholarly acquaintance with the evidences of the past. I have here worked up no more than a preliminary and unsophisticated exploration of a theme that really calls for serious, professional treatment.

Fortunately, the obvious primary source lay immediately at hand. In the stockroom of our physics department library were all the old bound volumes of one of the world's most famous and influential journals—*Nature*. This "weekly illustrated journal of science" was first published on 7 November 1869, and appeared thereafter at the rate of two volumes a year. Since each volume amounts to many hundreds of pages of double-column print, there was clearly more than enough material for my purpose. Indeed, I could only hope to sample this mine of information at random, leafing my way through every tenth volume, from volume 1 to volume 41, thus covering the period from about 1870 to about 1890 in five-year intervals.

This is an inadequate "data base" on which to construct an edifice of generalizations. It is also significantly biased by editorial policy: *Nature* was not a passive medium for scientific information and opinion. From the first, it proclaimed an intention "to urge the claims of Science to a more general recognition in Education and in Daily Life." It is difficult to guess how far such intentions were from the center of scientific opinion of the day. It was also, at that time, written almost entirely for British subscribers—at Home and in the Empire—and treated all "American and Continental" news as distinctively foreign.

But the 20,000 or so pages from which this ten percent sample was drawn are uniform in style and attitude. One immediately recognizes the high-minded tone of the late Victorian era of industrial progress, geographical expansion and cultural self-confidence. These voices may not be entirely representative of the world of science at the time of Alfred Nobel, but they are authentic of their day. They express a coherent view, as seen from a central point of vantage, and their words still echo in the thoughts of their successors, a century later.

A Retrospective Definition of a Theme

"Social responsibility in science" is an ill-defined notion, which has become a catch-phrase. But it is a convenient term for a cluster of issues, attitudes and concerns about the relationship between science and society at large. It stands for an outward orientation, away from the production and accumulation of knowledge towards the deliberate use (and abuse) of scientific method, or specific scientific techniques, to meet acknowledged human needs. It shifts the focus of attention away from the "internal" activities of the scientific community, from actual research discoveries, and from their essentially unpredictable consequences, towards the "external" social context of science.

"Social responsibility" thus implies an attitude of scientists towards societal problems and needs. Even in our own day, when no one doubts the effectiveness of scientific research as a conscious instrument of social purpose and power, this attitude is not paramount in the everyday thoughts of the working scientist. The sheer technical labor of experiment, observation, theorizing, communication, etc., keeps broader questions of ends and means at a distance. Even now, for example, *Nature* devotes only a few pages of each issue to social and political affairs, before getting down to the real meat of the latest discoveries in molecular biology or cosmology, or plate tectonics.

Nevertheless, around the periphery of the strictly academic domain, there is now an immensely active region of discussion about "science, technology and society." Topics relevant to this theme are expounded in books, reported in weekly magazines, taught about in lectures, and gravely considered by government committees. We recognize them under such headings as "science policy," "technology assessment," "the economics of research," "science and the media" or "rights and responsibilities of scientists," not to mention the wider social and political problems of war, energy, the environment, population, economic development and so on. These are the characteristic items in the agenda of the modern "socially responsible scientist." This is the theme, thus defined retrospectively, to be explored in the writings of the past.

What I have done, then, is to skim through those five old volumes of *Nature*, looking for articles, editorials, letters, or brief news items that might now be considered relevant to the theme of social responsibility in science. All communications referring solely to the content of science— claims of new observations, theoretical controversies, general accounts of recent discoveries, and so on—were ignored, along with interminable reports of the proceedings of various learned societies, British, Imperial, and foreign. A large number of biographical memoirs had to be passed over, together with various items of "internal" news, such as appointments, promotions and prizes awarded to individual scientists. Some of this discarded material is, of course, fascinating in its own right, and might yield relevant items to more thorough scrutiny; but the whole theme is too coarse-grained to pass through a finer sieve of definition and selection. Occasional items may also have been missed among the book reviews, which were not looked at carefully.

Perhaps we should greet this selected material in the same spirit as, say, a naturalist living at that time in New Zealand, receiving news of science in England through this single channel, six months in arrears, and sending back his own occasional reports of the meetings of the local philosophical society, out there in the colonies. It was not so much a matter of being minutely informed about events which one could not influence, as of keeping in touch with current attitudes and opinions which one felt one should be sharing.

A Significant Concern

We tend to think of natural scientists in the late nineteenth century as a highly academic community disconnected from social or technological issues, and concerned, almost exclusively, with "internal" questions, whether cognitive or communal. It was surprising, then, to find that my notes referred to a total of about 300 pages out of the 3074 pages of the five-volume sample. Admittedly, *Nature* was trying to speak to a wider public than it does today, and sought to fulfil some of the functions that are now undertaken by other general science magazines such as *New Scientist*, or by more specialized periodicals such as, say, the *School Science Review*. But even if questions of social relevance were not central to the interests of British scientists at that time, they were certainly not insignificantly marginal. The theme of social responsibility seems to have deeper and more solid roots in Victorian science than is often supposed.

In assessing the thrust of this concern, we must, of course, make due allowance for the immense change that has taken place both "within" science and "outside" it. The enormous growth of scientific knowledge and practical technique scarcely needs emphasis. For example, in the 1880s, the world itself was still being explored: the scientific literature is full of reports of geographical, anthropological and botanical discoveries, with all their speculative economic and political implications. The weight of many factors in the interaction between science and society has inevitably been altered by the growth in the scale and scope of research; for example, the cost of supporting science in those days was far below the threshold of national economic concern. Nineteenth-century society itself was not only very different politically and economically; it also had a very different cultural style. For example, it was almost romantically optimistic about its future. This comes out clearly in *Nature*'s "advertisement" of editorial policy. The intention was to make a major feature of

> articles written by men eminent in Science on subjects connected with the various points of contact of Natural knowledge with practical affairs, the public health and material progress; and on the advancement of Science and its educational and civilizing functions (1870; **1**, 323).

Nevertheless, in spite of far-reaching historical changes of content, scope and social context, the initial impression is of continuity of attitudes and themes rather than of a glimpse of an entirely unfamiliar world. This impression derives to some extent from the pleasure of recognizing such hardy perennials as expressions of dissatisfaction with the constitution of London University (1884; **31**, 145) or from a splendid debate at Oxford on vivisection (1885; **31**, 453); but it survives more serious analysis, as one recognizes the primitive ancestors of most of our present obsessions. One can already detect a general notion of socially responsible science, although the components are combined in somewhat different proportions and oriented towards different immediate goals.

But this general impression of thematic continuity may not stand up to more critical analysis. Let us take it simply as a tentative hypothesis, to be tested by more detailed consideration of particular topics.

Science Education

The broadest interface between nineteenth-century science and nineteenth-century society was through formal education. *Nature* proclaimed its intention to publish

> records of all efforts made for the encouragement of Natural knowledge in our Colleges and Schools and notices of aids in Science teaching (1870; **1**, 34).

About one-third of the 300 pages noted in my sample survey come under this heading.

That interest goes beyond the internal affairs of a community, whose members were often (though not exclusively) employed as teachers in universities, colleges and schools. I do not count the many specific announcements of academic appointments such as:

> Mr. J. J. Thomson has been elected to fill the post of Cavendish Professor of Experimental Physics in the University of Cambridge, in succession to Lord Rayleigh. A numerously signed requisition to Sir Wm. Thomson [Lord Kelvin] to become a candidate was declined (1884; **31**, 179);

or incidental chit-chat, such as:

> Cambridge was *en fête* on Monday. Peterhouse, the oldest collegiate institution in the University, was celebrating the six-hundredth anniversary of its foundation. It was stated at the dinner that one-third of the present Fellows were Fellows of the Royal Society (1884; **31**, 179).

It is zeal for science education as a general social influence that is so striking.

This mission was not without support from the outside: in 1875 (**11**, 241), for example, no less a political eminence than the Marquis of Salisbury was reported as speaking publicly in favor of scientific education. But the general impression is of an enlightened minority group, well persuaded that

> the only principle on which a satisfactory course of education can be constructed is, that it is essential for the well-being of every man and woman

that he and she should start in life with a well-trained mind and a fair knowledge of the principles and the main facts of everyday life (1874; **11**, 21),

fighting to change a situation where

In most cases where science has been admitted into our schools it has been only on sufferance as a kind of interloper for which any odd corner is good enough (1874; **11**, 21).

At the highest level, this implied an attempt to breach the entrenched positions of the classics and mathematics in the universities and socially prestigious secondary schools. The tone of this attack was not very confident, relying upon apologetic arguments such as that science was not so difficult as classics—hence suitable for weaker pupils (1869; **1**, 18)—and could be developed without damaging traditional standards of classical scholarship (1869; **1**, 25). The 1870s, in fact, saw the establishment of experimental science as a formal course of study at Cambridge, although there are reports of the unwillingness of the Cambridge colleges to support teaching posts in science (1870; **1**, 586), of the weakness of the new Natural Sciences Tripos in comparison with the old-established Mathematical Tripos (1875; **11**, 132), and various debates over new regulations and curricular structure (1879–1880; **21**, 26, 86, 125). Even at the end of the period, in 1890, there is bitter complaint (**41**, 25, 265) that "science men" are practically excluded from fair competition in the examinations for entry in the Indian Civil Service, the most esteemed branch of the Imperial bureaucracy. We observe, indeed, the initial phrases of the "Two Cultures" dichotomy, which still keeps the majority of our mandarins untouched by formal scientific education from the age of sixteen.

They apparently managed these matters better elsewhere, especially in Germany, At least that was one of the most important arguments for more science education in Britain. The comparisons are odious: the New Statutes (of 1879) "will scarcely place Cambridge on a level, so far as teaching power goes with a second rate German university" (**21**, 125). According to H.E. Roscoe (1869; **1**, 157), the German university system, although state-supported, was intellectually free, held in high esteem, non-sectarian, cheap for the student, well staffed, scientifically thorough, connected with industry, and generally without defects. An account of the new University of Strasbourg (1885; **31**, 557)—admittedly cultural window dressing for the German annexation of Alsace-Lorraine after the Franco-Prussian War—is as admiring and envious of enlightened prodigality as, say, a European description of the Bell Telephone Laboratories in the 1950s, or a present-day American report on an automated Japanese car factory. French education is not set up as an ideal model, but there are numerous favorable references to the much larger scale of expenditure on technical education than in England (1879; **21**, 139).

These comparisons apply not only to higher education in science; they also apply to technical training and to science in general education:

> ...plenty of time remains in German schools for the teaching of science, which forms so important a part of education throughout that country, and which gives the German a starting point in life so very much superior to that which the average Englishman has, even when educated at our public schools and universities (1874; **11**, 21).

Indeed, the scientific community, through such notably "responsible" scientists as T. H. Huxley, was obviously much involved in the movement to establish a system of compulsory primary education in Britain. Every effort was to be taken to incorporate elementary science in the standard curriculum of mass education—as Henry Armstrong put it:

> ...the advantages to be derived from even the most elementary acquaintance with what may be termed the science of daily life are so manifold that, if once understood by the public, the claims of science to a place in the ordinary school course must meet with universal recognition (1884; **31**, 19).

There was also the challenge of founding new institutions for technical education, where scientific subjects would be taught at an intermediate level for the skilled craftsmen and professional experts needed in advanced industry. This was a grave weakness of Britain compared with its commercial competitors, especially Germany, with its new system of polytechnic institutions (1870; **1**, 475), and France, with a long tradition of state-supported hautes écoles (1879; **21**, 139).

In its twentieth-anniversary number, *Nature* had good cause to celebrate the progress that had been achieved in science education:

> Twenty years ago England was in the birth-throes of a national system of primary instruction. This year has seen the State recognition of the necessity of a secondary and essentially a scientific system of education and the Technical Instruction Act marks an era in the scientific annals of the nation...A race is thus springing up which has sufficient knowledge of science to enforce due recognition of its importance and public opinion can now, far more than in the past, be relied on to support its demands (1889; **51**, 1).

Unfortunately, this momentum for reform was not maintained; a century later, we still regret and suffer from the continuing weakness and inadequacy of scientific and technical education in Britain, by comparison with what is achieved in other countries.

But I do not think we should blame our Victorian scientific forebears for these deficiencies. They backed up their rhetoric with both realistic and inventive detail. Thc administrative threads of such uptight institutions as the Universities of Cambridge (1987; **11**, 125) and London (1885; **31**, 145, 159, 352) are carefully unknotted for reform. Thoughtful plans for a whole system of "polytechnics" in London are propounded (1890; **41**, 242, 481). The implications of various clauses of the "Education Code" of the Technical Instruction Act are laboriously spelled out (1890; **41**, 385, 505). The accumulated wealth of the Guilds of the City of London is fastened on as appropriate endowment for the new College of Science to train science teachers (1880; **21**, 221).

This involvement in educational reform is not (as so often nowadays) limited to framing general objectives, demanding more financial resources and proposing new institutional arrangements. Henry Armstrong explained in detail his new method of teaching chemistry, right down to the description of an experiment to determine the composition of air (1884; **31**, 19). This was followed by a lively correspondence from other science teachers, including the familiar complaint that pedagogic innovations have to be abandoned because:

> Science masters are [entirely] at the mercy of examiners both of University examiners, periodically examining a school, and of examiners for open scholarships (1884; **31**, 28).

Armstrong was not alone in advocating a new approach to science teaching. Even today, we emphasize the value of experiments in science education, perhaps echoing C.K. Wead, an American physics professor, who wrote that

> ...discovery, with its necessary companions, self-reliance, independent thought, shrewdness of judgement—the very qualities which make a successful man of the world—are all developed by experimental science instead of the too frequent opposite effect which make anxious business fathers dread too much schooling for the sons who will have to follow them (1885; **31**, 578).

But laboratory teaching was expensive. W. J. Harrison's "New Method for the Teaching of Science in Public Elementary Schools" in Birmingham was to take demonstration apparatus from school to school in a handcart. As a result: "Among the boys the half-timers then muster strongly, often getting leave to come in for that lesson only, and sitting with bare arms and rolled-up aprons, just as they run from their work" (1884; **31**, 175). And although we sympathize with C.W. Quin's insistence on "the absolute necessity of beginning scientific training at a very early age," we are still trying to prove his assertion that "children of 8 or 9 are [not] too young for systematic science teaching" (1869; **1**, 209).

The modern scientist tends to regard science teaching as a specialized profession, for which he has little direct concern. One cannot imagine that modern readers of *Nature* or *New Scientist* would be interested in general accounts of educations problems, such as the "Report of the Commissioners of Education in the United States for the year 1882–83," drawing attention to the persistence of conditions such as:

> The variation in different States of the expenditure on education...is still exemplified in the fact that Massachusetts pays fifteen times the amount per head that Alabama pays.

and:

> We have not need to enlarge again here upon the United States difficulty; the education of the Negro. The burning question of course is, Who is to pay for it? (1885; **31**, 435).

The educational role of science was evidently felt with greater responsibility in the late nineteenth century than it is today.

Popularization of Science

In Victorian times, there does not seem to have been a clear demarcation between formal instruction and public lecturing about science. The first volume of *Nature* is enthusiastic about "Lectures for Women" (1870; **1**, 488) and, for working men, to "Help Them Get Things Right (1869; **1**, 71). Those who did the "arduous work of teaching," free, at the Working Men's College (1870; **1**, 511) were among the elite of the science community. In 1875, for example, Professor Roscoe lectured publicly in Glasgow to 3000 people (at thruppence per head) on "The History of the Chemical Elements" (**11**, 233). On the other hand, some of the attempts to popularize science were as inept and infuriating as those that can be found in the modern mass media. M. Lichtenstein wrote to the Editor about attending one of the well-endowed Gresham Lectures—given, it seems by a classical scholar with no knowledge of science:

> Choking with Indignation, I left the building, never having heard, in all my life, either in sermon or lecture, so many false statements uttered in the space of half an hour (1874; **11**, 28).

Nevertheless, such references to the popularization of science are quite brief. Apart from an account of steps towards the foundation for the science museums at South Kensington (1890; **41**, 409), I could find no further reference to this subject in later volumes. It is true that *Nature* itself was not a specialized scientific periodical. Its editorial policy statement distinguished between "Those portions of the Paper more especially devoted to the discussion of matters interesting to the public at large" and those portions "more especially interesting to Scientific men." But that public (like modern subscribers to the *Scientific American*, say) is evidently drawn from a narrow stratum of educated, technically oriented people, already far better informed about Science than the mass of ordinary people. I wonder whether the scientific community of Nobel's time was any more concerned about the mass popularization of science than scientists are today. This sample even suggests a decline in that particular manifestation of social responsibility from the 1870; to the 1890s; but the evidence is not sufficient to support any definite conclusion on this point.

Scientific Technology

Nowadays, "science" is almost synonymous with "technology." At least, one should say, they are symbiotic. The most abstrusely academic research is connected with practical affairs by a few overlapping links of experimentation, inventive application, industrial development and commercial manufacture. The modern scientist is expected to exercise social responsibility very largely through the mechanisms that give direction to the immense social forces of scientific technology.

Our Victorian forebears were certainly as interested in technological innovation as we are. About seventy-five pages of my sample were about recent or prospective advances in various fields of industry and engineering, both civil and military. This was the new age of electricity, celebrated, for example, in a whole series of articles on the "Progress of the Telegraph" (1875; **11**, 392, 450). The facilities of the new Merchants' Telephone Exchange in New York are described enthusiastically:

> If having established the connection the employé is obliged to withdraw his telephone, the communication between Edward and John is secret. If while these two are in conversation No 42, James, wishes to correspond with John, for example, the employé may join in the conversation of the two interlocutors just like a servant announcing a visitor. If required, conversation may be established between the three subscribers (1880; **21**, 495).

It is reported that Victoria Station is being lit with sixty electric lights driven from a twenty-horsepower motor (1879; **21**, 162); and there is an interesting discussion of the relative merits of various schemes of gas or electric lighting for Paris (1880; **21**, 282). In the same volume, there are several brief reports from America about Edison's new carbon-filament lamp (1879; **21**, 187,215), including the information that the shares of his company had recently gone up from $20 to $3500 (1879; **21**, 261). This was evidently the trigger for an intensely sarcastic editorial (1880; **21**, 341):

> What is then the nature of the inventions thus heralded before the world? Regarded quietly, and without prejudice, from a scientific standpoint, what is the value for the discoveries which can thus play havoc on the Stock Exchange?

But the attack goes beyond the "reckless and amusing statements made by newspaper correspondents and interviewers," to assertions that "Mr. Edison is thirty-five years behind the time in his invention," and regrets that "Mr. Edison does not devote some time to learn what has been already done in this field." If only it were possible, then or now, to

> Let the public opinion insist that the inventor should be allowed to pursue his way unhampered by the officious interference of the unprincipled speculators whom his soul abhors, or by the irresponsible unscientific reporter who is only one degree less reprehensible for the part he plays.

The attractions of the superstar technological project were also evident. The recently opened St. Gotthard Tunnel called up the same sentiments ("The greatest work hitherto attempted by man," (1880; **21**, 581) as a present-day description of the flight test of the space shuttle. An admirable series of articles (1869, 1870; **1**, 160, 303, 631) analyzes various proposals for "The Projected Channel Railways," rather as one might nowadays analyze schemes for collecting solar power by stationary earth satellites. Unfortunately, history has not yet vindicated their confident judgement:

> That a permanent railway across the English Channel will be built, we doubt not: we are equally confident that Messrs Bateman and Revy's scheme of a sunken iron tube on the sea-bed is a practical solution of the problem. No less an authority than the Emperor Napoleon III, after mature consideration of the scheme, wrote to say: 'C'est le seul réalisable,' and as the design is one that belongs essentially to England, His Majesty's opinion acquires enhanced value and importance (1870; **1**, 631).

Science and War

What is surprising, however, is the obvious fascination of scientists with military technology. We tend to assume that the liaison between science and war (with all the social responsibilities thereby incurred by scientists) was kept fairly quiet until, say, the First World War, and did not become a matter of great public concern until after Hiroshima. But the articles in *Nature* go far beyond portentous frivolities, such as the suggestion that

> ...germs of small-pox and similar malignant diseases [should be collected] in cotton or other dust-substances and [loaded into] shells. We should then hear of an enemy dislodged from his position by a volley of typhus or a few rounds of Asiatic cholera (1870; **1**, 562),

or mere news items such as that observation balloons were to be used by the Expeditionary Force going to the Soudan (1885; **31**, 368). There were detailed technical reports on current military innovations, such as the use of torpedoes in the American Civil War (1870; **1**, 656) or Hiram Maxim's latest invention:

> A gun which loads and fires itself is certainly a novelty, and presents many interesting features and possibilities to anyone who takes an interest in implements of warfare (1885; **31**, 414).

The series of articles by Sir Frederick Abel, FRS (of Her Majesty's Chemical Department at Woolwich Arsenal) on "Smokeless Explosives" (1890; **41**, 328,352) are of special interest. They make clear the great tactical advantages of smokeless powder in land and sea battles and report the considerable efforts of the major European powers to seize these advantages for themselves. Secrecy is an essential factor:

> As in the case of mélinite, the fabulously destructive effects of which were much vaunted at about the same time, the secret of the precise nature of the smokeless powder [used in the new Lebel rifle] was so well preserved by the French authorities, that surmises could only be made on the subject even by those most conversant with these matters.

Meanwhile, of course

> In Germany, the subject of smokeless powder for small arms and artillery was being steadily pursued in secret, and a small-arm powder giving excellent results in regard to ballistic properties and uniformity, was elaborated at the Rottweil powder-works...

In these circumstances, it is scarcely surprising that Alfred Nobel applied his inventive skill with blasting agents to the production of smokeless powder. We may be sure that others would soon have taken the same path to the far more efficient and destructive weaponry that this invention obviously made possible. Incidently, Abel took pains to combat the widespread fallacy that smokeless powder was also *noiseless*. We, too, on occasion find it

"...somewhat difficult to conceive that, in these comparatively enlightened days—an acquaintance with the first principles of physical science having for many years past constituted a preliminary condition of admission to the training establishments of the future warrior—the physical impossibility of such fairy tales which appear to be considered necessary in France for the delusion of the ordinary public, would not at once have been obvious."

The positive concern of late Victorian science with military technology is further evidenced by the controversy in the pages of *Nature* over the design of battleships. In the half-century since Trafalgar, naval architecture and gunnery had changed out of all recognition—but the heavily armored iron ships had never been tested in battle against the high-explosive shells of their opposite numbers. Some of the phrases of the editorial of Thursday, 26 February 1885 (31, 381–384) could be brought forward a hundred years and applied to strategic nuclear weapons. A Mr. Barnaby, the Director of Naval Construction, had at one time computed the "relative efficiency" of warships by an algebraic formula compounded of factors such as "weight of armor per ton of ship's measurement," "the height of battery port-sills above load water-line," etc., and then later asserted the entirely different principle that "the fairest available measure of [fighting] power" is the "displacement or total weight" of the ship. The editorial points out the complete inconsistency of these two principles, and demolishes all such simple numerical indicators.

"The fighting power of a ship is... composed of several diverse and independent elements; and there is nothing approaching to a consensus of professional opinion as to the relative importance of these elements. To assume that they all vary together with the ship's dimensions, or with her weight in tons, is in the highest degree delusive and absurd...
"...It would be extremely difficult to devise any simple standard by which the popular mind may be fairly impressed with the relative powers of our own and foreign navies; while for purposes of exact comparison or of technical discussion no such standard could be regarded as absolute."

One can approve the scientific rationality of such contributions to public affairs—but does it match to a higher ideal of social responsibility? It was the Prime Minister, Lord Salisbury, addressing the Institution of Electrical Engineers, on the subject of the electric telegraph—"that small discovery, worked out by a few distinguished men in their laboratories upon experiments of an apparently trivial character, on matter and instruments not in the first instance, of a very recondite description..."—who drew attention to its most significant social consequence:

> I would ask you to think of what is the most conspicuous feature in the politics of our time, the one which occupies the thoughts of every Statesman, and which places the whole future of the whole civilized world in a condition of doubt and question. It is the existence of those gigantic armies held in leash by the various Governments of the world, whose tremendous power may be a guarantee for the happiness of mankind and the maintenance of civilization, but who, on the other hand, hold in their hands powers of destruction which are almost equal to the task of levelling civilization to the ground (1889; **41**, 21).

Scientists and Inventors

There is no doubt that nineteenth-century scientists were generally interested in technological progress beyond its purely technical aspects. They took upon themselves much of the credit for the transformation of industry—for example, the progress made in the previous forty years in the "construction of self-acting machinery" that would surely make living cheaper (1870; **1**, 432). The scientific engineers of the Institute of Electrical Engineers would have applauded Lord Salisbury when he looked far into the future:

> If it ever does happen that in the house of the artisan you can turn on power as now you can turn on gas—and there is nothing in the essence of the problem, nothing in the facts of the science, as we know them, that should prevent such a consumption taking place—if ever that distribution of power should be so organized, you will then see men and women able to pursue in their own homes many of the industries which now require their aggregation at the factory. You may, above all, see women and children pursue these industries without that disruption of families which is one of the most unhappy results of the present requirements of industry (1889; **41**, 21).

They were not unaware of the social consequences of industrial change. There is a report that beet sugar is ruining the cane-sugar industry of Cuba (1875; **11**, 314). There is mention of the introduction of labor-saving machinery, by which it seems "the Swiss toolmakers annihilated the English watch toolmakers some years ago' (1880; **21**, 397). This comes in a fascinating article by Silvanus Thompson on the reports made by a number of skilled artisans who were sent over to Paris for the Exhibition of 1878—most of whom came away "fully alive to all the advantage which accrue to an industry from the extension of labor-saving appliances, and from the dissemination of higher technical knowledge," and presumably confident that craftsmen displaced by such appliances would soon find alternative employment. Industrial prosperity was an important national goal, to which, for example, reform of the patent system would undoubtedly contribute (1874; **11**, 141).

Nevertheless, it is not without significance that a review of the first twenty years of *Nature* made practically no direct reference to industrial or technological developments (1889; **41**, 1). It was all very well for George Gore to argue that "employment for workmen in this country can be increased" by "encouragement of experimental scientific research" (1869; **1**, 623). There was still a long and irreversible chain of cause and effect from a scientific discovery, through its application as an invention, to the final manufacture that provides employment; and the scientific community did not take any responsibility for what happened further down that chain. This aloofness is all too evident in a comment on Edison:

> We are doing no injustice to Mr. Edison's splendid genius when we say that it is to the character of the inventor not to that of the scientific thinker, that he aspires (1880; **31**, 341).

Except, perhaps, in relation to military technology, there seems to be no notion of a systematic activity of "research and development" that would bring science into close interaction with technical problems of direct social significance.

Science and the Environment

Scientists nowadays are urged to take responsible attitudes towards the environment. Every agenda for social action or education on science, technology and society lays great emphasis on issues such as environmental pollution, ecological balance, conservation of resources, industrial health and safety, and so on. These issues are taken up so earnestly that it would seem that no one had ever thought of them before.

This is a grave misconception. The scientists of a hundred years ago were as much concerned as we are to make the world a cleaner, healthier, safer place, where mankind and all other forms of life could prosper. The evidence is quite clear throughout my sample. The thirty-five or so pages on these issues are unequivocally on the same side as our angels: there are no apologetics for dirt, disease, poverty, waste or exploitation of labor.

A long report on scientific aspects of an "International Health Exhibition" organized by the Society of Arts (1884; **31**, 138–143) illustrates the wide range of issues in which scientists were closely involved in social action. For example, there is a call for more stringent legislation, similar in character to that at present in operation in the United States and France, on the adulteration of food and drugs, on the grounds that

> These are questions largely affecting the health of the whole nation, and especially affecting the welfare of the poor, who suffer most by the substitution of worthless, inferior, or adulterated articles in the fabrication of apparently cheap, but often very dear because worthless, articles of food.

There were tests on the cooking efficiency and relative smokiness of kitchen stoves, conducted by members of the "Smoke Abatement Institution." A new code of public sanitation was proposed. The author of the report, Mr. Ernest Hart, had

> for several years as Chairman of the National Health Society...occupied myself with collecting the facts and figures which demonstrate the urgent necessity of improved legislation for the safeguarding of the sanitary construction of our houses, and the improved education and registration of those builders and plumbers to whom we intrust that construction.

Since *Nature* is not a medical journal, there is little about human diseases as such, but a strong case is made for the foundation of an Institute of Public Health to foster higher education and research in sanitary science.

Water pollution was clearly a major concern. In the very first volume of *Nature* (1870; **1**, 578), there was a polemic on "The Abuse of Water," which could scarcely have been improved on, scientifically, humanely, economically or rhetorically, for the next hundred years. It was followed up with positive proposals, such as General Scott's paper on "Suggestions for Dealing with the Sewage of London" (1874; **21**, 133). It would be unfair to blame the scientists and engineers for the failure to clean up the Thames and other rivers until a much later day.

The risks of accidental explosions were as salient then as now. F. A. Abel makes no bones about the causes of bursting steam boilers: "Very few explosions in 1873 have been due to the neglect of the attendants, but by far the greater number to that of the boiler owners or the makers," and calls for statutory inspection (1875; **11**, 436). However, he blames poor working practices, often abetted by the miners, for the frightful colliery explosions that were then so frequent. He returned to the same theme in 1885 (**31**, 469, 493, 518), with a series of articles on explosions caused by the vapors of non-explosive liquids such as paint thinners. Improvements in railway safety are also commended (1874; **11**, 173: 1884; **31**, 84), and there is a tough article on ensuring stability in the design of ships (1880; **21**, 485).

The biological environment is seen mainly through the eyes of the farmer or fisherman—the dangers of the spread of plant diseases such as phylloxera and of pests such as the Colorado beetle and apple blood louse from the New World (1875; **11**, 394: 1880; **21**, 356) are set off against the remarkable successes of Louis Pasteur against threats to the wine, beer, silkworm, cattle and poultry industries (1884; **3**, 138–143). Presumably, the Royal Agricultural Society really deserved the scorn poured on its efforts to foster research on potato disease (1874; **11**, 67, 109). The socio-economic origins of scientific ecology are evident in the work of the United States Fish Commission, founded in 1871, when the Congress "has its attention directed to the alarming decrease in its east coast food fishes" (1884; **31**, 128: 1885; **31**, 294). These articles emphasize that, although "The Fish Commission's work in its original conception was really the solution of practical economic problems, and it has in the main adhered to this idea," it had also been permitted to deviate "from the rather uninteresting study of the shallow fishing-grounds to the rich field of [basic] deep-sea research." We, on the other hand, might now value the fact that "the natural history work of the Fish Commission has, of necessity, been mainly of a systemic character, dealing with species and their distribution more than with problems of anatomy, embryology and history." It is noteworthy, nevertheless, that there is a continuing demand for similar practical scientific support for British Fisheries (1870; **1**, 243: 1890; **41**, 497).

In some ways, the most surprising items in the whole of my sample were concerned with energy. For all his academic and social eminence, Lord Rayleigh shows no disdain for practical needs when he devotes a major part of a lecture on "The Dissipation of Energy" to "The prevention of unnecessary dissipation [which] is the guide to economy of fuel in industrial operations" (1875; **11**, 454). There is a gift for the sociology of knowledge in Balfour Stewart's metaphorical interpretation of the energy concept:

> ...in the social world we have what may justly be termed two kinds of energy, namely:
>
> 1. Actual or personal energy [i.e., force of character, persistence, etc.].
>
> 2. Energy derived from position [i.e., social standing, family fortune, etc.] (1870; **1**, 647).

But Sydney Lupton on "The Coal Question (1885; **31**, 242) is uncanny. He estimates the amount of accessible coal in Great Britain, extrapolates exponentially from past consumption trends, and calculates that the total supply will be exhausted by "about AD 1990." Ignoring oil, he accepts that tides, winds and waterfalls could be exploited, but points out that since we have no monopoly on these sources "we should compete with our neighbors rather at a disadvantage." The cost of importing coal would be economically insupportable. Soon

> By the scarcity of our coal our pre-eminence in cheapness of manufactures becomes a thing of the past, the means of paying for imported food will gradually cease, and the pressure of population, together with the increased cost of the necessities of life, by emigration, by an increased death-rate, and by a reduced birth-rate, will change the England of to-day into a country like the England of 1780.

Even his remedies give one a sense of déjà vu:

> After discussing and rejecting the expedience of limiting or taxing the output or export of coal, on the ground that any such measure would impose a serious burden upon our manufactures and commerce...

he supports the traditional monetarist policy of Stanley Jevons—make a serious effort to pay off the National Debt! I wonder whether the readers of *Nature* were relieved by an editorial by George Gore, a month later (1885; **31**, 357), insisting that all could be well if only governments and people understood that

> The practical value of new scientific knowledge as a source of wealth and progress is incomparably greater than that of all the coal deposits, petroleum springs, and gold-fields of the earth.

Science and the State

Social responsibility *in* science cannot really be disentangled from social responsibility *for* science. Modern science is so totally dependent upon state support that we regard government science policy as one of the principal means by which science is made to respond to social needs. Scientists themselves are held to bear great social responsibility through the influence they are able to exert on such policy.

A century ago, that particular moral burden could scarcely be envisaged. In Victorian Britain, both the scientific community and the central government machine were much smaller, much weaker and much more aloof from one another than they are today. They shared very few common institutions. In all those pages of *Nature*, there are few specific references to government research organizations as such. If an informal survey of government expenditure on science is to be trusted (1870; **1**, 589), these consisted of little more than several astronomical observatories, botanical gardens and museums, together with the Ordnance Survey, the Hydrographic Department of the Admiralty, and a few other small sections attached to military establishments such as Woolwich Arsenal. There was simply no government science worth having a policy for.

In this respect—as was frequently remarked at the time—Britain was exceptional among advanced industrial nations. There are occasional admiring references to countries such as France (e.g., 1870; **1**, 587), where science was much more obviously incorporated within the powerful bureaucratic apparatus of the State. Surprisingly enough, the United States Federal Government is also seen to be much more involved in science than is the British Government, perhaps through the necessity of creating a coherent scientific organization to survey the vast continental territories opening up in the West. There is a familiar sound to the report that Congress had at last created the United States Geological Survey, to put an end to "the chronic feuds to which so many independent United States Government Surveys with rival objects and officers gave rise" (1880; **21**, 197).

There seems nowhere to have been a really strong connection between science and politics. In an age of fierce national rivalry, the fact that each nation was very proud of its "own" scientists and their achievements for the benefit of their countrymen was not turned into an operational instrument. There are no reports of constraints on travel or of communication between scientists. And yet there is a decided absence of positive international collaboration on scientific projects—witness the seventy stations set up across the Pacific by astronomers from six countries to observe the transit of Venus (1874; **11**, 102), without any overall coordination of effort. The important conference in Washington that finally agreed on Greenwich as the Prime Meridian and standard for universal time was *inter*national, not in any sense *trans*national (1884; **31**, 7, 82): the delegates, "in some cases scientific men, in others the ambassadors accredited to the United States, were instructed by their respective Governments specifically for the settlement of [these] questions"—and, in the end, France abstained. We may observe the same clashes of national pride in settling a contemporary version of the same issue—the harmonization of the days of "summer time" in the European air travel region!

But the scientists themselves were not indifferent to the vast resources that could be tapped from the State—presumably without significant loss of independence in research. In its very first volume, *Nature* gave full editorial support to the report of the British Association recommending an inquiry into two questions:

I. Does there exist in the United Kingdom of Great Britain and Ireland sufficient provision for the vigorous prosecution of Physical Research?

II. If not, what further provision is needed? and what measures should be taken to secure it? (1869; **1**, 127).

It was confidently argued that a Royal Commission would show such deficiencies in the organization and magnitude of government support for science that a great reform of the system would be inevitable:

...one of the important results of the analysis will be bringing to light the scattered character of our scientific efforts: almost every department of the State having charge of some scientific institution—the Admiralty of one, the War Office of another, the Board of Trade of a third, and so on, a dispersion which is absolutely prohibitive of harmonious systems of progressive improvement, of efficient superintendence, of economy in expenditure, and of definite responsibility...

We may hope as another most important result, that a central administration of scientific affairs will be shown to be necessary. In all other civilized countries a Minister of State is charged with this duty. It seems absolutely impossible to organize or maintain in an efficient state anything like a harmonious scientific system, without a dominant authority presiding over the whole (1870; **1**, 589).

As was pointed out at a meeting at the Society of Arts,

Mr. Chadwick, on one occasion, desiring to ask in the House of Commons a question regarding some scientific matter, found that it affected four different departments, and should therefore elicit a quadruple reply, the horrors of which he evaded by most informally putting the question to the Premier himself (1870; **1**, 575).

Alas, this campaign can scarcely have been successful—for the present situation could be described in just the same terms. There has not yet been what the mid-Victorian reformers thought to be "probable"—i.e., "a total re-arrangement of the internal organization and the official distribution of our scientific institutions, with a view to concentrated superintendence and responsibility"—perhaps for the very reason that this "will also involve a revision of scientific staffs and salaries, with all the attendant questions of patronage, promotions, distinctions, privileges and pensions" (1870; **1**, 589). The whole subject is scarcely mentioned at all in later years. Even the material fruits of this agitation were not prolific. Looking

back over the first twenty years of its existence, *Nature's* report is cool
(1889; **41**, 1):

> The support afforded by the Governments of Western Europe to scientific
> investigation has been markedly increased within the period which we sur-
> vey. France has largely extended her subsidies to scientific research, whilst
> Germany has made use of a large part of her increased Imperial revenue to
> improve the arrangements for similar objects in her Universities. The Brit-
> ish Government has shown a decided inclination in the same direction: the
> grant to the Royal Society for the promotion of scientific research has been
> increased from £1000 to £4000 a year; whilst subsidies have been voted for
> the Marine laboratory at Plymouth, to the Committee of Solar Physics, to
> the Meteorological Council, and quite recently to the University Colleges
> throughout the country...

Science as a Profession

Much more information than can be gleaned from those old volumes of
Nature would be needed to enter deeper into this important historical issue. It does seem, however, that the demand for better "provision" for research was not primarily directed towards finding the funds for major
pieces of research apparatus, or to setting up larger, more sophisticated research institutions. There were complaints, of course, at the way in which
funds for "Large appliances" such as telescopes have come from private
benefactors rather than from public funds (1869; **1**, 263: 1870; **1**, 316, 375,
409: 1874; **11**, 62: 1879; **21**, 19, 47: etc.). This was not yet the age of Big
Science, with all that that can imply for science, for scientists and for society.
The real call was for better endowment of research as a profession, with
many more scientists being supported directly in paid employment.

The scientific community of the day was not so small, after all. Leone
Levi estimated that there were about 120 learned societies, with an aggregate of 60,000 members. Allowing for overlapping membership,

> we arrive at the interesting fact that there are in the United Kingdom,
> 45,000 men representing the scientific world, or in the proportion of fifteen
> in every ten thousand of the entire population; the "upper ten thousand" of
> the aristocracy of learning being thus three times as many as the "upper ten
> thousand" of the aristocracy of wealth (1869; **1**, 99).

But, despite the value of scientific research to industry and employment,

> there is in this country no recognized payment for the labors of scientific
> discovery, and no provision for the support of men who investigate science.

...In consequence of the peculiar nature of the occupation, its hopelessness as a source of emolument to the investigator, the great skill and extreme self-denial required, and frequently danger incurred in its pursuit, and the consequent great difficulty of achieving success in it, scarcely one person in one million of the population of England is exclusively devoted to it, although a much greater proportion occupy a small amount of their time in its advancement (1870; **1**, 624).

Why was this deficiency not remedied? On what grounds had journalists previously regarded "the proposal to endow scientific research as a visionary and wild scheme," so that it was still necessary to argue very cautiously "to show that it is practicable, by means of a judicious application of precarious salaries, to train up a class of scientific investigators, and that it is a safe investment to give endowments to young men before they have reached eminence in their studies?" (1874; **11**, 1).

Nature was evidently ahead of public opinion, even among scientists, on this point. Two letters to the Editor in 1870 suggest some of the objections that must have been widely held. A.R. Wallace (**1**, 288) argued forcefully that

the State has no moral right to apply funds raised by the taxation of all its members to any purpose which is not directly available for the benefit of all,

hence,

if our men of science want more complete laboratories, or finer telescopes, or more expensive apparatus of any kind, who but our scientific associations and the large and wealthy class now interested in science should supply the want.

Thus,

if we once admit the right of the Government to support institutions for the benefit of any class of students or amateurs, however large and respectable, we adopt a principle which will enable us to offer but a feeble resistance to the claims of less and less extensive interests whenever they happen to become fashion.

This letter was soundly rebuffed, of course, in an editorial in the same issue (**1**, 279); but recent political events in Britain and the United States show that it expressed a viewpoint that can still find substantial favor in the public mind. There must also have been many older scientists who would associate themselves with the "garrulous old man" who signed himself "in sicco" (**1**, 431) and asked, "In what way can Government most beneficially interfere with the spontaneous energy of original scientific laborers," and then spelled out all the theoretical objections to paying research workers either at piece rates "by results" or beforehand for work they were expected to do. It is an elegant bit of rhetoric, not without applicability to the modern scientific profession. Certainly, conditions in our own academic profession are not so far from those envisaged in the draft scheme, where

> no candidate is to establish his claim to a permanent endowment until he has previously served an apprenticeship of some ten years, during which he must furnish continual proofs of his aptitude and diligence, and will receive regular payment by results amounting to a continuous salary if his work is satisfactory (1874; **11**, 2).

In other words, science at the time of Alfred Nobel was to be regarded primarily as a *vocation*, from which the scientist was to be rewarded as much in personal satisfaction or public acclaim as in material gain. To pay for research was to corrupt it, not only bringing it down from the transcendental plane of truth for its own sake, but also reducing it to "hack work" whose intrinsic quality was then suspect. So strong was this sentiment, even outside science, that *Nature* had to combat a proposal that scientists would happily examine patents for free, even though lawyers, of course, would expect to be paid fees for their part in the same work (1874; **11**, 191).

Scientists in Society

Not being directly dependent on the production of research results to make a living, the nineteenth-century British scientist had all the freedom he needed to exercise social responsibility. But this spirit of enthusiastic amateurism disconnected science from other social realities and gave to research a superficial air of dilettantism which could easily slip into a frivolous indifference to the social consequences of what was being sought or accidently found. It has to be admitted that the world of science, a century ago, was far more self-sufficient, far more inward looking and parochial than it can afford to be today.

Yet the scientific community was by no means an isolated, disregarded, powerless sectarian group of "outsiders." *Nature* speaks in the confident voice of the professional upper-middle class, well aware of its superiority to the masses, in education, culture and material well-being, and not compelled to kowtow to the aristocracy or the plutocracy. The guest of honor at its ceremonial occasions might well be the Prime Minister of the day, and the leading scientific notables evidently had personal access to the centers of political and economic power. Although the formal links between science and the government were limited, there were informal connections—especially through higher education—through which scientific opinion could have its influence within the Establishment. This opinion was fragmented and practically unconcerned with the major political and social issues of the time. It could easily be doped into complacency about the ills of poverty and social degradation, or of militarism and imperialism, by a mild dose of social Darwinism (1869; **1**, 183). But it was not crudely reactionary, nor vulgarly arrogant, nor shrill with demoralized radicalism.

For all the frustrations that they suffered in their civilizing mission, the mid-Victorian scientists were quite confident that this mission was on its way to success. In science they had the *practical* means of improving society:

> It is an important national question, 'By what means can employment for workmen in this country be increased?' My reply is 'By encouragement of experimental scientific research.'

They also had the *spiritual* means of achieving progress (1870; **1**, 623). In their enlightenment, they could ask a leading politician to "set himself to rescue politics from its present degraded position as a mere theater for party strife, and to elevate it into something like a science of national life and progress" (1880; **21**, 295), thus indicting their very real feeling that the scientific attitude is the ultimate in social responsibility.

Has this attitude changed at all, in the scientific community, in the past century? Might not a modern editorial in *Nature* go on in the same vein as:

> There are one or two eminent men of science in parliament, but no one of either party ever seems to think of looking at any measure or any line of conduct apart from party bias, and solely as a matter for scientific consideration...And we should advise those of our public men who are really desirous to discover the science of statesmanship, and to guide their public conduct by its principles, to leave the method of agitation alone for a period, and take to calm but rigid scientific research in their own department, and we are sure the results will surprise even themselves.

A vain hope, alas, now as then!

The little historical investigation that is reported here is too weakly documented to prove any unusual hypothesis about the past. The whole question of how scientists in the time of Alfred Nobel saw themselves as social actors begs innumerable other questions of social psychology, of politics, of cultural history, and of the influence of science itself on human affairs. The particular archive from which I have sampled is certainly highly relevant and worth detailed study for evidence on just such questions, but needs to be supplemented from many other sources. But the working hypothesis of thematic continuity was not seriously disconfirmed, even on issues such as science and war, for which I had confidently expected to find a decisive change of tone. One could even pick up the traces of many particular controversies, such as the place of science in the primary school, or the value of numerical indicators of the fighting power of weapons, which seems to us to be so distinctly modern.

This selection of articles for attention, and of passages within them, has been a purely personal responsibility. Somebody else might read those volumes quite differently. For me, at least, it has been a valuable experience of the long time-scale of change in the inner ideology of a culture that seems to have been utterly transformed outwardly. I can clearly hear the characteristic thoughts and emotions, the enthusiasms and distastes, the concerns and indifferences, of my very own academic colleagues in the words of those nineteenth-century scientific worthies, drawn away from their laboratories, for a little while to discuss education and employment, health and safety, peace and war, and other such issues of the "outside" world. That could almost be heartening, for it suggests that *our* pessimism may be no more justified than *their* abounding optimism: if so few of their confident hopes for progress came to full fruition, need we be quite so dismal about *our* fears for the future, which could be equally without foundation?

Rights and Responsibilities in Research

I t is generally agreed that the practice of research in science and technology generates quite strong claims to the exercise of particular human rights, and also imposes substantial responsibilities of those involved. Any student of political sociology would expect to find some trade off, or balance, between these two moral categories. But there does not seem to be a rational calculus for computing the relative weights of the different components so that, in fact, the balance is struck at different points in the range in different cultures at different times. The question is—what relationship between rights and responsibilities for scientists and technologists is optimal for a developing country such as India?

The very words, science and technology, refer back to a cultural tradition that grew up as various advanced countries passed through their respective industrial and scientific revolutions. In this tradition one can separate two distinct strands—those of "Academia" and of "Industria." In their most extreme forms, these two ideological strands take very different stances towards the relative rights and responsibilities of what we might call "scientists," on the one hand, and "technologists," on the other. As we shall see in a moment this dichotomy should not be over-emphasized; but it provides a convenient basis for further discussion.

In Academia the tradition is that the scientists have many rights and few responsibilities. Freedom in teaching and scholarship, academic tenure, public recognition of research contributions, promotion solely on the basis of published work, encouragement to travel and protection from political harassment, are regarded as appropriate conditions for their work. Indeed, it is easy to argue that these rights are necessary if they are to be able to respect the fundamental norms of the scientific community—community-spirit, universalism, disinterestedness, originality, and skepticism, as listed by R. K. Merton. In fact, it is very instructive to link the sociology of science with political theory by showing that the modern international code of human rights, as defined in various international legal instruments, provides almost complete coverage for the traditional practices of academic science.* Academia does not ask for more rights for its members than ought to be available for all people—but it does have a strong stake in making sure that these rights are respected in practice.

And yet, traditionally, scientists in Academia are not charged with heavy responsibilities. They must, of course, do their teaching and be sure to keep up their list of publications, but their research is supposed to be quite "pure," simply knowledge gained for its own sake, and not answerable to the prior conditions of a plan, nor in the way it might turn out in practice. That, at any rate, is the idealized extreme of the academic ideology.

In Industria, on the other hand, responsibilities are heavy and rights are restricted. The technologist is expected to apply his knowledge to the service of his company, his country, or, nowadays, to less tangible human needs. If something goes wrong then he, or she, must bear the blame. Technology is knowledge for use, not just for its own sake.

By tradition, the technologist in industrial, or military, or governmental, research and development has very little freedom within the confines of his own job. The human rights that he enjoys as a citizen are carefully drawn in such a way that they do not interfere with the power of his employers over the initiation, prosecution and publication of his research. Commercial and military secrecy acts as a constraint in his contacts with his colleagues at home and abroad, and all questions of promotion are decided within the organization for which he works, without direct reference to the opinion of the community at large. The technologist in Industria is little more than a hired hand, and is certainly not in any position to follow the norms of academic science.

Many of you must now be protesting against this grossly exaggerated picture of completely contrasting archetypes. And, in reality, in the most advanced countries, Academia and Industria have become so mingled that the distinction between the "scientists" and the "technologists" is scarcely meaningful. One could argue that the supposed distinction has always been fictitious, that Academia, for example, has always really been serving the purposes of Industria, that the scientists were all, in principle, technologists and their allegiance to "pure" research an ideological myth.

---

*Scholarly Freedom and Human Rights (Council for Science and Society, 3/4, St. Andrews Hill, London EC4).

Merging of Industria and Academia

The real point is, however, that whether or not this myth has historical substance it is certainly not consistent with the contemporary situation. One can see quite clearly, nowadays, how Industria and Academia merge into one another in a variety of ways as follows:

(i) A very high proportion of the knowledge that is generated in Academia is potentially applicable. This is recognized by the financial support given to academic science—often to meet strictly technological objectives. Thus Academia must be much more "responsible" than the myth would suggest. At the same time, a great deal of "useless" knowledge is generated in Industria and made public in the academic style;

(ii) In times of national need—especially war—academic science is brought completely under control and the rights of the "scientists" are narrowly constricted. These conditions have persisted into the era of heavily armed truce that we now call peace; secrecy and other significant constraints on the characteristic cosmopolitanism of the academic tradition have greatly reduced the effective rights of the members of Academia;

(iii) A considerable proportion of the human effort that goes into "academic" research is now aggregated into big teams working with the big facilities of Big Science. A scientist working in such a team must subordinate himself to the program of the team in just the same way as if he were a technologist carrying out applied research and development;

(iv) On the other hand, because the academic style of science is often very fruitful technologically, Industria tries to woo the scientists by offering some of the benefits of Academia—e.g., more freedom and less responsibility than the technological realities seem to demand. This tendency is supported by the movement towards more direct recognition of human rights—as it might be, for example, in explicitly guaranteeing the property rights of technologists in their personal inventions, or in protecting them from reprisals if they put the public good above their contractual duties to their employers; and

(v) Within this tradition, also, one must include technological practitioners such as physicians and consulting engineers who learnt to protect their individual rights by forming themselves into professional associations, etc., which have considerable influence within Industria.

In most advanced countries, therefore, there is a dynamic balance between rights and responsibilities that is much the same for both scientists and technologists. A large body of QSEs (Qualified Scientists and Engineers) are employed by large bureaucratic organizations devoted to "R&D" (i.e., Research and Development), oriented towards missions of short term or long term technological significance. Each individual QSE is in duty bound to work within the program of the organization and must, therefore, bear at least some minimal responsibility for the outcome of that program. Generally speaking, he, or she, is also in a reasonable position to exercise the rights of a fairly well educated and responsible citizen in that particular country. Within this generally industrialized R&D system, however, there remain strong traditions of loyalties to a more transcendental conception of knowledge and to styles of research and invention that are much more individualistic in the exercise of both rights and responsibilities.

This stark description of the state of the research profession in most advanced countries is, of course, grossly over-simplified. But it does, I think within the limit of a short paragraph, correctly state the typical situation and, despite enormous contrasts in political ideology, economic organization, and social structure, between capitalist and socialist countries—e.g., the United States and the Soviet Union—this description applies more or less equally to them all. It is only when we look more closely at such matters as respect for particular human rights, relations between centralized governmental organs and private corporations, the degree of autonomy allowed to trade unions and professional associations, the social status accorded to the technical expert or scholar, etc., etc., that we can really see the differences from one advanced country to another.

In the Developing Countries

Now, what is the position in most developing countries? It seems to me that the traditional dichotomy between Academia and Industria is much more apparent than it is in advanced countries. As evidenced by this Seminar, the situation is changing rapidly, but at least until a few years ago the myth of an "irresponsible" claim to the right to pursue pure knowledge without serious interference still held sway in Academia. True, the academic scientists were poorly paid and heavily loaded with teaching, so that their freedom to participate in the advancement of universal knowledge was largely a sham, but their rights in this respect were seldom formally constrained by censorship, prohibitions on travel, etc. In these very same countries, however, the "technologists" in government research laboratories and in the employ of overseas industrial corporations were in the state characteristic of "Industria"—supposedly "responsible" for a variety of socially significant, or profitable, projects but, in fact, so subordinate in status and rights that the research that they carried out was seldom very effectual. Not only were these various communities cut off from one another, intellectually and socially but they also seemed quite distinct in their internal workings and external relations, as if "scientists"and "technologists" belonged to quite distinct estates of society.

It would be possible, I think, to trace these distinctions back to the various channels through which European science and technology actually came into each developing country. That is to say, the myth (or tradition") of the two distinct strands was zealously preserved and clothed in actuality—indeed, almost in caricature—in the course of this transfer process. Was this, as some historians would claim, a deliberate act of dividing to rule, or was it just a naive interpretation of current philosophies of science, which tended to emphasize a division of labor between the pure academic researcher and the applied industrial developer? Was this sharp dichotomy between Academia and Industria in the Third World inherent in the social and political situation, or was it, simply, a deep frozen version of a social situation that had long been outdated in the countries of its origin?

Historical questions like that are very difficult to resolve. What seems to me quite clear is that many of the factors that have drawn Academia and Industria together in the advanced countries are much weaker and less effective in bridging this gap in the Third World. Going back, for example, to the list of the various forces pushing "scientists" and "technologists" into the same general category of "QSE" we can readily check this argument in detail. Thus, for example, the meager results of research under poverty-stricken conditions, in both Academia and Industria, have not made them very tempting to the powers that be, whether for application or for national prestige. Again, the capabilities of Academia in wars fought with imported modern weapons have scarcely been worth enlisting. Since the material resources are quite inadequate for Big Science, there had been little occasion to "industrialize" academic research; although, of course, Academia has not been spared the clumsy and wasteful bureaucratization that is characteristic of so many facets of life in developing countries. It would be wrong to suggest that the general background of "human rights" in developing countries is worse, on the average, than it is in advanced countries; but it is quite obvious that professional organizations, such as learned societies, technological institutions, technical trade unions etc., are much less able to protect their members against the arbitrary powers of governments or large corporations. Thus in almost every respect, the forces that draw Academia and Industria together are much weaker in developing countries than they are in advanced countries.

In the last decade it is true that the myth of the separation of responsibilities from rights in the research world has been challenged, in the name of making research more relevant. In many ways, this is a very healthy trend towards a more realistic appraisal of the place of science in society, but the immediate effect on the rights/responsibility relationship has not been altogether beneficial. The main trend has been towards a loss of valuable rights in the name of responsibility—but not always with tangible results. We must all be aware of several disfunctions, such as excessive demands on research workers for short term achievements, barriers to international travel, degradation of the infrastructure of communications amongst scientists around the world, and many, many, follies arising from over-optimistic, over-zealous, and bureaucratic, planning of research. What is happening is that Academia in the developing countries is being pushed quite strongly into the subordinate estate characteristic in Industria, without any significant improvement in the condition of the latter.

But there has also been a significant change in attitudes towards the rights and responsibilities of scientists and technologists in advanced countries in recent years. For various historical reasons, this question has become much more salient in the scientific community in the past decade and is being debated much more thoughtfully than in the past. That is to say, instead of just taking for granted the traditional myth of the division of labor between scientists and technologists, as a matter of basic principle, there is now much closer attention to the realities of the R&D system as it now exists in all industrialized countries. There is certainly not the time here to review the arguments that are now brought to bear on such complex and subtle issues as the civil rights of research workers, responsibility for the unforeseen consequences of research and development, the balance between internal and external criteria in science policy, the balance between autonomy and accountability in funded research by academic institutions, etc., etc.

Need for Academic Freedom

In all these discussions, however, one returns to some basic principles which must continue to be emphasized. One of these is that the individual rights that we refer to generally under the heading of "academic freedom" are pre-requisite for a personal capacity to carry out research in a fully responsible manner. This is particularly true in developing countries, where there is only a very limited supply of highly competent research managers who can oversee and direct the activities of large numbers of scientists and technologists to prescribed goals. In other words, the very poor performance of Industria in most developing counties is not likely to be cured by imposing yet firmer authority onto people whose personal rights within the research system are already very limited; the right direction might be to recognize that a looser structure in which individuals can actually realize the responsibilities that are being thrust upon them may prove much more fruitful. On the other side of the equation, it is now more widely appreciated in the advanced countries that academic freedom and support for basic research is not an absolute right, but is earned by Academia in return for the long term, unforeseen, social benefits of research and for willingness to participate in shorter term technological development whenever this is appropriate. In other words, the proper response of Academia to the demand that it should become more responsible in its research activities is not to retreat behind the walls of the ivory tower of knowledge for its own sake, nor yet to surrender to every call for immediate relevance but, at least, to think out the potentially applicable aspects of the research being undertaken and to justify these as far as is reasonable in the corridors of power.

It may seem that these principles are not very novel, nor very spectacular and perhaps not even consistent; but then the whole relationship between individual rights and responsibilities is not new and has always seemed stubbornly contradictory to crude political rationality. The wisest course is not to try to "reconcile" these opposites, or to weaken them both by playing one off against the other; it is consciously to strengthen both sides—the rights of research workers on the one hand, their responsibilities, on the other—and to get the fullest possible benefit from the dynamic balance between them.

The fact is that wherever we look in all countries, nowadays, the place of science, and of science-based technology, in the polity, is still highly ambivalent. In some ways it is regarded as the instrument by which the state, or other powerful social institutions such as commercial corporations, meet the demand, or the needs, for which they are held responsible by society. In this instrumental mode, science is highly technical, rationally planned and managed, and fragmented corporately or nationally. But science is also seen to be an organ of transcendental knowledge, with loyalties to humanity at large and an ideology of universalism in which all knowledge is public and every significant community has world wide membership. This transcendental mode is not so much in favor nowadays as it used to be, and no longer appears to be justifiable simply by reference to knowledge as a universal good, more to be valued than any other component of the human condition. Nevertheless, it could be gross folly to try to establish science within the polity solely in the instrumental mode. It may well be that, paradoxically, the best way for science to serve society—in all sorts of countries, from the most advanced to the least developed—is to keep faith with its myth of universality and thus to maintain the creative tension between what is known and what may yet be discovered, between the imaginative and critical faculties of the mind, between the autonomy and the social accountability of the scientific community and its institutions, between basic science and technology—and above all, between the rights and responsibilities of scientists and technologists.

Science Education for the Real World

The question is whether one learns enough about science just by learning *science*. Every page of *New Scientist* illustrates the theme of "STS"—the intimate relationships between science, technology and society. Is this connection so clear and obvious that it doesn't bear teaching about?

The main stream of education in the natural sciences, medicine and engineering flows on, apparently unperturbed by this question. The professors and heads of science departments ward off the threat of doing something about it by showing that there is no room for it in the timetable ("We've already had to drop geometrical optics so as to give them a bit of laser logic") or by regretting that they have no one to teach it ("Not a *sociologist*; not in *my* department") or by uttering an incantation "Sheer Prometheanism, you know") or, as a last resort, by putting a humanistic gloss of history and philosophy of science on a few consenting intercultural transvestites.

But the movement for STS education will not go away. It has gained a foothold in many universities and in most polytechnics, and is beginning to appear sporadically in the schools. It is still a marginal enterprise, sustained by little ragged bands of enthusiasts rather than by people out to make a success of their own careers. Nevertheless, according to recent reports, STS courses of one sort or another now occupy about 1,000 student years of tertiary education in Britain, and probably comparable proportions in several other countries such as the United States, Holland and Australia. Although the current phase of austerity in education is unfavorable to what many sympathetic science teachers still regard as a luxury, the STS movement is steadily growing in coherence and influence.

Can one really describe as a "movement" such a diverse activity, ranging from sixth form general studies to higher degrees covering a miscellany of topics from history, philosophy, economics and politics, and taught from many different logical and ethical points of view? For most of those who are involved in it, STS education must seem as varied in its goals, scope and methods as science itself. Nevertheless, there is an underlying unit to this movement that does not always appear on the surface. This applies particularly to its fundamental goals. The motive force for STS education has two main strands, which are now merging into a single braid.

The first strand was originally very radical, politically and ethically. In some ways, science is conventional and conservative, and strongly entrenched within the "Establishment:" it is thus a natural target for revolutionary zeal. From its birth in the late 1960s, the STS theme has always been a vehicle for the idealism and enthusiasm of many good people who would like to make a much better world. Like every movement for *educational* reform it was soon loaded with *political* and *moral* responsibilities that it could scarcely bear. According to this view, the crimes and follies of contemporary society—Third World poverty, exploitation of the workers, degradation of the natural environment, escalating horrors of war, etc.—are revealed through the scientific technology that made such iniquities possible. Perhaps it even created them?

But experience has shown that students of chemistry or agriculture are not easily converted into militant radicals. They rarely catch fire from sophisticated variants of such slogans as "Down with capitalism," "Back to nature," or even "Burn the books." Many of those who began to teach about science and society in this spirit have been tamed by student indifference, and have become soberly committed to STS education in its own right, as a source of general enlightenment for students of the wide range of political and religious opinions tolerated in a democratic society.

The other strand was initially much more conventional. Thoughtful science teachers have long been concerned at the narrowly specialized education that their students receive, from O-level onwards. The historical and philosophical aspects of science can be made into a bridge to the "other culture," where humanistic skills and interests can be fostered. But modern history and philosophy of science are not fenced off from sociology, social psychology, and the other sciences of human behavior. The morally uplifting life story of Charles Darwin leads back to the origins of the theory of evolution in the political and economic writings of Thomas Malthus and on to the perversion of biological theory to support racism and imperialism. As one enters the general territory of STS, one inevitably becomes more and more aware of the many-sided political and moral issues in which science is deeply involved. Although they seldom turn completely against science itself, those who began to teach about science with vaguely liberal cultural intentions soon found themselves taking a more critical attitude to their subject.

But all these "liberalized radicals" and "radicalized liberals" in the STS movement combine their forces against the "scientism" and "technocracy" implicit in the way that the sciences are usually taught. Most courses of physics and biology, medicine and engineering, right up to the frontiers of the professional practice and advanced research, give the impression that scientific knowledge is unquestionably true, all-embracing in scope, and uniquely valuable to mankind. They suggest that the problems of the world can be solved (if at all) only by scientific experts using the scientific method. Research is represented as a dedicated, selfless pursuit, monkishly intense and aloof, and yet generating endless benefits—a religious vocation rather than a practical profession.

This bias in science education is not intended. It is generated by what is left out rather than by what is actually taught. The laws of thermodynamics and the genetic code are innocent in themselves, but their significance depends on how they are to be used in the real world. To correct the bias one must learn that what passes for scientific truth is usually uncertain and always incomplete, that the pursuit of knowledge "for its own sake" is not an absolute good, that the major and minor issues of life are so value-laden and imperfectly specified that they cannot be resolved by a purely scientific approach, and that scientists are as fallible and corruptible as any of the other technical personnel of modern society. All thoughtful research workers and teachers know this perfectly well; the fundamental goal of the STS movement is to make this knowledge public.

This process of demystification is like a voyage of exploration: it can be begun only from where the students happen to be. For those who are already well on the way to careers in science and technology, it is natural to start from the "interior" of the research world (Figure 1) and to work outwards. An STS course for undergraduate physicists might first examine, for example, the logic of research and discovery, on which contemporary philosophers such as Karl Popper and Thomas Kuhn have cast a cold light of doubt. This leads on to the working habits of the scientific community; are the high-minded "norms" of science suggested by Robert Merton realized in practice; should they be displaced by the less reverent formulations of other schools of sociology, taking account of the more mundane interests of scientists, such as getting on in the world or keeping their jobs? The student may then discover that the autonomy of this community is much exaggerated. To slake its insatiable thirst for research

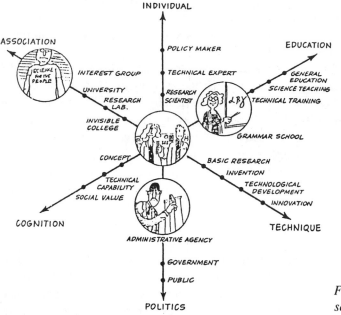

Figure 1. Science seen from within.

funds, academic science ingratiates itself with the state, and become subject to the various government organs for science policy. Indeed, it is now obvious that the basic research itself has become so highly organized and technical (think of high energy physics, or the latest methods in molecular biology) that it can scarcely be distinguished from the R&D carried out in mission-oriented government and industrial laboratories. In other words, the student learns that what the world at large calls science is not just the coy community of aspiring Nobel laureates but a sprawling unruly sector of society, shot through with political, economic, cultural and personal interests. He or she may thus come to realize that Hiroshima was not an historical aberration due to the physicists having "known sin," but was just as typical a product of science as the theory of black holes or the discovery of penicillin.

On the other hand, for the vast majority of people who are not intending to make direct use of science in their working lives, it is more appropriate to take up the STS theme from the standpoint of everyday life (Figure 2). Such a course might, for example, be based historically on the growing influence of science in industry, engineering, medicine, agriculture, warfare, and so on, over the past few centuries. This can be linked to the great issues of the present world *problématique*—nuclear weapons, population, energy, food, industrialization, and so on. Science is obvi-

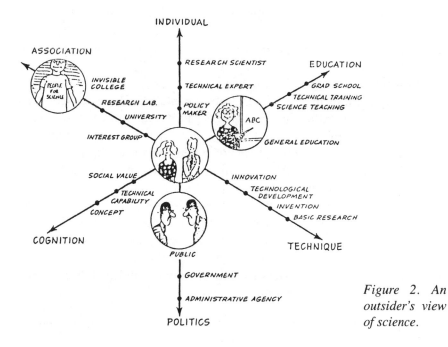

Figure 2. An outsider's view of science.

ously one of the causes of these troubles; how far can it be applied to resolving them?

Many rather naive STS courses end at this point. But if we really want to understand the way in which science is affecting the quality of life over the whole Earth (for good or ill; that may be a matter of opinion) we need to study the research process itself in a little more detail: what do scientists really do; who gives them their orders; how valid are their discoveries; and how are they put to use? In other words, we must eventually focus attention on the "internal" sociology, philosophy and psychology of science, where so much of the action surprisingly originates. The student may thus come to realize that the theory of black holes, apparently quite remote from all human affairs, may not be unconnected with events as worldly as the destruction of Hiroshima or the discovery of penicillin.

These are only two of the many different ways in which the STS theme can be explored. It is not like a conventional science, with a standard curriculum to be followed from level to level up the educational pyramid. For example, a good course designed to "integrate" the various components of a "combined science" degree can be constructed on many different lines, touching upon many different subjects. Although there is tacit agreement on general objectives and on the range of topics that might be considered relevant, STS education has not yet crystallized into a regular academic discipline. The is still plenty of scope for the independent-minded teacher with the urge to influence, inform, or enlighten his or her pupils, by all sorts of means towards all sorts of ends. Scientists dislike such educational anarchy; for many of us in the STS movement, a diversity of opinions is a sign of strength and vitality.

The original impetus for STS education came from high academia and from radical political circles outside the educational system. Whether or not it was supposed to be linked with political action, it was mainly conceived in the rather highbrow terms of advanced social theory. Much of that theory was as remote from real life as the science that it purported to criticize. It is fascinating to observe how it has been domesticated by educational experience and brought into closer harmony with its own "social context"—that is, the simple needs, capabilities and aspirations of those to whom it is addressed. What we are beginning to realize is that Science, Technology and Society are not distant abstractions: STS education is about tangible features of everyday life, worthy of informed discussion in school and college by every future citizen.

INTO THE SOCIAL DIMENSION

Science is an *institution*. Precisely what this proposition signifies is a matter of earnest sociological debate. It certainly says more than that scientists consciously relate to one another as members of a community, or that scientific progress is shaped by societal influences. It means, for example, that scientific activity can be described as an interlocking set of social *practices*, carried out regularly by individual *actors* performing typical *roles*. To complete our study we need to consider the social dimension of these roles.

An institution is characterized by its mission. The trouble is that science has two self-proclaimed missions which are by no means equivalent: it aims to produce knowledge, and it aims to be useful. The effect of emphasis one or the other of these missions is to produce contrasting conceptions of science in the public mind. Up to this point, we have followed scientists themselves in favoring a "discovery" model, which concentrates on the mechanisms by which research results are generated and validated. But most people talk about science in terms of an "instrumental" model, which stresses the means by which science and technology are directed and exploited for practical purposes. Of course these distinctive missions are closely connected. In reality these contrasting conceptions are just different aspects of a very complex institution. But many people (including some self-assured metascientists!) misconceive science at vital points by referring exclusively to one or the other model.

For the scientist at the laboratory bench, grand missions and general models are largely irrelevant. The whole vast institution manifests itself locally as a social scene within which to play out a personal role. How, then, is this role constituted? At one extreme, there are social theorists who would argue that the script is so strongly determined by structural principles, functional norms, established hierarchies, etc. that even the most creative researcher is largely the unconscious puppet of uncontrollable forces. At the other extreme. there are metascientists who define the institutional environment as a jungle of perils, where every action is designed to maximize the perceived interests of the actor. But these apparently contrary views are merely dialectical truisms about social life in general, and provide no basis for further conclusions.

Science is, after all, a very distinctive institution, whose dual mission is concerned with the real world. Scientists would agree that they are socially bound by the established principles of their profession, and also that they are socially driven to compete with one another for recognition. They would insist, however, that their actions are largely determined by nature itself, through the necessities of logic and the contingencies of experience. The truly individual element in their roles comes through freedom in deciding what research they should do, and how it should be undertaken. "Problem choice" is thus what a physicist might describe as a triple point in the phase diagram of the scientific enterprise. Many significant features of science are revealed by the interaction of personal, social and intellectual considerations at this much neglected site for metascientific investigation.

Another vital function is the process by which the results of research are put into the hands of the scientific community. My very first venture into the field of science studies was *Public Knowledge* (Cambridge University Press: 1967), where I showed that the whole scientific enterprise was centered socially on its communication system. Here again, a subtle balance is maintained between the selfish personal interests of researchers seeking to get their work accepted for publication, and their shared communal interest in ensuring that the knowledge claims in the archives are as reliable and convincing as humanly possible. An analysis of the background to the sardonic careerist slogan "publish or perish" thus reveals a great deal about the workings of science as an institution, showing how social practices and individual attitudes slowly evolve and adapt to scientific progress and societal change.

These are two of the key elements in the scientific enterprise. But how do they articulate with one another, and with other elements in the system, such as the supply of material resource the recruitment of researchers, or the practical exploitation of knowledge? For example, what ensures that innumerable individual problem choices, as in "What Are the Options," add up to a major movement towards new scientific goals, as in "Reconstructing the Reality of Scientific Growth" and "What Shall We Look into Now?" What features of the communication system discussed in "Publish or Perish" serve to guard the frontiers against the inroads of parascience, self deception, or fraud?

The answer is that science works on the market principle of mutually beneficial exchanges of goods and services between independent customers and vendors. Some of the items exchanged are abstract and intangible, such as "recognition," and some of the transactions are notional or metaphorical. There are several distinct marketplaces, dealing in research projects, discovery claims, scientific reputations, academic posts, intellectual property, etc. But these are all geared together, by customary practices and attitudes, into a complex system, where all the grassroot functions are performed without top-down control.

This implies much more than the familiar dictum that the advance of science is impossible to plan, or that it is foolish to try to organize research on conventional bureaucratic lines. The political ideologues who go on about making science more competitive seem not to know that this has been the name of the game in research since modern science emerged as a social institution in 17th century Europe. Indeed, crude attempts to boost *commercial* entrepreneurship in basic science simply interfere with the well-established forces favoring *intellectual* enterprise.

The spirit of the market is intensely individualistic, in keeping with the stereotype of the lonely seeker after truth. The plain fact is, however, that this extreme individualism is being challenged by scientific and technological progress, which make it more and more difficult to seek after truth on one's own—even within a narrowly specialized field. Serious research nowadays usually requires elaborate apparatus, sophisticated facilities, diverse expertise, and the coordinated efforts of many hands and minds. Scientists have to work in teams, hunt in packs, and share in the use of many expensive instruments. In many fields, internationally competitive results can only be achieved by exceeding a certain "critical mass" of human and material resources. Science is being "collectivized." in the sense that individual researchers are forced to work closely together, in smaller or larger groups, to achieve a collective effect.

Science is also being "collectivized" in another sense of the word: as science becomes more influential in our whole culture, it comes more under the control of the grand collectives of society, such as government agencies and large industrial corporations. This is not a new tendency, but it has speeded up recently as the resources required for research reach an appreciable fraction of the national income. The consequences for science of thus approaching its "limit to growth" are diverse, pervasive, and far-reaching. This is a very large and important topic in its own right, which I discuss in detail in *Prometheus Bound: Science in a Dynamic Steady State* (Cambridge University Press: 1994).

For these various reasons, science is passing through a period of unprecedented structural change. For the moment, this does not invalidate the market principle, since research groups behave like small firms, replacing individuals as the competing units. But when these firms become quite large, as they do in "Big Science" fields such as experimental particle physics and space research, they begin to develop an internal structure which is much more like a bureaucracy than a free market. This, in turn modifies many of the traditional practices and norms of the scientific enterprise, such as freedom in the choice of research problems, or the recognition of individual achievement on the basis of published contributions. In other words, many of the apparently stable features of a professional research career may, within a few years, seem entirely outdated.

Like most spectacularly successful institutions, science is being transformed internally under the weight of its own achievements. For good or ill, the subject of our investigation—the relationship between the personal and social aspects of the scientific life—is changing under our eyes, and taking on new, unpredictable forms. Although this only becomes apparent in these last few chapters of the book, it is the message towards which all the earlier chapters lead.

This conclusion is reinforced by a consideration of what it means to be a physicist. These remarks were intended originally for a lecture in memory of Cecil Powell, who won a Nobel Prize for the discovery of the pi meson in 1950, and who died in 1969 on the day of his retirement from the senior physics chair at Bristol, where he had spent most of his research career. Cecil Powell was a charming and sympathetic man, who had been one of the leading figures in the Pugwash movement, and other international scientific enterprises in the Cold War era. This lecture was to have been given in Prague, at the 1984 Meeting of the European Physical Society. But in the end I publicly declined to take part in a scientific event to which my friend Frantisek Janouch—an outspoken Czechoslovak physicist then domiciled in Sweden—had been refused entry by the Czechoslovak government. This episode illustrates perfectly the moral dilemmas of seeking the truth, whether alone or in good company, in an all too real world.

Conceptions of Science

Old Models and New Perspectives

In the past half century, science has moved from the periphery of social affairs into the center. Public policies for research and development in basic science, and in a wide range of science-based technologies, are of perceptible weight in national and international politics. Technical experts with scientific qualifications have become major actors in public affairs.[1] Attitudes towards scientific innovations and their possible effects on the quality of life determine the fate of political parties and influence the direction of economic development in many countries.[2]

The conceptions that people have of science and/or its associate technologies may thus exert a powerful influence on social action. Since science is a social institution, it is a suitable object for study within the social sciences. Knowledge derived from this source eventually contributes to the public conception of science, and sooner or later manifests itself in political forms. In other words, to make science policy, one must apply some sort of science theory. The rationale of practical R&D is based, consciously or unconsciously, on a metascientific paradigm which is not derived so much from immediate practical experience as from the cultural milieu of the day.

This paradigm represents the current answer to the general question "What *is* science?" Over the centuries, many fine scholars have devoted their careers to this enquiry, in narrow depth or in broad sweep. But the wisdom embodied in their various learned answers is not immediately effective in society at large: it is what science is *believed* to be, by academically unsophisticated people like business executives, members of parliament, judges, scientific research workers, physicians and most other taxpayers, that really counts. These beliefs may not be well-founded in principle: they are, nevertheless, the grounds on which decisions about science are based in practice.

The general public does not, in fact, have a coherent conception of science. The *instrumental* image that views science-and-technology as simply an organized means for solving practical problems by rational methods seems to have little in common with the *discovery* image in which science is regarded as essentially the exploration of the material domain by specially skilled and motivated individuals: Considered as distinct aspects of a single human activity, these contrasting conceptions are not necessarily inconsistent with one another. But without a unifying theoretical scheme, there is no rationale for reconciling the policies to which they would logically lead—on the one hand the subordination of all forms of research to economic or political opportunism; on the other hand, the public support of academic science as an autonomous activity, regardless of any likely utility. In extreme forms, these policies are in flagrant contradiction. On the one hand, for example, the scientific worker is regarded as the servant of any state or industrial corporation that can command or hire his or her expertise: on the other hand, the savant is supposed to belong to a world wide community whose activities transcend national frontiers and commercial competition.[3]

The intellectual origins of each of these contrasting conceptions of science lie far back in history. The instrumental conception can certainly be traced back through Leibniz to Francis Bacon: the discovery conception has even deeper roots in Western thought. In their respective spheres of political economy and philosophy they have been ceaselessly discussed and elaborated, to the extent that they are both embedded almost beyond question in contemporary opinion. Although the policy contradictions that they imply have long been apparent,[4] neither of these traditional doctrines can be shown to be essentially fallacious within its particular sphere of operation, so that these contradictions are popularly regarded as demarcation disputes between "pure" science and "applied" science, without more fundamental significance.

But recent developments in academic metascience—i.e., in the philosophy, sociology, social psychology, economics, political theory, etc., of science and technology[5]—have shown up the inadequacy of these two popular images of science. These developments take account of, and throw light upon many new aspects of this subtle and complex human activity. For example, the internal sociology of science has demonstrated the influence of social relations within the scientific community on the research process—a process whose evolution is also profoundly affected by the external forces studied in the more general sociology of knowledge.

These new perspectives in science studies have not yet found their way into popular understanding. They are not even very familiar to those people directly involved in making science policy, such as politicians, higher civil servants, scientific notables, industrial managers and science journalists. This is evident from the language they use. In such circles, there might be an occasional reference to Popperian falsification, Kuhnian paradigms or Weinberg's criteria for scientific choice: more sophisticated metascientific concepts, such as Mertonian norms, invisible colleges, or co-citation networks would seldom be mentioned.[6]

The Emerging Synthesis

Such a gap between theory and practice is perfectly normal, and can be attributed to a variety of factors. Many of the new concepts are still regarded as controversial within the disciplines where they have arisen. The transfer of novel paradigms from various fields of metascience is hindered by the communication barrier between the social sciences and the natural sciences. Fragmented information is less persuasive, and interdisciplinary perspectives are subject to more skepticism, than data and theories lying within a single traditional academic category. Several decades may be needed for ideas to percolate from the separate departments of academia into common knowledge.

Nevertheless—and here we come to my central theme—these new developments are not as antithetic and chaotic as the sophisticated controversies surrounding them tend to suggest.[7] Despite the multiplication and divergence of specific responses to the question "what is science and how does it work?," there is an underlying convergence of scholarly opinion that a satisfactory answer to this question must be broadly based on insights derived from a wide range of academic disciplines. Diverse philosophical, sociological, psychological and political conceptions of science are now to be recognized as particular aspects of a science/technology complex, which cannot be described satisfactorily within a single disciplinary framework of thought.[8]

This convergence may only be apparent to the eye of hope. It would be going too far to claim that a complete and coherent model of science is emerging from the academic dust storms. Although there might well be general agreement on a list of the significant features of such a model, there is no consensus on their relative weights and interactions. The particular model described here is purely schematic—a design sketch, not a blueprint. The important point is, however, that the elementary instrumental and discovery conceptions of science can be reconciled and connected within a more general theoretical synthesis of this kind, which provides a reference framework for such burning questions of principle as the epistemological status of scientific knowledge, the ethical responsibilities of scientists, and the cultural and ideological roles of science and technology, etc.

Our concern here goes beyond the formulation or validation of the new metascientific paradigms: it is to connect well-founded theory with social practice. If this new transdisciplinary conception of science were more widely diffused outside academia, it could have a powerful influence on the place of science and technology in society. A number of puzzling policy issues could be dealt with more satisfactorily if there were better public understanding of such contributory factors as the limited validity of scientific expertise, the weak and uncertain linkage between basic science and technological innovation, the effects of constraints of scientific communication, the extent to which value judgements enter research activity, or the mismatch between scientific and journalistic media of communication. In the past, it has been almost impossible to weigh up factors such as these, operating in the conceptual no-man's-land between the traditional spheres of influence of the instrumental and discovery image; this is the area that we can now begin to map.

The Instrumental Model of Science

The significance of recent developments in metascience can best be appreciated against the background of a brief account of the two traditional points of view. The *utilitarian* conception of science as a major source of applicable knowledge dates back to the 17th century. In recent years, this conception has become more explicitly instrumental. The main characteristic of science is now supposed to be that it can be deliberately directed towards achieving predetermined goals. The institutional structures, budgetary allocations and other detailed features of *R&D systems* are generated and determined largely by these goals. These systems are regarded as instruments by which nations or commercial corporations can solve the problems with which they happen to be faced.[9]

From this point of view, the fundamental issue concerning science would appear to be the list of problems to which it might be applied. These problems arise in political, commercial, industrial, military or other social domains: the function of science policy is to establish their relative urgencies and possibilities, and to direct the research by which they might be solved. This naturally implies that the necessary personnel should be trained and deployed, and that appropriate financial resources and material facilities should be made available. But the essence of an R&D system is conceived to be the administrative apparatus by which priorities negotiated in the political or commercial forum are transformed into effective programs of scientific research and technological development.[10]

This conception of science is unequivocally technological, in that the objective of research, and the value of its results, are to be judged by standards of practical use. But it is not necessarily scientistic or technocratic. It does not imply, for example, that all social and material problems could in principle be solved by the application of science, or that only "scientific" solutions to such problems are valid. It leaves plenty of room for ethical and ideological manoeuvre in the political domain, both in defining needs and assessing the quality of the means by which they might be met.[11] Nor does it give undue social authority to the scientists and technologists who man the system. Indeed, it is recognized that science can be a very imperfect instrument, which takes considerable skill to use. Many industrial corporations, for example, are often justifiably disappointed in the performance of their in-house R&D divisions, whilst the national R&D systems of many countries are administratively fragmented, and ill-matched to national economic and social goals. In many features such as these, the instrumental model of science is not unrealistic, and can even be shared by people with conflicting socio-political opinions ranging from technological optimism to anti-scientific radicalism.[12]

This model does not, however, include a detailed analysis of *how* scientific research gets the results we think we want. An R&D system is treated as a "black box" that solves problems by an activating principle called "scientific method." Although this method is not supposed to be entirely unproblematic, all questions concerning the reliability of the results that it gives, the role of individual scientists within the system, the social mechanisms by which they interact to produce scientific knowledge, etc., are regarded as belonging to certain technical disciplines such as philosophy, and hence of no more significance in general social policy than purely scientific questions such as the validity of the laws of thermodynamics or the role of viruses in cancer of the lung. From time to time, the answers to such questions may, of course, turn out to be relevant to the use of science as a social instrument, but they are considered to be of little immediate concern to the political decision makers or to the general public, and are almost entirely ignored in political, economical and managerial studies of R&D systems.

In particular, the instrumental conception of science can make nothing of *basic* research—that is research that is not directed towards the solution of any particular practical problem. The possibility that the results of such research might eventually turn out to be useful is not, of course, discounted, but there is no rationale for undertaking it in the first place. Strictly speaking of an R&D system designed as a "black box" without internal sources of action can have no motive for prosecuting such investigations, no mechanisms for deciding the direction they might take, and no criteria for evaluating their outcome.

This is not merely a theoretical inadequacy: in the long run, it makes nonsense of the instrumental model as a whole. Fundamental scientific knowledge obtained by basic research has the potential of completely transforming all our technological capabilities, so that it eventually dominates all that can be achieved in the shorter term by mission-oriented R&D.[13] A model of science that seems to have no conception of the place of basic research in the scheme of things is clearly inadequate as a guide to social practice.

The Discovery Model of Science

To fill this gap basic research is conventionally rationalized by reference to the *discovery model* of science. This is a metascientific model requiring the dimensions of description and analysis. On the one hand, there is the *psychological* dimension of the individual scientist, exploring the natural world, reporting observations and hypothesizing about them: on the other hand, there is the *philosophical* dimension of the body of knowledge obtained in this way. The essential characteristics of science (as distinct from other forms of organized knowledge) are supposed to reside in the validity criteria through which the personal experiences and insights of the research worker are filtered for incorporation in the scientific corpus.

Although this model of science has long been accepted in principle, and has been studied in great depth and detail, it is by no means fully worked out. Philosophers generally agree, for example, that there must be a rationale for the method (or methods) used by scientists in their research—but there is no consensus on the definition or ranking of the diverse criteria for scientific validity. Is scientific knowledge primarily validated by objective observation, or empirical evidence, or hypothetical simplicity, or systematic orderliness, or what? There is plenty of room in the philosophy of science for a wide range of views concerning the nature of scientific method and its actual applications in the research process.[14]

By focussing on these fundamental epistemological issues, the discovery model of science offers to every research worker a valuable critique of his or her technical procedures. This critique is often as relevant in technological development as it is in more academic investigations. It also suggests a rationale for basic research. The generally accepted criteria for scientific validity are not merely standards for assessing "truth"; they are also closely linked with more transcendental criteria such as "explanatory power," "fundamental generality," "theoretical elegance," etc., which indicate the relative "importance" of particular scientific results.[15] Thus, for example, Einstein's relativistic mass-energy equation is no more true, in principle, than the chemical formula for common salt—but it is of far greater scientific importance. The objective of basic research is to obtain knowledge that is "important" according to these criteria—another rich topic for philosophical, psychological and sociological study.

This model is also realistic in explaining the close similarity, from the point of view of the individual scientist, of pure and applied research. This follows quite simply if one includes amongst the indicators of "importance" a substantial component for practical utility. The chemical formula for penicillin for example, was much sought after, not because it seemed of fundamental epistemological significance but because of the extraordinary medical value of this substance.[16] At the tactical level, at the laboratory bench, there is no serious inconsistency between the intellectual concept of a scientific method and the instrumental model of an R&D system where it is put into daily practice.

But these two conceptions of science are in fundamental contradiction in their attitude towards the likely results of research. The instrumental value of R&D cannot be dissociated from a high degree of predictability in the outcome of a particular experiment, prototype trial, field test etc. The overall rationality of a system for solving prescribed problems implies a well-founded expectation that a reasonably high proportion of useful solutions will in due course be achieved. The discovery model, on the other hand, emphasizes the essential unpredictability of research: it is the *un*expected discovery that is most important. This is the principle at the heart of the falsifiability criterion for science theories—that conjectures must always be subject to the uncertainties of subsequent refutation.[17] The design of a crucial experiment is, of course, a form of planning in basic science, but does not form part of a long sequence of steps leading in anticipation to more distant goals.

These two popular conceptions of science cannot be reconciled as they stand. It is not very convincing to assert that the uncertainties in the results of particular investigations can simply be averaged out, leaving the research as a whole with a predictable trend towards its planned objectives. Yet these uncertainties could only be suppressed by abandoning the scientific method itself in each particular investigation. How does it come about, in practice, that scientists are able to collaborate in their research, and achieve a high degree of coordination in moving towards desired goals, without sacrificing this essential element of individual choice and unpredictable outcome in their work? This question can only answered by going outside the two-dimensional discovery model as a description of the "method" of science, even at its most academic; as modern science studies have emphasized, the *sociological* dimension of science is as fundamental as its psychological and philosophical aspects.

The Sociological Dimension of Academic Science

The traditional discovery model of science takes little account of the social interactions between scientists. Until fairly recently, the fact that research is essentially a communal activity was almost completely ignored. In the past thirty years, however, the sociology of science has grown into a major metascientific discipline, coequal with philosophy.

The *internal sociology* of science has drawn attention to the fact that research scientists do not behave simply as individuals making separate contributions to knowledge.[18] Even in the traditional academic style of research, their behavior is strongly constrained by norms that are peculiar to the scientific community.[19] These norms—*communalism, universalism, disinterestedness, originality, skepticism,* etc.—are not explicitly laid down, but they are implicit in the functioning of the public institutions of science, such as learned societies, conferences, scholarly journals, etc. Opinions differ concerning the status of various normative systems and the means by which they are enforced, but there is general agreement that the social organization of the scientific community ensures that original research is encouraged, and yet receives adequate critical assessment.[20] This is the social framework within which each scientist actually carries out research.

As indicated in Figure 1, this sociological dimension of science cannot be separated from its other aspects. This applies not only to psychological aspects such as personal creativity and motivation: it also has a profound

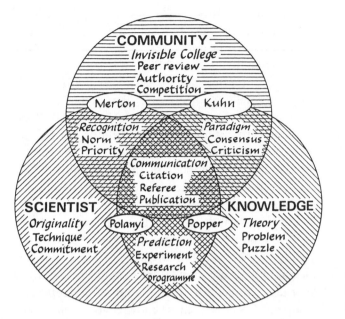

Figure 1. Three dimensions of academic science.

effect on the nature and quality of scientific knowledge.[21] The criteria by which research results are assessed as "valid" or "important" are not philosophically invariant but are strongly influenced by the current climate of thought within the scientific community. The individual dialectic of "creativity" versus "criticism" has its social counterpart in the contrast between *revolutionary* and *normal* modes of research, relative to the theoretical *paradigms* of the day.[22] In practice, the credibility and reliability of scientific knowledge are as much dependent on such sociological factors as on experimental precision and theoretical rigor.[23]

This model thus provides a framework within which to describe many familiar features of what we might call *academic* science. At the center, where all three dimensions overlap, we find the *communication system* of science, with its specialized primary journals, referees, citations, abstract journals, etc.[24] We can observe the way in which the scientific community condenses locally into informal *invisible colleges*, concentrated cognitively around active *research programs*, as indicated by nodal regions in the citation network.[25] The model can also be modified, in the spirit of social psychology, to make allowance for such functional imperfections as the steep stratification of status, and the excessive esteem given to intellectual "authority" in scientific affairs.[26] These and many other insights into the ways of academia could be valuable contributions from contemporary metascientific theory to practical policy for basic research.

In this light, the "predictability" antithesis between the simple discovery and instrumentation models can be largely resolved. On the one hand, basic research is not just random exploration, and its discoveries are seldom entirely unexpected; although the possibility of chancing upon a revolutionary anomaly can never be discounted, basic research is mainly conducted in the spirit of "normal" science, under the guidance of accepted theoretical paradigms. On the other hand, technological R&D would be superfluous if there were not a significant likelihood of unforeseeable failure in the outcome of every test: indeed, it is of the utmost practical importance that all dangerous modes of failure should be adequately explored by just such tests.[27] "Predictability" is not, therefore, a criterion by which technological development can be distinguished in principle from basic academic research. At the tactical level, both are planned and executed in much the same manner—to solve technical puzzles, to check up empirically on theoretical predictions, to survey half-explored fields of phenomena, to measure significant parameters—and just sometimes to formulate or test a conjecture with more profound repercussions. Over the whole range of R&D activities, knowledge is sought and established by a variety of methods involving the same broad mixture of formal analytical procedures, accepted conceptual paradigms, and individual human judgement.

This convergence applies also to the longer-term strategic objectives of research. Thus, on the one hand, the instrumental model exaggerates the likelihood that R&D will reach its more distant ostensible goals (such as, for example, energy supply from controlled nuclear fusion); on the other hand, the discovery model takes insufficient account of the socially mediated orientation of basic science towards the solution of specific problems, such as the origin of the universe or the mechanisms of biological growth and form. In this respect, also, "pure" science is not to be distinguished fundamentally from its more applied variants.

The academic and instrumental conceptions of science seem to imply large differences in the way that research work ought to be organized. On the one hand, the Mertonian norms give the impression of life in a highly democratic community, approximating to the ideal of an ancient Greek city-state.[28] On the other hand, an effective R&D system must surely adopt the rational, hierarchical administrative structure typical of a commercial, industrial or governmental organization.[29]

In practice, however, this theoretical difference in organizational structure is not clearly observed. The two styles tend to converge. Academic science is usually much more stratified and bureaucratic than this anarchical model would suggest; a successful R&D system must make adequate allowance internally for such egalitarian norms as communalism and universalism. It is an empirical observation, with which contemporary metascience has not come fully to grips, that the same "scientific method" seems to function quite satisfactorily over a wide range of organizational styles between these two extremes.[30]

But there remains a real distinction of principle in the way that the objectives of research are set, and the results evaluated. A fundamental feature of our "three-dimensional" conception of academic science is that it is self-contained. Plans, policies, and rewards are established internally, at an appropriate level of authority within the scientific community. By contrast, a technological R&D system always "belongs" to some larger social institution such as a commercial enterprise or the state; its objectives are laid down by external authorities, to whom, eventually its research workers are answerable for what they achieve.[31]

Until, say, the Second World War, this theoretical distinction between "pure" and "applied" science was grounded in reality. The autonomy of the academic community in the pursuit of basic research was a fundamental tenet of the open society.[32] But our uneasy awareness nowadays that basic science cannot be hived off from the general R&D system suggests that the conventional theoretical justification for this separation needs to be reconsidered. In other words, even the "three-dimensional" model of academic science is incomplete and out of date, because it fails to show the connections with society at large, from whence these "external" influences must come.

The Material Dimension of Academic Science

The most direct connection between basic science and "society" is through the material resources employed in research. Until the last few decades, these resources—laboratory equipment, technical services, etc.—could be treated as mere tools at the disposal of people, institutions and conceptual principles. Nowadays, however, *apparatus* must be considered a distinct dimension of all research, even in the academic domain.

This is measurable in financial terms. A modern research laboratory spends more in aggregate on instruments, workshops, animal houses, technical and secretarial staff, and expendable materials than it does on the salaries of its Ph.D.-level scientists. Not all research is "Big Science" in the most extreme sense, but the total cost of ordinary laboratory equipment such as mass spectrometers, electron microscopes, centrifuges and

mini-computers is not insignificant, even by comparison with the immense amounts that are spent on "indivisible" facilities such as particle
accelerators and radio telescopes.[33] These are not merely the material
background against which the scientific game is now played; they constitute an active factor in any realistic model of an R&D system.

The historical growth of this material dimension of research profoundly modifies basic science.[34] For example, the traditional norms of
the scientific community do not allow for substantial inequalities of
"communality" and "universality" between scientists with unequal access
to expensive research facilities. This goes far to explain the ambivalent
situation, within the world scientific community, of scientists working in
the less developed countries.[35] Again, the balance between "critical" and
"creative" talents is altered by the fact that large scale apparatus can only
be used by organized *teams*, where managerial authority is important, and
where the labor is divided into a number of technical tasks that need not
be very original in themselves.[36] Thus, the opportunities for scientists to
make individual contributions worthy of "recognition" are significantly
restricted. Even in the philosophical dimension, a research program designed to reproduce a doubtful experimental result or to refute a theoretical conjecture must accommodate itself to the availability of the necessary
apparatus or to the decision of a committee allocating "running time" on a
major research facility.[37]

Simply as an additional dimension of the *internal* sociology of science,
apparatus is a major factor transforming academic research into an activity that is almost indistinguishable from instrumental R&D. Whether discovery-oriented or use-oriented, whether responsible to scientific peers or
to external authority, the practice of research is effectively determined by
the technical resources at the disposal of the scientist. Theoretical paradigms and technological objectives converge and mutually reinforce one
another in the tangible forms of sophisticated instruments and specialized
personnel organized into research groups and teams.[38] Social relations between junior and senior members of invisible colleges, between graduate
students and professors, between research scientists and research managers, similarly converge on a characteristic administrative structure, loose
enough not to stifle the more persistent voices of originality and/or skepticism but continually striving for efficiency in the utilization of the diverse
material resources under its control. In other words, our conception of basic science must draw as much from the pragmatic art of "the management of R&D organizations" as from the conventional metascientific disciplines of philosophy, psychology and sociology.

But the most significant feature of this "four-dimensional" conception of basic research is that it has no place for the traditional model of academic science as a self-contained autonomous social institution. This model now lies wide open to external influences along its material dimension. It cannot work without money. "He who pays the piper calls the tune."

The place of R&D in economic theory is still a highly controversial subject, but there is almost complete agreement that basic science can only be supported adequately from collective sources such as governmental agencies, educational institutions and other "public interest" corporate bodies. It is much too expensive now to be carried on the normal academic budget as a vocational ancillary to university teaching. Basic scientific knowledge is not a marketable commodity that can be exchanged for the resources needed to produce it.[39] In so far as it is made publicly available, it is a communal good, whose generation is a function of the social infrastructure, along with the legal system and public highways.

The provision and distribution of funds for basic research has thus become a legitimate component of *science policy*. Through the power of the purse, academic science is subject to external forces that can decisively limit its autonomy, not only in the means by which knowledge is pursued but in the ends to which this pursuit is directed. Although the transcendental criteria by which science is valued "for its own sake" are not discredited as a source of personal inspiration,[40] at the level of social decision they are almost invariably subordinated to instrumental criteria by which knowledge is valued "for the good of humanity," "for the good of society," "for the good of the state," or "for the good of the company."

This is not, of course, to imply that basic research should now be put on a much shorter rein. On the contrary, this would be a short-sighted policy with grave negative consequences. Although one may be confident that basic research will, in the long run, yield large social benefits, the precise forms of such benefits cannot be anticipated and they would melt away if hotly pursued. The creation of a wide body of knowledge without specific applications is thus a more rational intermediate objective. Under such circumstances, it is a shrewd policy to exploit the psychological/sociological/philosophical mechanism of norms, rewards and sanctions that used to drive academic science so effectively and to set up a very loose managerial structure, in which, for example, quasi-independent research groups can compete for grants awarded by panels of their scientific peers.[41] Thus, for most academic scientists, freedom for research is not so much a fundamental right as a condition permitted by enlightened administrative expediency, capable only of being exercised within prescribed limits.

Basic science thus differs from other forms of R&D only in the degree of generality of its goals, of unpredictability of its outcome, of remoteness of its time horizons and of independence of its practitioners. It can no longer be realistically represented by the two-dimensional "discovery" model, or by the three-dimensional "academic" model as if it were a fundamentally different kind of activity from applied science or technology.

The implication of modern science studies is that basic science must be treated as part of a unified R&D activity whose goals and criteria are ultimately instrumental rather than transcendental. As we shall see, this paradigm shift in metascience is complemented by a historical change in the nature of "technology," which has moved towards "science" from the other direction. This opens up further channels by which social influences permeate R&D systems and the basic science that they now incorporate.[42]

Scientific Technology

A direct confrontation between the discovery and instrumental models of science could be avoided in the past by regarding them as descriptions of two distinct phases in the creation and use of knowledge. There was supposed to be a natural division of labor between the pure scientist who made discoveries and the applied scientist who used them to solve practical problems. In addition, it was recognized that certain technical arts, such as gunnery or surgery, provided examples of phenomena about which a great deal might be discovered by fundamental research. One wing of the history of science has tended to emphasize the former process by celebrating the appearance of *science-based technologies*, such as electronics or radiation therapy, which could never have developed at all without fundamental research on electromagnetic theory or nuclear physics. The other wing lays stress on *technology-based sciences*, such as thermodynamics or physiology, whose main impetus has come from the needs of technical practitioners such as engineers or physicians.[43]

In most fields of material activity, this separation between the pursuit of knowledge and the perfection of technique is no longer meaningful. Discovery and invention, research and craftsmanship, are melded into a unified *scientific technology*, rooted equally in theory and in practice. Solid state electronics, for example, extends without any sharply defined disciplinary boundaries from quantum theory to production engineering; the pharmaceutical industry is as deeply involved in the latest advances in molecular biology as in empirical clinical trials of its products. To make progress in an advanced technology such as nuclear power production, there must be a continual interchange of research results, design concepts, test data, theoretical analyses, instruments, etc., between the basic research laboratories, the design offices, the test departments and the construction workshops. In these circumstances, the theoretical division of labor between pure and applied research is of little practical significance within the enterprise as a whole.

This interweaving of "discovery" and "instrumental" modes of investigation provides a convincing argument for the cultivation of basic science even in the most pragmatic R&D organization. By treating apparently useless knowledge seriously, "for its own sake," and improving upon it according to the accepted criteria of scientific validity and importance, the organization gains the means to perform its mission more effectively.[44] Conversely—as many academic scientists are coming to realize—the puzzling behavior of many technical artifacts, such as chemical catalysts, has as much power as "natural" phenomena to arouse intellectual curiosity or to exemplify the most fundamental scientific principles.

It is misleading, therefore, to represent basic research as a distinct activity, *shielded* from social influences by a great mass of practical technology. On the contrary, basic scientific knowledge acquires social relevance because it is connected to society through its role in technological development and industrial innovation.[45] Technological decisions on issues such as energy supply policy cannot be distanced from their political and economic consequences, but they usually turn out to be inseparable from basic scientific questions, such as the effect of atmospheric carbon dioxide on the climate, or the likelihood, within the next million years or so, of water percolating into deep beds of rock.[46] Such questions, in their turn, can best be answered by scientific experts—that is by individuals with recognized authority in the research community. But in this dimension also, although there is a wide range of employment and experience between say, the academic professor of pure mathematics and the managing director of a firm of consultant engineers, the middle ground is occupied by a large body of people who are as much "technological scientists" as they are "scientific technologists."[47]

A Unified Conception of Science

Starting from the old-fashioned "discovery" conception of science, we have built up a model with many of the outward features of the "instrumental" conception. But the R&D system represented in Figure 2—let us just call is *Science*, with a capital S—is no longer a "black box." Within it

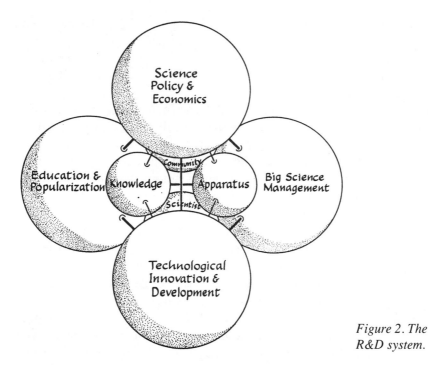

*Figure 2. The
R&D system.*

we can recognize four distinct dimensions—psychological, philosophical, sociological and material. These dimensions are characteristic of all forms of Science, from the most academically abstruse to the most practically empirical.

What we have learnt from contemporary science studies is that we cannot expect to understand how such a system really works unless we take specific account of these various categories—people, ideas, communities, and apparatus. Each dimension is of significance in the research process, and the interactions between them are complex and subtle. Just how these interactions manifest themselves in practice is not yet clearly understood, except in the case of traditional academic science which has been the main subject of study by the standard metascientific disciplines.

In addition to these four *internal dimensions*, our unified conception of Science has four major *external aspects*, through which it is connected with society at large. Each of these aspects has also become the subject matter of a distinct discipline. Science policy now seems to be as much concerned with industrial investment and technological development as with the government financing of academic research: it is essentially a combination of economics and politics.[48] *Research management* deals with the detailed implementation of such policies, and with the incorporation of Big Science apparatus into R&D organizations: it could be regarded as a specialized branch of business management or of public administration. The study of *technological innovation* is a somewhat more specialized subject, concerned with the processes by which discoveries and inventions are transformed into useful goods and services; it combines design and production engineering, industrial economics and commerce, often with a strong historical emphasis.[49]

Finally, we must not forget that Science is connected with Society through science education and science journalism. Through this *mass communication* interface, schematic information about scientific and technological knowledge and devices flows into the public consciousness. These images are incorporated into cultural institutions in the form of philosophical, ethical or political ideologies, and are eventually reflected back as instrumental demands upon Science—that it should generate endless material progress, for example, or that it should strengthen the nation state. This aspect of Science is now a familiar theme in metaphysics, ethics, political philosophy, sociology of knowledge and other more humanistic disciplines.[50]

Once again, it would be a gross oversimplification to treat this as an exhaustive list of quite distinct aspects of this multi-facetted social institution. But perhaps this is enough to establish my main point—that it is possible to set up a representation of science that unifies, without obvious contradiction, a diversity of facts and theories from a wide range of science studies. This unified conception of Science is seldom made explicit, despite the fact that it is the implicit basis of modern metascientific scholarship in many different departments and faculties of academia, and is the unacknowledged source of many valuable theoretical and practical ideas.

Misconceptions Leading to Dysfunctions

How might social practice benefit from these new developments in social theory? Would the R&D system function better if scientific notables, company executives, civil servants, politicians, journalists, and others fully grasped this new conception of Science? Our claims must be modest. We certainly do not have a well-tested working model of the research process, ready to install in any social environment to do any job required of it. The "conception" sketched out here is little more than an abstract framework of incommensurable categories ("dimensions" and "aspects") whose dynamical interactions are still unexplored or highly controversial.

Some features of Science, such as the norms of academic research or the planning of government R&D, are deeply researched or widely discussed; others of equal importance, such as the education of research workers in Big Science, have scarcely been considered in the metascience literature. This literature contains many valuable insights—often instructively antithetic—for the thoughtful science policy maker, but few reliable maps, blueprints or recipes for action.

It is important to emphasize, moreover, that neither the instrumental nor the discovery conceptions of science are entirely superseded by these new investigations. They still retain their value as "aspects" of Science, as seen from particular viewpoints. For many practical political and economic purposes, it is often sufficient to treat R&D as an instrument of policy, capable of producing the required results to order. From the standpoint of the individual research worker, the metaphysical ideal of personal discovery is often an entirely satisfactory ideology. It is only when we need to look more closely at how Science works that we must have a more comprehensive model, of which these traditional images are only partial views observed from contrary directions. As in all social theory, the elementary collective and individual representations of reality are much too simple for precise understanding, but they retain their value as heuristic devices over a wide range of social action.

What can be shown quite clearly however, are the serious consequences of trying to act entirely on the basis of such over-simplified models. To illustrate this, let us consider (very schematically, of course) the following examples of severe social dysfunctions of Science that can be traced to prevalent misconceptions about Science.

The "War on Cancer."[51] In the early 1970s, the U.S. Congress and the Nixon Administration combined in an attempt to break through to a cure for cancer by very heavy research funding. As predicted by most thoughtful observers, this brute-force attack on a major human problem was a costly failure. This was because it was planned from above entirely in the "instrumental" mode, without any reference to the internal dimensions of the biomedical research system. Financial and managerial decisions were pushed through in the spirit of technological development, when the state of available knowledge indicated the need for much more research directed towards the discovery of basic mechanisms. In particular, the shortage of trained scientists and the lack of well-founded paradigms in certain fields (e.g., the Special Cancer Virus Program) could not be compensated for by ambitious management wielding large material resources. This whole episode illustrates the dangers of failing to match the social objectives of Science to the realities and capabilities of an R&D system.

Science and National Development.[52] The fiasco of UNCSTD (the
United Nations Conference on Science and Technology for Development
in Vienna, in 1979) was a typical episode in a long and disheartening experience. In almost all developing countries, science policy since the Second World War has swung wildly from one over-simplified model of science to another. In the 1950s and '60s, there was a happy belief that
support of academic or quasi-academic research would "trickle down"
into economic benefits, without regard for the whole range of technological institutions, such as engineering schools and industrial R&D laboratories, that are needed to bring these benefits to society. When this was realized, there was a complete change towards a strictly instrumental policy,
insisting that all research should have immediate social relevance, regardless of long term strategic needs. Present day debate now concentrates on
the economic and management aspects of national R&D systems, with little reference to the realities of industrial innovation and practically no
consideration of the internal dimensions—scientists, apparatus and communal institutions—along whose axes such systems must be made to
work.[53]

Human rights. The scientific community has responded weakly and in-
effectually to the oppression of individual scientists, and to the inhuman
misuse of scientific techniques in a number of countries. There has been
little recognition of the inseparable connection between the traditional
norms of basic science and the political and legal liberties embodied in
the international and national codes of human rights.[54] In many cases, the
natural sympathy of individuals and learned societies for the plight of scientific colleagues has been inhibited by appeals to the supposed traditions[55] of apolitical transnationalism (i.e., to very naive versions of the
"discovery" conception of science) without any analysis of the sociological function of such traditions (i.e., the norms of communalism and universalism) or of the extent to which these traditions were being completely flouted, in the spirit of instrumental nationalism, by the offending
governments.

These are just a few of the issues where superficial examination suggests that unwise decisions have been taken by responsible people who
have simply misunderstood the basic realities of Science in its social context. One could mention many other current issues worth detailed study in
this light: the weight to be given to scientific advice or testimony in public
affairs; the means and extent to which scientists should take responsibility
for the social consequences of research; the mismatch between formal scientific theories and the prevailing scientific "world view;" the functioning
of the traditional learned societies as representative institutions; the social

assessment of technological innovations; and so on and so on.

It is easy to say, in defense of human frailty, that mistakes and misconceptions are woven into the fabric of human affairs. That is not to be doubted. Nor should the influence of abstract social theory be exaggerated: policy is usually made by shrewd people whose personal experience and intuition keeps them in touch with realities that are not dreamt of in their philosophies. Most of those who manage R&D systems quickly learn to fill the gap between the two traditional conceptions of Science with a variety of practical expedients that probably work much better than any theory-based blueprints. Indeed, they might well regard my argument as no more than a summary of what life has taught them.[56]

Nevertheless, it is dangerous to allow theory to lag too far behind practice. If the social study of science is to claim any social value (i.e., if the metascientific disciplines do not entirely repudiate an "instrumental" evaluation of themselves, taking refuge in the plea that they are only doing research on Science "for its own sake") then it must take some responsibility for not having communicated the gist of its discoveries to those who might make use of it. We cannot tell the world how its business ought to be run, but there might at least be a salutary lesson to learn by casting a cold eye on what well-intentioned people believe about social institutions when they make serious mistakes in trying to manage or change them.

References

1. Like many other statements in this paper, this scarcely needs to be backed up with a formal citation. A useful reference from the voluminous literature on this subject is *Advice and Dissent: Scientists in the Political Arena* by J. Primack and F. von Hippel (New York, 1974).
2. L. Winner, in *Daedalus,* 109, 121–136 (Winter 1980) makes the point by asking "Do Artifacts have Politics?"
3. J. M. Ziman, *Minerva.* 16, 42–72 (1978) "Solidarity within the Republic of Science."
4. This was the underlying issue in the debates on planning and freedom in science—see, e.g., W. McGucken. *Minerva,* 16, 42–72 (1978) "On Freedom and Planning in Science: The Society for Freedom in Science, 1940–46."
5. For comprehensive reviews of all these developments, see *Science, Technology and Society: A Cross Disciplinary Perspective*, edited by I. Spiegel-Rösing and D. de Solla Price (London: 1978).
6. This applies even to such high-brow publications as the issues of *Daedalus* devoted to "Limits of Scientific Inquiry" (Spring 1978) and "Modern Technology: Problem or Opportunity" (Winter 1980).
7. As, for example, between "Popperian" and "Kuhnian" views on scientific epistemology, or on the status of the "Mertonian norms" in the internal sociology of science.
8. The importance of taking this comprehensive view in science education is emphasized in

Teaching and Learning about Science and Society (Cambridge University Press: 1980) where many of the arguments of the present paper are set out at length.

9. This is the point of view adopted without question in most official publications, such as, for example, the various surveys of national research systems published by OECD.

10. As typified, for example, in the "Rothschild Report" advocating reforms in the "Framework for Government Research and Development" (London: HMSO Cmnd 4814, 1971), and in almost all the public debate that it occasioned.

11. This applies to almost all that is said about environmental protection, technology assessment, etc.

12. Including almost all those writing from a Marxist or neo-Marxist standpoint.

13. The standard historical cases are electromagnetism and nuclear physics.

14. Two of the most influential books in the recent swing away from epistemological positivism—*The Logic of Scientific Discovery* by K.R. Popper (English translation, London 1959) and *Patterns of Discovery* by N.R. Hanson (Cambridge, 1958)—indicate in their titles that they lie entirely within the framework of the "discovery" model.

15. The more closely one investigates the psychological dimension in research, the larger loom the non-logical criteria discussed by G. Holton in *Thematic Origins of Scientific Thought* (Cambridge, Mass. 1973).

16. This case is discussed briefly in *The Force of Knowledge* by J. M. Ziman (Cambridge, 1976) pp 188–192.

17. *Conjectures and Refutations*, by K.R. Popper (London 1963).

18. *The Scientific Community* by W.O. Hagstrom (New York 1965).

19. *The Sociology of Science* by R.K. Merton (Chicago, 1973), pp 267–278 "The Normative Structure of Science."

20. A functional interpretation of these norms and institutions as a means of arriving freely at a rational consensus was put forward in *Public Knowledge,* by J.M. Ziman (Cambridge, 1968), following ideas from *Personal Knowledge* by M. Polanyi (London 1958).

21. Never more eloquently expressed than by Ludwik Fleck in his neglected masterpiece *The Genesis and Development of a Scientific Fact.* (1955: English translation, Chicago 1979).

22. *The Structure of Scientific Revolutions* by T.S. Kuhn (Chicago 1962).

23. *Reliable Knowledge: An Exploration of the Grounds for Belief in Science* by J.M. Ziman (Cambridge, 1978).

24. *Communication in Science* by A. J. Meadows (London, 1974).

25. *Citation Indexing* by E. Garfield (New York, 1979).

26. R.K. Merton, *ibid*, pp 439–459 "The Matthew Effect in Science."

27. Despite all such endeavors, unpredictable accidents will always occur: see *The Acceptability of Risks* by the Council for Science and Society (London 1977).

28. Nowhere better stated than by M. Polanyi in *Personal Knowledge*, and, more briefly in *Minerva*, 1, 53–73 (1962) "The Republic of Science: Its Political and Economic Theory."

29. The political character of organized science is discussed in *Science and Politics* by J.-J. Salomon (London, 1973).

30. This is taken for granted in *Scientific Productivity: The Effectiveness of Research Groups in Six Countries,* edited by F.M. Andrews (Cambridge, 1979) which reports the results of an immense sociological investigation, sponsored by UNESCO, covering a wide range of academic and industrial research.

31. The distinction between "internal" and "external" criteria of scientific choice is made by A. Weinberg in *Reflections on Big Science* (Cambridge, Mass, 1967).

32. And was largely respected even in countries, such as the Soviet Union, ruled by quite a different social philosophy—see, e.g., *Soviet Science* by Zh. A. Medvedev (New York, 1978).

33. See, for example, *The Force of Knowledge*, pp 210–239.

34. The implications of increasing instrumental sophistication are analyzed by N. Rescher in *Scientific Progress* (Pittsburgh, 1978).

35. Science Development: *The Building of Science in Less Developed Countries* by M.J. Moravcsik (Bloomington Ind. 1975).

36. The important distinction between research and technical work is made by J.R. Ravetz in *Scientific Knowledge and its Social Problems* (Oxford, 1971).
37. Reliable Knowledge pp 60–64.
38. This convergence is not yet complete. The UNESCO project (note 30) began with the assumption that the social structure of research units was the same in all types of R&D: they were apparently slightly disconcerted to discover that there seem to be such typological distinctions between, say, academic and industrial research that "The possibility of identifying a single structural model capable of assisting science policy makers is...open to question" (p. 393). What is revealing is not the fact but the hypothesis that it disconfirmed.
39. J.M Ziman, *Minerva*, 10, 384–388 (1972) "Can Scientific Knowledge Be an Economic Category."
40. Most eloquently expressed by E. Shils, *Minerva*, 17, 129–177 (1979) "'Render unto Caesar...'. Government, Society and the Universities in their Reciprocal Rights and Duties."
41. Even in the Soviet Union, where the official philosophy is that all science is the instrument of the state, the Academy of Sciences is still allowed considerable autonomy in the lines of research to be pursued by basic scientists.
42. By referring to R&D systems in the plural, I avoid the implication that science is a single social institution, and suggest rather that research is a characteristic social process that may manifest itself in a variety of distinct corporate forms.
43. *The Force of Knowledge* pp 146–179.
44. Characteristically, when Britain set out to produce nuclear weapons for itself, after the Second World War, one of the major components of the project, the Harwell laboratory, was established as near as possible on "academic" lines—see *Independence and Deterrence: Britain and Atomic Energy* by M. Gowing (London, 1974).
45. *The Force of Knowledge* pp 180–209.
46. *Deciding about Energy Policy* by the Council for Science and Society (London, 1979).
47. Although the Royal Society stretches between these extremes, it is heavily weighted towards the more academic scientific disciplines. If it were to follow the U.S. example and set up a separate "National Academy of Engineering," this would not be because science and technology had drifted further apart but because this traditional institution could not accommodate itself to the shift of science towards greater involvement in technology in recent decades.
48. This may be verified by a glance at almost any program of "Science Policy Research."
49. See e.g., *The Sources of Invention* by J. Jewkes et al. (London, 1969) *Wealth from Knowledge* by J. Langrish et al. (London, 1972).
50. A recent collection of typical papers on such matters is to be found in *Theory of Knowledge and Science Policy* edited by W. Callebaut, M. de Mey, R. Pinxten and F. Vandamme (Ghent, 1979).
51. *The Patchwork Mouse* by J. Hixson (New York, 1976).
52. I have not yet found a satisfactory analytical account of this subject, on which there is, of course, a vast bulk of primary material in the form of government reports and journalistic commentaries. Perhaps the main reason why such an account has not yet been written is that it would need these new conceptions of Science to make sense!
53. J.M. Ziman, *New Scientist*, 21 Sept. 1978, 650–651 "Research as if relevance matters."
54. *Scholarly Freedom and Human Rights* by the Council for Science and Society and the British Institute of Human Rights (London, 1977).
55. The historical realities of transnational science are detailed in *Les Scientifiques et la Paix*, by B. Schroeder-Gudhus.
56. See, e.g., the presidential address to the American Physical Society by L.M. Branscomb, *Physics Today*, April 1980, 42–50.

What Are the Options?

S cience evolves through innumerable particular decisions of individual scientists to undertake specific investigations. Since the outcome of a completed investigation is an historical fact which cannot be changed, and the outcome of a proposed investigation cannot, from the very definition of research, be foreseen, conscious intention can be related only to a question to be answered or a problem to be solved. The rationality of the process of research, i.e., the extent to which it is an activity capable of well-ordered public or introspective explication, evinces itself primarily at the point where the scientist has to decide the question: "What research shall I do now?"[1]

This question arises in implicit form—hourly, daily, weekly, yearly, or even from decade to decade—in the life of every scientist. But mostly it is decided in relation to some larger plan of which the investigation under consideration is merely a contributory part. The scientific life is like "everyday" life in that its actions are normally governed by a hierarchy of plans the interconnected consequences and goals of which extend to every-widening horizons.[2] Thus, today I decide to discover why the electron microscope is not giving a steady image. I do this in order to continue observations of the dislocation structure in an alloy specimen, which is one of a batch of 20 that are being deliberately subjected to various mechanical stresses, so that I may test a conjecture concerning creep in turbine blades and thus contribute to the design of a new jet engine being undertaken for the air force by my company. Although I have not worked out a detailed program of action in advance, the words I use show that I shall be able to decide the direction of each successive step by the logic of the current situation, within this larger framework of intentions. Unlike the lecturer who plans to explain the indeterminacy principle to the stage I students in room G42 between 12:00 p.m. and 12:50 p.m. on Thursday 16

June, 1983, I do not yet know what research I shall be attempting that morning—but I feel sure that, when it comes to it, the alternative will be clear enough and the decision between them will often look like a foregone conclusion.

Nevertheless, the alternatives and the freedom to choose from them are not illusory. If there is any human activity in which human beings can be said to exercise some genuine freedom of action, it lies within research. In any particular case, the range of alternatives may be restricted, but the choice cannot be predetermined by reference to a formula, nor preempted by a prescribed code of behavior. Some degree of personal judgement and initiative—for example in assessing the credibility of competing theories—is constitutive in all science.[3] The individuality and historical uniqueness of each decision to take up this or that investigation, within the permissible range, cannot be gainsaid.

But what are the alternatives, and how are they limited? To what range or "realistic" possibilities do scientists restrict themselves when they consciously consider their research objectives? This question involves so many intangible and idiosyncratic factors that it can only be posed schematically. Yet it touches upon a crucial point in the social structure of scientific activity—the means by which individual scientists are kept, at least to some extent, under social control. This mechanism of social control has so much "slack" in it that individuals normally accommodate themselves unthinkingly to its constraints. The "next" short-term goals within the hierarchy of plans generally present themselves in an open-ended range without harshly "unreasonable" limitations. But we cannot suppose that there is eternal harmony between the inner life of the individual and the social milieu in which he or she is living. The guiding hand cannot remain entirely invisible. There are occasions when what we loosely call the "social relations of science " make themselves manifest in the form of partial but powerful determinants of personal research plans.

Such an occasion might be, for example, when a major goal of research has been achieved ("Now, at last, I understand the cause of creep in turbine blades!"), so that new plans must be made. It is when each of us then sits down and tries to think out "What research shall I do now?"—not for the next week, not even for the year, but for the years ahead—that we may become uneasily aware of the alternatives that are really open to us. The considerations that we turn over in our minds may seem highly idiosyncratic, and yet they relate inevitably to our circumstances as social beings. They thus throw light upon the mechanisms by which society influences the orientation of scientific research.

This theme can only be explored hermeneutically by the interpretation of introspective accounts of crucial individual decisions. These are essential items of evidence. We must also assume that the same considerations continue to apply at less significant branching points along each personal "research trail."[4] At any given moment, the whole hierarchy of plans by which our actions are given meaning could presumably be called into consciousness for possible revision. Generally speaking, however, these plans interlock so tightly that their longer term goals are accepted "intuitively" as if they were no longer susceptible to conscious reflection. We take it for granted, in everyday life, that an intention that reveals itself to introspection in a moment of hesitation preceding deliberate action continues to inform that action as it proceeds. By fastening upon the few moments when we ask ourselves, in so many words, "What research shall I do now?", we thus hope to capture some of the underlying rationale of the scientific life. But there is no way of knowing what "unconscious," "irrational" factors may also be at work upon even our most carefully considered decisions.

Instrumentalized Scientists

Apart from its biographical and anecdotal instances, the general question of the research plans of individual scientists seems to have been neglected by sociologists of science.[5] This neglect may be attributed to the dominance of two excessively simplified conceptions of science[6] the inconsistency of which would show up all too clearly if the question were carefully explored. The instrumental model of science and the model of science as discovery impinge upon one another, and overlap to contradiction, through the social influences on the research plans of the individual. Within the framework of each of these models, this influence is made to appear either unproblematically strong or else insignificantly weak. Inadvertently, incompatible interpretations are being offered of the same human event.

The instrumental model treats personal research plans as unproblematic in principle because it assumes that all research is oriented to objectives that are ultimately determined by "external" powers and policies. Within this model, therefore, the individual scientist is instrumentalized— i.e., subordinated to the success of the mission as a whole for which he or she is engaged in research. Whenever the question of research alternatives arises, it can always be referred to specific goals that are beyond the competence of the individual scientists to evaluate.

This is not to say that all short term "tactical" research goals are pre-scribed in advance. This would be logically inconsistent with the unfore-seeable outcome of any research. The instrumental model does not have to define a research worker as a mere technician whose work is laid down on a daily or weekly basis. But whenever alternatives are open, the deci-sion between them is supposedly taken on strictly technical grounds: "I must now do the research that is most likely to solve the problem in hand, which has been defined as 'my' problem by the research organization to which I belong."

This model is exemplified by "mission-oriented" scientific organiza-tions with a bureaucratic hierarchy of "authorities" whose ranges of per-sonal discretion broaden out towards successively more distant research horizons. At any given level of authority, however, they pre-empt the re-search options of the subordinate individuals, whose hierarchy of personal research plans is thus truncated at a corresponding level of generality. Thus, for example, as a scientific officer, I am permitted to undertake my microscopic investigation of alloys in my own way, to answer specific questions suggested by the head of my research group— a senior scien-tific officer—who is free to undertake or manage a variety of other metal-lurgical investigations directed towards the explanation of creep—and so forth. At every level, however, the alternatives are supposed to be either unquestionably prescribed from above—hence, eventually deriving from authorities outside the process of research altogether, in the center of power in society—or can be rationally related to such goals by the "nest-ing" of research plans down to the work of each day.

Although it is not an entirely unrealistic description of some of the ways in which research is organized,[7] the instrumental conception of sci-ence is obviously incomplete, in that it has no place for basic research.[8] It can, of course, be argued that this apparent deficiency is an illusion. There is a substantial metascientific literature based upon the proposition that scientists are never really free to construct their own individual research plans since the alternatives available to them are strictly limited by social, economic, and political factors.[9] According to this argument, even the "purest," most basic research is motivated instrumentally towards prob-lems derived from interests such as the interest of a social class in domi-nating society. It is conceded, however, that this influence is often deeply hidden from the actors themselves, by what seem to be more immediate personal or cognitive considerations. This is as may be. But if this influ-ence is to be discerned at all, it is most likely to manifest itself at the mo-ment when new research plans are in the making. Even if we are of the opinion that the conception of science as discovery is essentially a "false ideology," we may study the choice of research goals within that ideology to reveal the linkages with more potent interests. In other words, to assume in advance that science is totally controlled by social interests would beg the question of this study, which is not so much about the reality of such influences as about their magnitude, direction and effects.

Scientists as Discoverers

In the traditional conception of science as discovery, the making of every research plan is a unique historical event, the rationale of which is of no more metascientific significance than the way in which a particular pair of gas molecules interact is of significance for thermodynamics. According to this view the answer to the question "What research shall I do now?" is seldom founded on rational considerations and must not, in any case, be adhered to in practice. At the limit, it is said,[11] the "Dionysian" scientist "knows only the direction in which he wants to go out into the unknown; he has no idea what he is going to find there and how he is going to find it. Defining the unknown or writing down the subconscious is a contradiction in absurdum." Historical studies merely reveal that an investigation was instigated by some external event (e.g., the observation of a floppy-eared rabbit[12]) or by some internal mental episode (e.g., "What would a light wave look like to someone travelling with it at the same speed?"[13]). Although such reports are endlessly fascinating, they are held to be no more than superficial manifestations of mysterious forces of imagination and "creativity."

This appeal to inspired talent or good fortune or both as the prime factors in research should be clarified. It clearly does less than justice to the achievements of the "Apollonian" scientist who "clearly sees the future lines of his research" and "tends to develop established lines to perfection."[14]

The emphasis on serendipity is well deserved, for it strikes at the very heart of our subject. It is essential to distinguish between the fortuitous combination of circumstances opening the way to the intimation of a scientific discovery and the deliberate action needed to exploit this opportunity. Consider, for example a typical recent case.[15] Dr. J. Lovelock "happened" to be an expert in gas chromatography; he "happened" to be on holiday in Ireland; he "happened" to notice that the air was hazy when the wind blew from the European continent; he "happened" to wonder if air pollution could travel so far: what more can one say, except that he was observant and thoughtful, as any good scientist should be.

The significant step in a serendipitous discovery comes with the realization that "there might just possibly be something there," which requires a decision whether or not to investigate the matter further. Thus, as Dr. Lovelock reports, "So, in 1968, much to the disgust of my wife and family, I took with me a gas chromatograph with an electron capture detector. The purpose was to seek in the air, when it was hazy, the presence of some unequivocally man-made substance..." From this modest new research plan, and others which followed from it in the same spirit, there eventually grew a substantial new field of research, including, incidentally, the vexed and highly "instrumental" question of the effects of chlorofluorocarbons on the ozone layer in the upper atmosphere.

Without this decision to investigate further, the history of this "socially relevant" field of science would have been somewhat different. It was clearly a branching point both in an individual research trail and in the development of a scientific speciality.[16] But such a decision is not automatically in the "right" direction: it calls for a genuine choice between rationally identifiable alternatives. Some other personal project, such as a fishing expedition, may have to be deferred in order to undertake the research. A preconceived nesting sequence of research plans, for example, a major program of consultation for NASA, may have to be drastically altered. External constraints, for example, the availability of a grant to carry out the research, may seem to stand in the way. The opportunity might be allowed to slip away because it does not seem to be within the terms of one's appointment ("Shall I be allowed to analyze the data when I get home?"), or, more subtly, because it seems inconsistent with one's self-definition as a specialist within a particular discipline ("Isn't this really a question for a meteorologist?").

These considerations clearly involve a variety of considerations arising from obligations, commitments and norms which significantly affect the outcome. Dr. Lovelock knew that he could allocate time to the research because he was that rare person nowadays—an independent scientific practitioner; if he had been an employee of a "mission-oriented" research organization this might not have been permissible. On the other hand, the discovery was nearly stifled at birth for lack of financial support. Before the recent public concern with the quality of the environment, he probably could not have obtained a grant for such an "unfashionable" subject. Again, it required a certain independence of mind to discount the consequences of moving across the socially defined and maintained boundaries between scientific disciplines. These social factors are simply not taken account of at all in the traditional conception of a scientific discovery.

Serendipity is a familiar phenomenon along the trail of every individual research worker—and in every organized research program from the smallest to the largest.[17] But this phenomenon does not of itself make the elementary mode of discovery an adequate conception of science, even at its most basic. Nor does it justify the traditional corollary that "the objectives of the basic research enterprise are furthered by allowing the individual scientist complete freedom both to choose the subject matter of his investigations, and to draw the conclusions to which they lead, consistent with laws of logic and nature."[18] The most that it suggests is that the objectives of every research enterprise are furthered by allowing the individual scientist the maximum possible freedom to make a radical change of individual research plans in order to exploit the opportunity presented by an unanticipated discovery or intimation of a discovery.

Communality

These traditional conceptions of science are normally regarded as polar opposites, demarcating distinct regions of "pure" and "applied" research. Modern metascience, however, shows that they can be thought of without contradiction as different aspects, from quite different viewpoints, of a general class of systems of research and development, encompassing both basic science and technology.[19]

The conventional approach to the planning of research to meet socially prescribed objectives naturally starts from the "instrumental" viewpoint, looking at science from the "outside." But to appreciate the significance of personal research plans it is more instructive to start from the "inside" of this model, at first taking the point of view of the individual scientists[20] as "discoverer" and then working outwards to the social framework in which the model is embedded. This order of discussion does not imply that the "internal" sociology of science predominates in the final analysis. At each stage, as some new "external" social factor is introduced, we must reassess the weight and balance of the forces already considered, until we eventually see the subjective logic of the situation for the individual in an objective social perspective.

In the first place, it is obvious that even the most elementary notion of scientific discovery must be set against a background of existing knowledge, to which the scientist has access and to which it is desired to contribute. For the individual it may seem that scientific knowledge is something "objective" and apart from humanity; in reality it is a result of intellectual activities in social settings; it is a public phenomenon in the noetic domain.[21] "Curiosity" might reasonably be described as an unintended interaction between the mental domain of the observer and the material domain of the "external world." "Reflection" or theorizing is normally regarded as a process going on within the mental domain of the individual thinker—although it usually operates on mental "pictures" derived from noetic "maps."[22]

But one cannot be said to have made a scientific discovery until the state of public knowledge has been changed.[23] A personal research plan is usually treated by philosophers and historians of science as a private construction within the mental domain of the scientist, to be acted out subsequently by experiment or observation. But such a plan always has the intention, however vague or fantastic, of making a discovery relevant to some particular body of knowledge. It can only be founded upon rational beliefs about the current state of such knowledge—i.e., upon the result of a deliberate inspection of some of the "maps" in the noetic domain.

This appears, for example, in the conventional expression of purpose which introduces nearly every report on research. Phrases such as "The aims of this paper are to generalize...," "The first purpose of this paper is to examine more fully...," or, "It was, therefore, considered worthwhile to examine the validity of...,"[24] are not, of course, to be taken literally. The original goal of a research project is no more than an obsolete biographical or historical fact, of no current or future scientific significance.[25] Nevertheless, the peculiar rhetoric of a scientific publication makes the work seem an inevitable "step in the right direction," and hence an "objective" contribution to knowledge,[26] made by a properly informed, and hence technically competent person.

This element of communality in the research objectives of even the most sublime scientific genius is thus a corollary of the epistemological impersonality of the scientific method. Consider, for example, the typica' account of a "crucial" experiment.[27] It is not stated in the philosophical protocol whether the attempted "refutation" is to be made by the person who made the original "conjecture" or by some other person. Once "we" have publicly formulated a theory, then there is an onus on "us" to test it—it may be a century later! Indeed, the intellectual craftsmanship of an episode such as the Michelson-Morley experiment mainly lies in seeing that such a test is now possible and in constructing a rational plan for research with this goal in view.

There is more to this communal influence than is usually allowed for in the philosophical notion that knowledge grows by the successive unfolding of "research programs" which somehow attract the attention of individual scientists and draw to themselves their personal research plans.[28]

The community of "other" scientists not only validates the cognitive claims of the day: it is also the source of, so to speak, a list of "do-able" research projects, assessed with respect to their apparent difficulty and the potential value of their results.

This list is not, of course, brought up to date and published like a list of stock exchange prices. Along with related paradigmatic evaluations of "the current state of the art" and "what is well established" it belongs largely to the "tacit knowledge" of experienced scientists, and is always highly controversial.[29] Nevertheless, the occasional obiter dicta of a notable scientist on a question such as "What problems of physics and astrophysics are of special importance and interest at present?"[30] are usually given much attention in the scientific world. A bold sweep over such a wide and distant horizon probably has very little direct effect on the progress of science, but one should not altogether discount the influence of more specialized "survey lectures" and "invited review papers" which make up a major part of the program of scientific conferences. The mass of the audience genuinely want to know what the leading members of their "invisible college" consider to be "the kind of information which theorists want..." or "the various unknowns which beset the problem" or "[a method which] may still be fruitful" and expect them to "pick out the problems that still seem interesting" or "check that important issues have not been ignored" or assert the view that "we shall probably need to use both methods to obtain our final goal."[31] Such a conference is not merely an occasion for the exchange and collective criticism of the results of recent research; it is also a social institution through which the individual research worker can look at his or her personal research plans from a communal point of view.

Much more precisely, much less coherently, but with irresistible effect, the communal influence is brought to bear upon the research plans of the immature academic scientist in the course of his or her training. We often forget that the commonest occasion for asking the question "What research shall I do now?" is when the graduate student goes to a research supervisor for advice on a topic for his dissertation;[32] i.e., to be given a scientific problem which is both "do-able" and "worth doing." At a slightly more advanced level, the post-doctoral scientist naturally adopts the research methods, problems and objectives of the research-milieu in which he or she may be a temporary visitor.[33] In each case, access is being sought, through a more experienced "authority," to the current list of appropriate problems. One of the major difficulties of research in less developed countries is isolation from such advice: indeed, a leading scientist visiting such a country is often asked point-blank by local scientists to suggest suitable topics for their research, as if this were a simple question to which a real expert must surely know the "correct" answer.[34]

There is thus no lack of empirical evidence for the assertion that within any field of science, there is normally "a criterion for choosing problems that...can be assumed to have solutions." It may be too strong to insist that "to a great extent these are the only problems that the community will admit as scientific," but one can surely agree that "[they are the only ones it will] encourage its members to undertake,"[35] and that when "an anomaly comes to seem more than just another puzzle of normal science...[it] comes to be more generally recognized as such by the profession. More and more attention is devoted to it by more and more of the field's most eminent men. If it still continues to resist, as it usually does not, many of them may come to view its resolution as the subject matter of their discipline."[36]

This mechanism by which scientific paradigms direct the process of research is seldom sufficiently emphasized.[37] The "disciplinary matrix" is usually described as a constraint on the ex post facto legitimation of theoretical interpretations in the public, noetic, domain or as a limitation on the resources of creative imagination in the mental domain of each scientist of the day. And yet, from the point of view of the individual research worker, a quasi-consensual body of public belief is surely much more cogent and convincing in the advance assessment of puzzles "worth solving" and of anomalies "needing to be resolved" than it is when the unique results of an actual investigation must somehow be explained. In the history of science it is often much easier to understand how it was that certain apparently simple problems were not studied at all[38] than to reconstruct rationally all the arguments that seemed at the time to make a proposed theory or new discovery unconvincing. To put the point colloquially—anyone might be forgiven for not embarking with Columbus on that voyage that people said would get nowhere; whereas they must have been proper fools not to have accepted Galileo's argument that the earth goes round the sun! By slightly shifting the emphasis from retrospective to prospective paradigmatic influences in the process of research, we may make reasonable allowance for the weight of collective opinion on the growth of knowledge without having to rely too heavily on imponderable factors of credulity and irrationality amongst even the most thoughtful and well-informed persons.[39] Once more, it is at their moments of intentionality that scientists are to be seen by an empathic observer at their most rational and comprehensible.

The Importance of Being Difficult

Having enlarged the elementary model of science as discovery by the introduction of a "communal" dimension, we can now represent scientific research as the activity of a more or less self-contained community where cognitive, social and psychological factors strongly interact. This academic conception of science is still far from a complete model. It is unrealistic to dissociate the "internal" factors of a system of research and development from its "external" material, political and cultural aspects. Nevertheless, the next step along our general line of argument is to consider the situation of the individual research worker within some such idealized social structure.

The archetypal academic scientist is assumed to be able to make his or her personal research plans with the minimum of direct interference by other persona. The majority of scientists nowadays do not have anything like this degree of autonomy. But it is still an essential qualification for a senior appointment in an academic institution, let us say, as associate professor, reader, or full professor, to be scientifically "self-winding." By this I mean that he should have a proven ability to set his own research goals and to work diligently towards them. As we shall see, this freedom is more closely bounded than they would like to admit, but this is the pattern which many academic scientists try to find and in which they prefer to operate.

The traditional outlook of academic science is thus highly individualistic. It is presumed that every scientist in a particular field of research is well acquainted with the local cognitive landscape—seen, of course, through a blinkered or refracting paradigm— and is best left alone to make his or her own decisions on what to do next.[40] It is not to the point of our present investigation to ask whether or not laissez-faire is the best policy for the advancement of knowledge. For the moment we simply take for granted that the research plans of an academic scientist, however, enlightened and far-seeing, are ultimately made for the benefit of "number one." This is the only basis on which we can discuss the personal considerations which enter into a calculated decision to undertake this or that piece of research.

Because the outcome of research is always uncertain, the underlying rationale of any such decision is inevitably akin to a gambling strategy. A personal stake of time and effort is to be risked in the hope of "pay-off" in the form of intellectual satisfaction, material reward, or social esteem.

These factors are so imponderable and subjective that they ar seldom consciously weighed up. But if we seek by introspection to make sense of our situation, they must be linked to more objective factors drawn from a more public domain. A potential research project is represented to oneself schematically, as a question of a certain degree of importance, the answer to which might be obtainable with a certain degree of difficulty.[41] In a general way, we take it for granted in academic science that the difficulty of a scientific question is a measure of the psychic energy we must spend in order to get a solution the importance of which is correlated with the gratification it will eventually afford us.[42] Within the acceptable range of risks, the logic of the situation is to tackle a question with a good ratio of importance to difficulty.[43]

It must be emphasized, however, that the publicly recognized schedule of "timely" research projects provides only limited guidance in such a private calculation. The shrewdest attempt to assess their relative importance and difficulty cannot be expected to rank the alternatives from the point of view of the person who is to undertake them and to be rewarded for their results. From any given situation, different persons may quite reasonably decide to follow quite different research trails, consonant with what they judge to be their own capacities and their own ultimate interests within the larger structures of their own "life worlds."[44]

It is widely acknowledged, for example, that the tacit knowledge that goes into the choice of "good" research problems is one of the most enviable assets of the successful scientist.[45] This talent is more than a nose for the ripeness of a question for answering; it cannot easily be separated from personal skill in seeking that answer. Such skills are very variable and diverse. The "difficulty" of a scientific question is relative to the capacity of the person who is to set about answering it—that is, in a truly personal research plan, oneself.[46]

This mental calculation of the ratio of importance to difficulty for a scientific question is thus essentially notional. No objective meaning can be attached to formal algebraic manipulations of such subjective variables, the properties and values of which are only very vaguely defined within the mental domain of each one of us. Nevertheless, the dimensional characteristics of these variables can be expressed in the public language of rational discourse.

The academic scientist tends to assess the difficulty of a scientific question in units of time. Within the available material resources—which the academic conception of research apparently takes for granted—the prime consideration is: "How long might it take me to get some sort of answer?" Thus, for example, at a relatively low level of difficulty, a topic for a doctoral dissertation should be easy enough to occupy an inexperienced student for about three years—having been assessed at, say, a few months work, if it were done by the research supervisor in person. This scale has a natural upper limit which is a human lifetime. Within the individualistic ethos of academic science, a question that would not be expected to yield any significant result for the person investigating it must be counted as infinitely difficult, and therefore not a rational research objective.

The conventional minimal unit of importance would correspond to what might be barely publishable as a scientific paper. In principle, this is a universal standard invoked by the editors and referees of reputable scientific journals, and enjoined upon the examiners of doctoral dissertations. In practice, it is a level that varies considerably from journal to journal within any particular field, and is not strictly comparable from one discipline to another.[47] However, this variation in minimal acceptability for publication is much narrower than the observable range in the quality of published scientific papers. Measured by "citations," for example, answers to certain questions have proved to be worth hundreds or thousands of times more than most ordinary scientific results.[48] Our a priori estimates of the importance of unanswered questions must be ranked on a logarithmic scale if we are to accommodate this wide range of possible objectives of research.[49]

Generally speaking, in any conscientious survey of the alternatives for research, these notional attributes present themselves as positively correlated: the easy questions seem unimportant and the important questions seem difficult. To some extent, this correlation is a social construction; there is a natural tendency to acclaim a technically skillful answer to a question that has long seemed difficult even when the result is really of little intrinsic scientific interest.[50] There is also an unconscious filtering process in the alternatives that are surveyed: we do not seriously consider taking up unimportant research questions that look very difficult to answer anyway.[51] And if it does occasionally happen that a recent scientific "breakthrough" makes a large number of important questions suddenly look rather easy, we may be sure that these will soon be answered by competitors crowding into the field.

In a rational analysis of what research to do next, the academic scientist is thus faced with a variety of alternatives, ranging from apparently easy but probably unimportant puzzles[52] to obviously important but difficult problems. The social system of academic science guarantees that the eventual "pay-offs" will greatly exceed, in aggregate, the stakes that must be ventured, but the choice of a particular project is always something of a gamble. Under such circumstances, different persons adopt different personal research policies which can only be related to the imponderable temperamental factors that presumably govern their plans over the long run of a lifetime.

The safest policy, of course, is to go for "small profits and quick returns." The average Apollonian scientist[53] tries to make a modest living by solving a succession of puzzles within a specialized disciplinary matrix. The questions to be answered are not necessarily trivial, but they are chosen because they are expected to give publishable answers within a reasonably short time. By having several different projects going on at once,[54] so that various puzzles overlap and form a continuous "braid" of research activities, one can avoid any occasion for thinking about larger problems calling for longer-term research plans. The majority of scientific careers give the impression that "minimal risk" is their underlying policy. Once they have been set going in a particular direction, they seem to evolve continually in small steps, following the logic of the immediate situation within a clearly defined field of specialization.[55]

This unheroic policy[56] is often contrasted unfavorably[57] with the "achievement-orientation" of scientists who accept the risks involved in deliberately tackling larger, more difficult problems. The traditional accounts of discovery[58] enjoin the virtue of perseverance—exemplified, say, by Kepler in his long search for the laws of planetary motion or by the 20 years spent by Dr. Max Perutz in seeking to determine protein structures by X-ray diffraction. There are often long odds against such ventures. For every famous success there must be dozens or hundreds who have little to show for their efforts.[59] But the personal reward for a really important discovery is usually reckoned to be so large[60] that a whole scientific career is not a disproportionate stake to lay down for it.

In fact, the distinction between research policies which take small risks and those which take high risks is not so simple. What looks to the outsider like an unimportant technical question may be attacked as part of the approach to a major scientific problem. We do not always appreciate the hierarchy of individual research plans within which the solution of some particular puzzle is nested. The element of risk also can be greatly reduced by a tacit division of labor within the scientific community. A scientific discipline is often defined cognitively as a problem area. Once a significant question has been set up publicly as a suitable goal for research in that area, every result that claims to be an advance towards a possible answer can be considered a publishable contribution to knowledge. In a "revolutionary" phase of science, standards of "importance" are often tacitly lowered[61] so as not to discourage scientists from setting up research programs which might lead, in a series of easy steps and low risks, to an important result.

But the key factor in a personal research policy is not skillful cost-benefit, or stake/pay-off analysis of particular plans or programs: it is response to change. The scientific scene is never static. As new knowledge is gained, the odds on every research project are altered. The importance or difficulty of a particular scientific question can only be assessed in relation to the present moment, according to present understanding. Even one tiny scrap of new information may quite transform the situation.

This is why all active scientists are so concerned about access to the very latest results of research. It also explains the considerable contributions made by "Dionysian" scientists, with wide interests, who are continually on the alert for the intimations of discovery.[62] Their gift is to realize quickly that an important question is much easier to answer than anyone had previously suspected, or that a question which nobody had imagined worth formulating (e.g., why is it sometimes so hazy in the West of Ireland?) could be investigated with very little difficulty and might lead to important results. Their general research policy is not to tie themselves down for life to a single major problem, not to get caught in the web of routine solutions of puzzles, but to keep themselves as free as possible to exploit any serendipitous opportunity presented by experience or reflection.

The most dramatic effects of change are to be seen in the wake of a major scientific discovery. Once somebody has effectively solved an important problem of long standing, everybody in the field must reconsider his or her research program. Difficult projects that were directed towards possible solutions to the problem lose their rationale. The efforts which have been cautiously invested over a period of years to get recognized as an authority on the subject must be reassessed. The pathos of such a situation is one of the human costs of academic science—and a major source of psychological resistance to revolutionary new scientific ideas.

Less dramatic changes in the cognitive map of a conventional problem area may have a more insidious effect on a personal research program. One may become slowly aware that one's research plans are of diminishing relevance to live scientific issues—that they are related to questions which have been effectively answered, to techniques which no longer seem promising, or to theories which no longer seem credible. But these my be the very puzzles which one has learnt to solve without undue difficulty—and which still seem to be treated as important by one's scientific contemporaries. Should one make a drastic change of plan? There is no immediate reason why one should not stay safely at home as a recognized expert in a local disciplinary area, allowing one's research to evolve naturally out of work in progress towards the traditional objectives of one's "invisible college." In this dilemma, psychological and social factors rein force the inertial tendency towards undue persistence along personal re search trails that yield diminishing scientific returns.[63]

But a "low-risk" personal policy of specialized puzzle-solving may turn out to be a losing plan on a longer time-scale. A scientist who fails to reorient his or her individual research program before it becomes quite obsolete may eventually have to pay a price in demoralization and disappointment, as happens in the familiar case of the university teacher who "drops out" of research in middle age. This is one of the major traps of academic science, only to be avoided by relocating the intellectual site of one's research before it is too late.

Such radical changes of course are not uncommon along the research trails of relatively successful scientists. In many cases they seem to follow upon more or less chance events such as a serendipitous discovery which opens the way into quite a different field.[64] In other cases, a calculated decision is made to move out of an established discipline into a new field of research which is just opening up. This latter process is obviously essential for the advancement of science. A new multi-disciplinary problem-area can only be exploited and developed by specialists migrating into it from other fields.[65]

Any drastic reorientation or relocation of personal research plans obviously "costs" a great deal in the short run. An unproductive period of a few years learning about the new field must make every question there seem much more "difficult" than any of the puzzles that have been left behind.[66] Very few scientists can afford such a hiatus of productivity before achieving an established reputation in a recognized discipline. Nevertheless, the short-term risk is often made up, in the long run, by handsome rewards of authority and esteem for having contributed to an important subject at an early stage.[67]

The gravest danger in moving out of an established field of research is that one may fall into an area of deviant science,[68] which never develops into a recognized discipline. It often happens, for example, that a scientist of high standing takes up research on an enigma[69]—that is to say, a question the paradigmatic importance of which is not recognized by any orthodox scientific group. Although a question such as "Why do the continents fit so neatly together?"[70] or "Is telepathy experimentally detectable?"[71] may seem cogent enough as a goal for research, this judgement may not be shared by other scientists, who are then liable to belittle such a plan and call it irrational or pathological.[72] Personal research plans directed towards such questions are the seeds of new sciences: their germination is extraordinarily sensitive to the subtle distinction made by scientists between a respectable problem and an unacknowledged or disreputable enigma.

This discussion of the rationale of personal research policies in academic science is obviously very sketchy. Many other policies are justifiable (and observable), according to individual taste and inclination. But enough has been said to refute the conventional notion that any such discussion is pointless simply because scientific discovery is unpredictable. On the contrary, it is precisely through his or her personal response to this challenge of uncertainty that each research worker directly influences the evolution of knowledge. The considerations that enter into this response are not ineffably subjective: they are related not only to the scientific situation at the time, but also the social circumstances of the scientist.

The essence of the traditional laissez-faire conception of academic science is that scientific knowledge would evolve optimally if each individual scientist were investigating whatever question he or she considered most important. This individualistic outlook encourages specialization and perseverance. It celebrates the research worker who has successfully devoted the efforts of a lifetime to the solution of a particularly difficult problem. But for most academic scientists, the logic of the situation seems to favor very prudent personal policies, leading to the proliferation of routine, pot-boiling research and to undue persistence towards obsolete research objectives. At the crucial moment of considering research options the average scientist is not so bold as the laissez-faire theory seems to suppose[73] and is not given enough encouragement to adapt to scientific change. Even in its idealized conception, academic science is imperfectly matched to psychological realities, and thus exhibits characteristic defects as a social institution.

Apparatus: A Material Consideration

The three dimensions which are usually referred to in the study of academic science—the psychological, the cognitive and the social—are no longer adequate for a realistic conception of the process of research. Modern science is much affected by a more material determinant—the technical apparatus of research.[75] Even in basic research, with no "external" goals, the advancement of knowledge depends vitally on the provision of ever-increasing material resources and technological facilities.[76] The availability of appropriate apparatus is thus a major consideration in the research plans of every scientist.[77] This applies even to the traditional analytical sciences, such as economics and theoretical physics, where access to powerful computational facilities is now a precondition for most research projects.[78]

Instrumental innovation is an ambivalent factor in the rationale of research plans. From a communal point of view it is very positive, in that it can make important questions appear much easier to answer. An instrument of completely novel principle, such as the scanning electron microscope, or a conventional instrument of vastly improved sensitivity, such as a modern optical spectrometer, can be exploited like a conceptual discovery to open up and explore in detail a whole area of nature hitherto hidden.[79] This possibility is so attractive that scientists frequently make the construction or acquisition of more powerful apparatus the primary objective of their research plans, without detailed evaluation of the research problems to which this apparatus is to be applied. The traditional ethos of academic science awards public recognition for the development of research instrumentation, as if for its own sake.[80] Even in fundamental disciplines such as cosmology and molecular biology, the technology of investigation is regarded as a legitimate field for scientific endeavor.

From the point of view of the individual scientist, however, the increasing need for more sophisticated apparatus continually adds to the difficulty of most of the alternatives under consideration. This effect is much larger, and more profound, than it appears. Strictly speaking, the individualistic ethos of the academic conception of science implies that the "stake" which is ventured in a research project should include the whole effort needed to carry out the project, as if this effort were made by the investigator in person.[81] That is to say, an instrument costing x times his annual salary might be presumed to have taken at least x years of his own research time to build.[82]

It is absurd to suppose that this is in fact the basis on which scientists calculate the odds on various individual research projects. If this were the case, then the construction of a theory, however implausible, would almost always look a better bet than experimental research. But the absurdity derives from the unreality of the individualistic ethos that supposedly prevails in academic science. In practice, the logic of the situation impels scientists to collaborate—that is, to divide the labor, and thus share the "difficulty" of their projects with others.[83] They cooperate voluntarily in research teams, they apply for financial support from external sources such as governmental bodies, or they take up regular employment in a research organization. In other words, in order to meet the material needs of research, social institutions have grown up in and around academic science, which render obsolete its traditional outlook.

This radical transformation of the social system of science profoundly alters the rationale of most individual research plans. These effects will be discussed below. But even the quasi-academic scientist, entrusted by the system with the resources for "personal" research, i.e., instruments, technical facilities, assistants and so on, is not unaffected by the increasing technological sophistication and complexity of the apparatus needed for advanced research.

The obvious tendency is towards increasing inertia. Research projects which require the facilities of "big science" such as particle accelerators, oceanographic ships or satellite observatories, take much longer to complete.[84] A single experiment may take a number of years to plan and carry out.[85] Relative to the rate at which fundamental knowledge is accumulated for society, i.e., sub specie aeternitatis, this is nothing. But unusual nerve is needed to stake such a significant fraction of an individual lifetime for an uncertain outcome. The balance is thus shifted towards the minimal risk, which is recommended by an "Apollonian" policy in the personal choice of projects.

The effort required to assemble appropriate apparatus may also restrict the freedom to exploit a serendipitous opportunity.[86] Even when the apparatus is immediately on hand[87] it may be very disruptive of work in progress to switch a sophisticated instrumental program onto an unexpected problem.[88] And yet research continues to produce its surprises. The outcome of the most carefully planned research project, using the most advanced apparatus, cannot be foreseen. Even in high-energy physics "rarely has a new accelerator or storage ring had its expected goal turn out to be the eventual area of its most important impact."[89]

Whether through an opportunity appearing by chance or by deliberate intention, any move towards research in a new field or problem-domain is severely hampered by the ponderous structure of apparatus.[90] Apparatus is material; it cannot be created at will, or frivolously discarded. In the personal dimension, specialized skill in a particular instrumental technique such as electron microscopy takes long to acquire and becomes a valued professional asset. The research worker is not only captured cognitively by a particular disciplinary matrix of theoretical paradigms, and socially by self-definition as a member of a particular scientific community; he or she is also effectively trapped within a particular speciality by quite mundane considerations of getting together the necessary equipment and learning how to use it.

The inertia of advanced instrumentation and its accompanying methodology thus militates against mobility and flexibility in the personal research plans of the quasi-academic scientist. And yet research techniques are not immune from the characteristic modern phenomenon or rapid obsolescence by technological innovation. The tendency towards undue persistence with outdated methods is just as dangerous as the tendency to persist in the investigation of outdated problems. It is paradoxical that the research scientist—the pioneer of the advancing frontier of knowledge, and the progenitor of technological change—may be in more need than most other technical personnel of further education and of retraining for new professional roles, several times in the course of a career.

Participation by Subordination

The only way to get to one's first scientific research frontier is as a subordinate in an ongoing research project. The answer to the graduate student's question "What research shall I do?" is not really advice: it is a command. Social pressure to conform to the current paradigm becomes a logistic imperative. A serious research project using sophisticated apparatus can only be undertaken within the framework of the research plans of the research supervisor and his or her research group.[91]

Research training in a graduate school thus contradicts its own basic goal—to train young scientists to make original contributions to knowledge. The practical necessity of subordinating one's individual research plans to those of the group cannot be formally reconciled with the notion that one is engaged upon an investigation for which one is personally responsible both in conception and execution. This formal contradiction has to be overcome by deliberately devolving more and more responsibility on to the student in the course of the period of training so that the dissertation presents itself as the work of a "self-winding" scientist who can be trusted to make imaginative decisions on what research ought to be done next. This character trait is often a better indicator of competence in research than the scientific quality of the dissertation, which is so susceptible to local variations of comprehension and technique.[92]

But if this trait is to be fostered, it must relate to a genuine possibility. The notion of following one's personal plan of research must not seem entirely unrealistic. Otherwise, there can be no force to the wise dictum: "After graduate students have taken their Ph.D.s they must on no account continue with their Ph.D. work for the rest of their lives."[93]

Even in the academic world, however, the prospect of achieving personal autonomy in research must be deferred for a number of years. It has become the custom to prolong the training required for a research career somewhat beyond the traditional qualification of a Ph.D.[94] Post-doctoral employment for three or four years as a research assistant or research fellow is now considered a prerequisite for a permanent appointment.[95] Very few scientists are likely to achieve a "quasi-academic" position of independence in research before they are well into their thirties.[96] Indeed many scientists with high qualifications and technical expertise may have to be content with "unfaculty" appointments throughout their working lives, even though they work in universities.[97]

This trend has undoubtedly been accelerated by economic constraints on the growth of science,[98] but it cannot be regarded as a temporary phenomenon. It is a natural consequence of the increasing "collectivization" of the research process, under pressure to expand in the dimension of apparatus. This apparatus must be shared in use by groups of scientists who are thus forced to co-ordinate their research plans. Although this co-ordination can sometimes be arranged as a voluntary co-operation between more or less equal parties, it is very sensitive to the stratification of esteem in the scientific community. Younger, less experienced scientists are almost bound to subordinate their individual research plans to the goals of those who are held to be more senior in experience and reputation—and who usually control the institutional allocation of material resources for research. Voluntary submission to intellectual leadership in defining research programs is transformed into obedience to social authority.

This transformation is enhanced by team research. The increasing complexity of research work, both in its matérial and in the deployment of its personnel, requires a division of the labor of each project into separate technical tasks. Whatever degree of democracy is practiced in the management of the project as a whole, each individual participant is constrained personally to perform a specified role and thus gives up considerable independence in making further individual research plans. This limitation of choice by specialization in research is to be seen in the advertisements for post-doctoral appointments; although these are for research work in universities and other educational institutions, they call for persons with highly specialized expertise to take part in projects the goals and methods of which are prescribed in considerable detail.[99] Thus, at this stage in a career in research, a particular individual, with a particular curriculum vitae, appears to have only a very restricted range of alternatives from which to choose and has practically no possibility at all of moving into a significantly different scientific field.[100]

It must be emphasized that most young scientists do not seem to regard these limitations on their choice in research as unnatural or even uncongenial. Although the traditional conception of academic science praises bold research, its situational logic favors an individual policy of minimal risk, even to the extent of providing a justification for professional immobilisme.[101] A scientific worker who has come to define himself as a specialist of a particular kind can cogently argue that the expectations aroused by long years of training should be rewarded with appropriate employment.[102] Indeed, the doctrine of academic freedom may be enlisted against pressure to work in a new field; in extreme cases, a scientist may even prefer to give up research altogether rather than adopt a new scientific identity.[103] But there are probably very few who are so devoted to science that they would "prefer reasonable unemployment (while still maintaining professional contacts and future opportunities) to leaving science for another profession."[104] The great majority of scientists adapt their research plans realistically to their opportunities for employment.

Personal research plans are thus more closely interwoven with personal career plans than is assumed in the traditional conception of research as a process of individual "discovery."[105] Only a small proportion of present-day research workers are permitted, or are likely to be permitted, the degree of individual autonomy that is envisaged in the academic model of science. This applies even in universities. Whether they work in a higher educational institution, in a governmental laboratory, or in an industrial corporation, the majority of research workers are involved in research projects the goals of which they are not free to choose or to change. For them an instrumental conception of science is the ultimate reality.

This does not mean, however, that the answer to every question "What research should now be done?" is pre-empted by superior authority. A professional scientist cannot function satisfactorily as a technician, carrying out a prescribed schedule of tasks. The most bureaucratic research organization, dedicated to the most urgent specific "mission," must nevertheless rely heavily on individual initiative. Within the limits of personal autonomy allowed by the organization, the rationale of the process of research is remarkably similar to the pattern described by the conception of academic science. The exploitation of serendipity, reference to communal paradigms, the uncertain calculus of "importance" relative to "difficulty," the choice between short term "puzzles" and longer term "problems," the subtle distinction between the virtue of perseverance and the vice of "undue persistence" and so on—these considerations are just as relevant to discovering why the electron microscope is not giving a steady image this morning as they were in the discovery of the structure of DNA. They are constitutive of the complex cognitive, psychological and social activities which result in scientific knowledge, even in its most instrumental form. When people describe science as "problem-solving," this is what they have in mind.

For this reason, the conditions under which scientists may choose or change their research projects are of major significance in the management of research.[106] The high proportion of scientists in industrial laboratories who become expert problem-solvers can best be rewarded with independence in the choice of the problems they are committed to solve.[107] They get the same satisfaction as the archetypal academic scientist from "doing their own work," i.e., from following out individual research plans that have not been directly imposed upon them by executive authority. Unless they are given a specific allowance of elective time for such work,[108] they may even go to the extent of "bootlegging" research on projects which the management could not be persuaded to include in the official program.[109]

In their policies of recruitment and management, many "mission-oriented" research organizations interpret this demand by the members of their scientific staff for personal autonomy as a desire to do "pure" research, in the form of continued participation in "public science" after they have taken the Ph.D. degree.[110] But the theoretical antithesis of "pure" and "applied" science may not be so significant from an individual viewpoint.[111] The norm of independence[112] embodied in the individualistic outlook of academic science can be largely satisfied in practice within a much smaller domain of action than, for example, in a recognized scientific subspeciality. There are relatively few "natural philosophers" with transcendental research goals among the tens of thousands of persons entering the scientific profession. A considerable number are willing enough to become "private scientists"[113] mainly concerned with "success"[114] as understood by their immediate associates within the domain where their responsibilities are evident.

But it is not sufficient to replace public recognition by the scientific community by more tangible private rewards. In a well-managed research organization, personal commitment to the mission of the organization generates its own psychological gratification.[115] That is to say, the individual scientist learns to follow a personal research policy in which the risks and rewards have been so redefined that the "importance" of a project is assessed in terms such as "contributing to the interests of the company I work for," rather than as "contributing to the theory of alloys."[116] As in academic science, such a calculation necessarily includes the cognitive and technical considerations, i.e., paradigms of knowledge, the "state of the art," available apparatus and so on, which define the specialized expertise and responsibility of the scientist in question. It is also necessary to appreciate more general factors relating to the mission of the organization, such as manufacturing capacities and commercial conditions. One of the functions of the management of research is to help scientists to gain this appreciation, e.g., by "broadening their experience by properly timed moves to neighboring fields of study."[117]

The various considerations which can arise in making personal research plans are vital to the art of research management. In some cases, the wise exercise of executive authority in the research process can be of much more benefit to the individual than complete freedom, for example, in putting a stop to "undue persistence" in a sterile line of enquiry.[118] Even in the delicate matter of individual autonomy as a professional scientist, the convergence of the academic and instrumental forms of the organization of research is not all to the disadvantage of the former.[119]

Collaboration

Organizational constraints limit personal autonomy in research. As science is "collectivized" and "industrialized"[120] this domain tends to shrink. The majority of scientists in subordinate posts now have little more autonomy than other professional technical workers such as doctors or engineers. Nevertheless, in large-scale industrial and governmental research organizations, many experienced scientists are allowed considerable personal discretion in planning their research. Within the limits of general institutional objectives, and the resources of matériel and personnel at their disposal, they may have as much effective autonomy in their research as any university professor. The social role of the quasi-academic scientist is not be to found only in universities. It is characteristic of leadership in research—as distinct from the management of research—throughout the entire system of research and development.[121]

This autonomy may be exercised very individualistically by following a personal policy in pursuit of purely personal rewards. But there is also the alternative of voluntary collaboration with colleagues, i.e., with other scientists of similar status and degrees of individual autonomy. A research project may thus have a significant social component which is neither strictly managerial nor communal in a broader sense.

The traditional ethos of academic science is not unfavorable to collaborative research. It is well understood that "two heads are better than one," especially when they bring together complementary skills or experience or points of view. As in such famous partnerships as that of Crick and Watson, the subtle psychological interaction of well-matched temperaments in close communication may be far more creative than either could be alone. Although academic science is highly competitive, it is not a zero-sum game, and there is far from a linear relationship between difficulty and importance. Even a narrowly selfish interest may be advanced by forming a coalition with one or more well-chosen colleagues in the hope of sharing more than the proportionate rewards.[122] Indeed, a narrow calculation of "selfish interest" is regarded as unenlightened in science. Any proposal for voluntary co-operation in research is naturally thought well of as a manifestation of altruism. For all its individualism, science is an enterprise aiming at the advancement of knowledge. The communal norms of academic science are justified functionally by the way they generate and validate this collective consensus.[123] The autonomy of the individual scientist is not an absolute right: as in society at large, it is justified by its contribution to the good of the enterprise as a whole. The commitment of the individual to the advancement of the body of knowledge with which he is concerned, like personal identification with the fortunes of one's corporate employers, is a significant consideration which often transcends any calculable personal gain from a research plan.

Despite these rational arguments in favor of voluntary collaboration in research, the historical evidence suggests that the psychological balance in academic science tends to be against co-operation and towards competitive individualism.[124] Until about 50 years ago the majority of primary scientific publications in all fields of research were by single authors.[125] The increased proportion of papers with plural authorship is almost certainly a manifestation of collaboration enforced by the increasing sophistication of instruments and techniques. Quite apart from the economic incentives to share the use of large apparatus, many research projects nowadays are essentially interdisciplinary: they can only be undertaken if the labor is divided among two or more scientists with complementary knowledge and skill. Quasi-academic scientists who are unwilling to enter into such projects on an equal footing with professional colleagues are in danger of closing off too many promising alternatives in their individual research plans. This applies with particular force in "mission-oriented" research organizations, where willingness to co-operate "voluntarily" with other scientists is usually regarded by the management as a virtue.

The pressure on quasi-academic scientists to merge their individual research plans in a co-operative venture is observable in an extreme form in high-energy physics, where the apparatus is so immense and complex that experimental research can only be done by a large team of fully qualified scientists with a wide range of experience in research and with very specialized skills. In other fields of "big science," such as space research, such teams are normally set up and controlled by executive officials within the managerial structure of a single organization. But the academic tradition is so strong among high-energy physicists that they have devised a means of working together without nominally sacrificing the individual autonomy which scientists prize so highly. At the Center Européen de la Recherche Nucléaire (CERN), for example, major experiments are undertaken by semi-permanent "collaborations" made up of hundreds of scientists and technicians drawn from as many as a dozen different universities in four or five different countries.[126]

Why are high-energy physicists willing to participate in such projects, knowing that their own personal research plans must be subordinated for long years to the imperatives of the group? Is sheer altruism a sufficient motive? Is it true that "the essential reason for the success of international scientific collaboration is that the members have a common purpose—to do research in science—and a common characteristic—scientific curiosity," and that "these common attitudes give a drive to achieve scientific results which transcends all personal differences?"[127] Does the weighing of importance against difficulty make any sense when the reward of even the most successful investigation has to be divided among so many individuals seeking personal recognition? Or are the participants really getting their satisfaction from self-actualization as technical experts, contributing their professional skills to a novel technological device? Is the participatory democracy of such groups capable of reconciling large scale "mission-oriented" research with every research worker's need for a recognized domain of individual autonomy? These questions pose themselves sharply when the answer to the question "What research shall I do now?" can only be "Join the team!"

Peer Review

The typical quasi-academic scientist has a characteristic domain of autonomy, within the limits set by terms of employment, resources of apparatus, established colleagues, specialized skills, recognized professional standing, and a reliable disciplinary paradigm. Research plans within this domain can be made fairly freely, and may be expected to yield a modest harvest of the personal benefits promised by the traditional ideology of academic science. In the longer run, however, this is a dangerous policy. To get out of the groove of undue persistence, to exploit a shift in the paradigm of the discipline or take advantage of a methodological innovation, to follow up a serendipitous discovery, to continue to participate in the dynamic progress of science, it is essential to undertake novel research projects that challenge the limits to this domain. In the course of the year, projects involving a new range of problems, new pieces of apparatus, new subordinate colleagues with new skills, are bound to enter into the research plans of every active quasi-academic scientist.

Access to these new resources then becomes a primary consideration in personal research plans. It is not sufficient to assess a project in terms of personal risks and rewards, and to be fully persuaded that it has a favorable ratio of importance to difficulty. It is also necessary to calculate the likelihood that these resources will in fact be made available by those who control them. To put it at its most mundane level: "I estimate that I could carry out this investigation at a cost of about £100,000 spread over the next 3 or 4 years. Who will provide me (strictly speaking, my university or research institute) with these funds for this purpose?" In a sense, this is no more than an additional term in the "difficulty" of the project. But it subtly shifts the fundamental question from "What research shall I do now?" to "What research can I persuade others to support my doing?"

Although greatly aggravated by the growing expense of research apparatus, this is not an entirely new consideration in personal research plans. In Germany in the nineteenth century, for example, it was very risky to undertake an original research project in an interdisciplinary field without the security of a university chair, and there was fierce competition to win the recognition needed for the acquisition of this essential resource.[128] It has always been easier to establish new fields of research in periods of educational expansion, when unconventional scholars could achieve quasi-academic autonomy in newly founded posts, often in newly founded institutions.[129] For this reason, there is concern lest the present period of stationary expenditures may affect scientific progress more adversely than would be suggested by a linear model of the research process.[130]

In most advanced countries, however, a system of project-grants has been established to support specific research projects which are to be undertaken under the leadership of individual, quasi-academic scientists.[131] The "principal investigator" writes a proposal to investigate a particular question, using particular methods, with the intention of attaining particular scientific objectives by a particular date. The proposal takes the form: "It is proposed to determine the superconducting transition temperature of poly(sulphur nitride), using a dilution refrigerator (to be purchased at a cost of £X,000) and employing a post-doctoral research assistant (at a salary of £Y,000 per annum) over a period of Z months." This proposal is submitted to an appropriate grant-awarding body, whether it be a research council, a foundation, a government department, or some other governmental institution, where it is accepted or rejected in competition with other proposals of a similar nature. If it is successful, the required funds are made available to the institution where the principal investigator holds an appointment, the apparatus is purchased, the post-doctoral assistant is appointed, and the research proceeds.

In outward form, this procedure does not transgress the norm of academic independence. Although there may be some negotiation over the details of the grant, such as the duration of the project, the standing of research assistants, travel costs and so on, the distance between the applicant and the grant-awarding body is carefully maintained. But it is not really a free bargain between equals, since the professional career of the applicant is ultimately shaped by the response of grant-awarding bodies to a series of such applications.[132] This is therefore a critical branching point in the unfolding of a personal research policy. At each such point, personal judgement or inclination must give way to an influence which is external to the individual concerned. From the point of view of the scientific community, a mistaken decision by a grant-awarding body not to support a particular research proposal might be even more damaging to the advancement of knowledge than, say, an editor's decision not to publish the results of the completed investigation.

The research-grant system, therefore, must be just as sensitive to the traditional academic norms as is the system of scientific communication.[133] It is not just a matter of making sure that research proposals are honestly considered according to practical criteria, and not approved for entirely inappropriate reasons such as the kinship of the applicant to a powerful politician. The proposal must stand or fall on its "intrinsic scientific merit," without reference to secondary features such as the reputation of the principal investigator, the status of the institution, or even the literary quality of the supporting documents.

Needless to say, this is a counsel of perfection. A system which seems to satisfy the norms of "universalism" on a statistical basis[134] may turn out, in the light of hindsight, to have made many serious misjudgments.[135] The difficulty of winning initial support for novel interdisciplinary projects, or at the early stages of following up a serendipitous discovery, is notorious.[136] But this merely demonstrates the fundamental uncertainty and unpredictability of the process of scientific discovery, where not all bright ideas are good, and not all surprises are happy ones.[137]

Intrinsic scientific merit is not a well-defined, measurable property. The qualities of "timeliness" and "promise" which are sought after by grant-awarding bodies[138] are no more than variation on the theme of "importance" against which the relative "difficulty" of the investigation must be calculated in monetary terms. These considerations cannot be weighed "objectively"; they can only be assessed by scientists with intimate understanding and experience of the subject in question. The decision can only be taken on the basis of "peer review," where these qualities are "widely interpreted and judged by the applicant's scientific peers,"[139] either sitting together in committee or reporting as referees, often anonymously. Again, the way in which this process works in practice is not our present interest. The essential point is that a research grant proposal is by its very nature an explicit justification of a personal research plan addressed impersonally to other scientists within a specific range of specialities.

This transformation of a "subjective" research plan into an "objective" research proposal favors consensual considerations above individual "hunches." In formulating such a proposal, a quasi-academic scientist is bound to emphasize topics and arguments which are likely to commend themselves to the group of scientific peers who will be asked to judge it. It is hard for the applicant to resist a tendency to limit the range of alternatives worth his personal consideration to projects which use "established" techniques to investigate "recognized" scientific questions. The peer-review system inevitably shifts the balance towards communal paradigms even within the mental domains of individual scientists, where creative imagination has its active source.

The applicant's anticipation of the responses of his peers is just as significant—and just as difficult to counteract—as the caution and orthodoxy to which most grant-awarding bodies naturally tend in their allocation of grants. This tendency is to be observed, for example, when the applicant deliberately attempts to ingratiate himself with the peer group by designing the proposal as closely as possible according to some fashionable model. An "unduly persistent" relationship can then develop between the two parties. Such conduct is strongly encouraged by recent economic pressures in universities, where the ability to obtain research grants appears to have become an essential qualification for a permanent appointment.[140] The same pressures drive quasi-academic scientists towards fields of research that are directly solicited by grant-awarding bodies in the form of "requests for proposals."[141] This procedure is reminiscent of the old practice of encouraging research by offering prizes for the solution of specified scientific problems,[142] with the significant difference that a prize could only be earned for an actual solution, while grants are now given for the promise of one.

The peer-review system has grown up to protect the competitive individualism of academic science. Ideally, each scientist takes the uncertainty of research as a personal risk: the labors of a lifetime are staked for the possible rewards of discovery. In the pursuit of resources and of occupational security, however, the quasi-academic scientist is tempted to move into an entrepreneurial role, offering to perform research on a contractual basis for an external body.[143] With this the prime value of research in academic science is abandoned for the sake of a place in an "instrumental" organization. Some of the burdens of the individual risk are thus shifted on the broader shoulders of corporate institutions, who effectively decide the question "What research shall I do now?" by the support they can offer to willing agents of their desires and plans.

The significant difference between a scientist with a quasi-academic status in a full-time research organization and a scientist in a university who is supported by grants, is not so much in the degree to which their long-term plans are subject to external authority. It is in the permeability of the barriers through which this authority is exercised. Research is never really strictly managed along bureaucratic lines. Plans, proposals, critical comments, novel ideas, personal predilections and group commitments pass informally up and down the hierarchies of control through various processes of face-to-face negotiation, assessment by colleagues and collective decision. The privileges of individual autonomy are foregone for the opportunity to collaborate with a congenial colleague or to participate in a promising major project. Paradoxically, a scientist in a university may feel more constrained than if he or she were employed in a large research organization[144] which is permeable and responsive to all such influences, from the bottom up as well as from the top down.

By contrast, the contemporary system of awarding grants on the basis of formal review of applications by "scientific peers" is only semi-permeable. Only research proposals may flow upwards, and only grant decisions may flow down. This protocol imposes on the relationship between applicants and reviewers an unrealistic rationale. Since a research proposal is designed to persuade a group of fellow-scientists, who are in principle anonymous, of the advantages to be gained from undertaking a certain action (i.e., the investigation in question), it has nearly the same rhetorical purpose as a primary scientific communication.[145] It should endeavor to compel assent by rigorous logic applied to well-attested instances arranged in sharply defined categories. Almost inevitably, both proposers and reviewers come to expect that it should be composed and assessed according to the same criteria as a conventional scientific paper,[146] in short that it should conform to the same canons of scientific rationality as the announcement and justification of a scientific discovery.

But a research plan is intentional, rather than justificatory. It contains an unavoidable component of genuine uncertainty, which breaks every chain of logical deduction and blurs the edges of every preconceived category. This does not mean that decisions in research are non-rational or irrational. Like any sane person conducting his or her everyday affairs, the scientist makes full use of a capacity:

> to calculate; to act deliberately; to project alternative plans of action; to se-
> lect before the actual fall of events the conditions under which he will fol-
> low one plan or another; to give priority in the selection of means to their
> technical efficacy; to be much concerned with predictability and desirous
> of "surprise in small amounts;" to prefer the analysis of alternatives and
> consequences prior to action in preference to improvisation; to be much
> concerned with questions of what is to be done and how it is to be done; to
> be aware of, to wish to, and to exercise choice; to be insistent upon "fine"
> as contrasted with "gross" structures in characterizations in the knowledge
> of situations that one considers valuable and realistic knowledge, and the
> rest.[147]

But serious contradictions arise in the planning of research,[148] as in
everyday life, if one seeks to live according to the additional characteristics of scientific rationality, e.g., "compatibility of end-means relationships with principles of formal logic," "semantic clarity and distinctness," "clarity and distinctness" "for its own sake" and "compatibility of
the definition of a situation with scientific knowledge."[149]

These contradictions are sensed intuitively by experienced actors in the
research process, who often insist that they "never make research plans"
or that they "do not know what research they will do tomorrow."[150] What
they mean is that they cannot live up the exalted standards of scientific rationality. They quite overlook the immense resource of everyday rationality which they apply, tacitly, as they glimpse and grope towards rewarding
goals in research. And when, in their turn, they review the research proposals of their peers, they are embarrassed to realize that the true rationale
of their decisions derives from considerations that cannot be formulated
and "rationalized" in scientific terms.

From Individual Discovery to Collective Instruments

The discovery and instrumental models of science seem to offer completely contradictory accounts of the process by which research is "undertaken," i.e., carried out with rational intent.[151] Our analysis of the rationale
of individual research plans shows, however, that these traditional conceptions of science are not logically incompatible but rather complementary.[152] They are, indeed, ideal-typical variants of the individualist
and collectivist versions of just this process of deciding what research
to undertake.[153]

The contradiction is resolved by reference to the scale of time and the institutional involvement or independence which is entailed in the plan over which a research plan is designed to run. Within a restricted domain, where, for example, individual plans can be freely adjusted to exploit serendipity, the concept of discovery is entirely meaningful—although absolute individualism is always restrained by collective influences such as the prevailing consensus regarding what is thought "important" and "difficult." As this domain is enlarged, research decisions are increasingly affected by institutional structures and arrangements—by the management of the organization, by voluntary collaboration, or by the existing system of providing financial support on the basis of judgement by peers. On the largest scale, only determination by institutional authority is visible; research appears to be an instrument used by society to achieve its diverse ends.

This last step from "collectivism" to "instrumentalism" is obviously valid for the science institutionalized in bureaucratic research organizations with specific technological tasks. We know from experience that "research and development" is a highly efficacious purposeful activity, which has been instrumental in achieving many desired objectives in industry, engineering warfare, medicine, and agriculture. And yet a considerable amount of individual autonomy is allowed to many scientists employed in such research. Unexpected discoveries also have disconcerting effects on the unfolding of research plans within such organizations. The art of the management of research is to harness this individualism and to profit from these discoveries.[154]

> There is much uncertainty in research. Nevertheless, by being confined to the laboratory's pathways, research is given a measure of certainty. The extent to which this is accomplished sets research in a predetermined direction. It provides an overall controlling system for the direction research will take. It even does more than this. It influences and controls the research results.[155]

A research result cannot be predetermined, nor can it be willed away after it has been found. It was only the act of seeking it that could have been frustrated by deliberate intent. Control over initiative in research is the means by which science is made a positive instrument of human purpose.

Does this apply to quasi-academic scientists, attempting sincerely to pursue knowledge "for its own sake?" Can they be said to be serving ends that are really laid down by social actors with other objectives than this pursuit?[156]

For the great majority of scientists in universities, a certain amount of subjection to an instrumental position cannot be avoided. Those in subordinate research posts, for example, have no choice; they must follow the plans laid down for them by their superiors. It make little difference whether the plans have ill-defined "basic" goals or more specific technological objectives: they have no voice in setting these goals, and they take no responsibility for these more distant ends. Paradoxically, the same abstention from responsibility for the decision of those who are in positions of institutional authority is often felt by the small number of individualists who manage to persist in "cottage industry" styles of research and who follow their own independent lines with minimal support from grant-awarding bodies. For both these groups, the system of grants for research is seen as "external." In the latter, they have no interest in it because it does not affect them; in the former, it is "external" because it cannot be affected by them.[157]

But the relatively privileged strata of quasi-academic scientists who participate in the making of decisions about which applicants are to be awarded financial support must take some responsibility for the institutional policies involved in such decisions.[158] In the assessment of a particular research-proposal, a group of scientific peers may wisely penetrate the rhetorical rationalization and fully appreciate, from their own experience and understanding, the personal considerations of curiosity and reflection which go into the making of a research-plan. Their own criteria of "difficulty" and "importance" may be very close to those of the applicant. But because they must exercise deliberate choice between different proposals competing for limited funds, they are bound to give more weight to collective factors than the traditional academic conception of science would suggest.

At a certain level, for example, this group of scientists who are asked to assess applications for financial support might see themselves as acting on behalf of the "invisible college" of a particular sub-discipline, trying to answer the question "What research shall we do now?" The peer-review system becomes the mechanism through which individual quasi-academic scientists collaborate voluntarily in order to maintain the autonomy of the scientific community over resources and research plans. The scientific community thus contrives to act according to its own "internal" norms, without reference to "external" social goals. In the same spirit, a group of leading scientists may initiate plans for a major new research instrument, such as a particle accelerator, to be shared in use by many independent investigators.[159] Indeed, they move naturally into the role of formulating a "proposal" for a research project, on a larger scale, addressed to a higher stratum of authority within the research funding system and soon find themselves adjusting their goals to the corresponding question "What research can we persuade others to support our doing?"

As such a proposal is transmitted upwards through the hierarchy of institutions which support scientific research, its rhetoric must be adapted to the stratum which will eventually decide its fate. The more costly the project, the more it must be considered in the light of criteria which go beyond the development of a single discipline. It is not capable of being justified by its "scientific" virtues alone. In the end, it will have to commend itself to powerful persons for whom academic science and the academic conception of science are not of absolute value and for whom the disinterested pursuit of knowledge is not their first concern. Hence the proposal is bound to shift its ground towards non-scientific criteria, justifying itself mainly in terms of the applicability of the knowledge to be sought; the "use" to which the results might be put; the industrial, economic and military "importance" of the problem to be solved. The same trend towards the instrumental justification of the financial support of research also occurs as budgets for grants in different fields are aggregated at higher organizational levels and choices must be made among many different fields.[160] At some stage there is an inevitable shift of emphasis from the "internal" criteria of "scientific merit" to the "external" criteria of "social and technological value."[161]

At the very top of the system, where funds for research and development are no more than line items in the operating budgets of government departments and industrial corporations, the criteria for decisions about "what research should be undertaken?"—generalized to the national level—cannot be separated from economic, industrial, military and political criteria. The legitimacy and potency of "science policy" cannot be determined entirely by reference to abstract knowledge; it also derives from the distribution of power in society at large.[162] Although the successive linkages of influence form the top down may not be as direct as in a formal bureaucratic organization, they are effective enough in the long run. Over large areas of quasi-academic science, research is "set in a predetermined direction" and its results "influenced and controlled," through the strings of the public purse, just as it is by the more explicit management of industrial research and development.[163]

It must be emphasized, however, that this general social process of collectivization and instrumentalization of science through control over research plans does not invalidate the conception of research as a personal process of discovery, often motivated by individual intellectual forces such as curiosity and reflection. Nor does it justify a conception of science which makes scientists into complete servants—even slaves— of "social need."[164] On the contrary, our study of the complex mixture of personal and social considerations, of "internal" and "external" factors, that enter into research plans and choices, shows that such a conception would be self-defeating and sterile. It is by their personal commitment to scientific research, their self-actualization through technical mastery of unsolved problems, their relative autonomy and independence within their own life-worlds, that scientists are most securely drawn into the service of larger social aspirations and demands. It is helpful and ethically proper that they should not be kept in ignorance of these ends, but not to the extent that these take priority over more immediate considerations of cognitive validity, paradigmatic significance or methodological efficacy in the process of discovery. "Relevance" is only one of the many criteria of the "importance" of a research project.

The scientist obviously cannot begin his investigation with assurance that his work will ultimately be important in satisfying a particular human need. But, some paths in the network are more probable than others...provided the right choices are made...and some end-results are more important than others. Thus even at the beginning of a scientific investigation it may be possible to perceive some of the ways in which it could be relevant to important human needs. I believe the perception of these decision-paths, however qualitative and even subjective they might be, should play a role in the research decisions of the scientist. I am not suggesting that he leave aside the traditional factors of intrinsic curiosity or intrinsic scientific interest. I am suggesting that a perception of the relevance to human need, however subjective and uncertain, should be added to the scientific ethic for choosing research problems.[165]

This perception is of the greatest importance for members of the scientific "Establishment" i.e., for those scientists who have risen into the higher strata of the institutions which conduct and support research, whether as managers of research organizations or as influences in the decisions of grant-awarding bodies. Generally speaking, they win this high status by outstanding performance in a narrow field of research. Nevertheless, they are soon called upon for decisive opinions on the research plans which transcend the boundaries of their personal experience and expertise, and must somehow assess the ratio of "importance" to "difficulty" for projects extending into disciplines where they are as ignorant as any layman.[166] At a certain level they can no longer confine themselves to internal considerations of "scientific" merit, but find themselves making judgements based upon "external" criteria of social value and need.

Many scientific notables shy away from this responsibility, justifying themselves by an exaggerated notion of the role of "scientific" rationality in practical affairs, or withdrawing into a self-contained academic individualism. Others are—not so many—drawn naturally into positions with influence far beyond the research-system, and are thus assimilated into the circle of the powerful in society. But to give a wise response to the central question of this paper, "What are the options for future research?" demands mediation between "internal" and "external" considerations and a balance between individualistic and collectivistic assessments of the "risks" and "rewards" in each proposed project. It demands of each member of the "establishment" something more than "social responsibility." There is also a need for intellectual imagination, not to instrumentalize other scientists, and to remember that the real question to be answered is "What should be undertaken so that they can give of their best?" It is the intentions of other people, as they enter into the structures of our own life-world, that are so difficult to grasp.

The final words on this subject must come for the heart of James Clerk Maxwell who, at the age of 23, wrote:

> He that would enjoy life and act with freedom must have the work of the day continually before his eyes. Not yesterday's work, lest he fall into despair, nor tomorrow's, lest he become a visionary,—not that which ends with the day, which is a worldly work, nor yet that only which remains to eternity, for by it he cannot shape his actions.

> Happy is the man who can recognize in the work of To-day a connected portion of the work of life, and an embodiment of the work of Eternity. The foundations of this confidence are unchangeable, for he has been made a partaker of Infinity. He strenuously works out his daily enterprises, because the present is given him for a possession.

> Thus ought Man to be an impersonation of the divine process of nature, and to show forth the union of the infinite with the finite, not slighting his temporal existence, remembering that in it only is individual action possible, nor yet shutting out from his view that which is eternal, knowing that Time is a mystery which man cannot endure to contemplate until eternal Truth enlighten it.[167]

References

1. This is not to say that the interpretation of research results is not rational—or at least highly rationalized. We are concerned here with research as social action, rather than with the validity of what this action produces.
2. "There may at times be a plan in the foreground of my consciousness that is determined by a governing interest. However, it is always surrounded with a horizon of meaning to which I can again explicatively advert. If I do so, I will discover that the governing interest is connected with other interests, that a goal that is to be actualized is a partial step towards the actualization of higher goals, that decisions have resulted from previous decisions. In daily

life, acts are components within a higher-order system of plans for a specific province within the life-world, for the day, for the year, for work and leisure—which in turn have their place in a more or less determined life-plan." Schütz, A. and Luckmann, T., *The Structures of the Life-World* (London: Heinemann, 1974), p.19.

3. Polanyi, M., *Personal Knowledge* (London: Routledge & Kegan Paul, 1958); Ziman, J.M., *Reliable Knowledge* (Cambridge: Cambridge University Press, 1979).

4. This term was introduced by Chubin, D.E. and Connolly, T., "Research Trails and Scientific Policies," *Sociology of the Sciences Yearbook,* VI (1980) to refer to the historical con sequences of the ensemble of "research plans" making up the biography of an individual research worker.

5. With the significant exception of Chubin, D.E. and Connolly, T., op cit. Recent studies of "Theory choice and problem choice in science" are reviewed by Zuckerman, H., *Sociological Inquiry*, XLVIII (1979), pp 67–95.

6. Ziman, J.M., "Conceptions of Science" *Realizing Social Science Knowledge*, B. Holzner at al. (eds.) (Vienna: Physica Verlag 1983), pp. 179–196.

7. In the extreme case, the perfectly instrumentalized scientist would be a slave. The prisoners of Alexander Solzhenitsyn's *The First Circle* (London: Collins, 1968) are successfully coerced to do research by fear of worse alternatives; the large numbers of scientists who work in secret military research laboratories in the Soviet Union are much better treated, but not much more free in their individual research plans. See Medvedev, Zh. A., *Soviet Science* (New York: Norton, 1978) and Popovsky, M., *Science in Chains* (London: Collins, 1980).

8. Ziman, J.M., "Conceptions of Science."

9. A characteristic expression of this viewpoint is the following: "Great men [such as Newton], no matter now notable their genius, in all spheres formulate and resolve those tasks which have been raised for accomplishment by the historical development of productive forces and production relationships." Hessen, Boris, "The Social and Economic Roots of Newton's Principia" in Bukharin, N.I., et al., *Science at the Cross Roads* (London: Kniga, 1931), p. 203.

10. In this I follow Hesse, M., *Revolutions and Reconstructions in the Philosophy of Science* (Brighton: Harvester Press, 1980), p. 49.

11. Szent-Gyorgi, A., "Dionysians and Apollonians," *Science* CLXXVI (2 June, 1972), p. 966. Many experienced scientists would echo his remark: "When I go home from my laboratory in the late afternoon, I often do not know what I am gong to do the next day. I expect to think that up during the night"—not reflecting, perhaps, on the actual invariance of their objectives under their own wider research plans.

12. Barber, R, and Fox, R.C., "The Case of the Floppy-Eared Rabbits: An Instance of Serendipity Gained and Serendipity Lost," *American Journal of Sociology*, XLIV (1956), pp. 128–136.

13. Hoffmann, B., *Albert Einstein: Creator and Rebel* (London: Hart-Davis, 1973), p. 28.

14. Szent-Gyorgi, A., op. cit.

15. Lovelock, J., "The Independent Practice of Science," *New Scientist*, LXXXII (6 September, 1979) pp. 714–717.

16. Chubin, D.E, and Connolly, T., op. cit.

17. Panofsky, W., "Needs versus Means in High-Energy Physics," *Physics Today*, XXXIII (June 1980), p. 29, makes this point for high-energy physics. As he shows from a table of examples, "rarely has a new accelerator or storage ring had its expected goal turn out to be the eventual area of its most important impact."

18. Marshak, R., "Basic Research in the University and Industrial Laboratory" *Science*, CLIV (23 December, 1966), p. 1,521.

19. Ziman, J.M., "Conceptions of Science."

20. From this point, words such as "scientists", "science," "research," etc. must be interpreted broadly to include much of what might be narrowly termed "technologist," "technology," "technological development," etc., without explicit qualification.

21. Ziman, J.M., "Conceptions of Science."

22. Ibid., p. 86.

23. Ibid., p. 70.

24. I quote these from four consecutive papers in an issue of a physics journal opened at random.

25. As pointed out by P. B. Medawar in *The Listener*, LXXI (12 September, 1963), this is one of the characteristically "fraudulent" features of most scientific papers.

26. Ziman, J.M., *Public Knowledge* (Cambridge: Cambridge University Press, 1968), p. 34.

27. See, for example, Popper, K.R., *Conjectures and Refutations* (London: Routledge & Kegan Paul, 1963), p. 112.

28. Lakatos, I., "Falsification and the Methodology of Scientific Research Programs" in Lakatos, I. and Musgrave, A. (eds.), *Criticism and the Growth of Knowledge* (Cambridge: Cambridge University Press, 1970), pp. 91–195, does not suggest that such programs are publicly adumbrated. On the contrary, by setting them up as "rational reconstructions," he clearly classes them with the ahistorical, a posteriori rationalizations which scientific authors claim for their papers, to which we have already referred.

30. Ginzburg, V.I., *Soviet Physics Uspekhi*, XIV (1979), pp. 21–39. This is a most interesting document for metascientific analysis, not only as a survey of the contemporary Zeitgeist of fundamental physics but also for its sociological self-consciousness: "...it is in general difficult to define in some consistent manner what is 'unimportant' and (or) 'not interesting' in science. Yet there is no doubt that a hierarchy of problems and tasks does exist. Its effects are felt in practice, it is reflected throughout scientific (and sometimes also in nonscientific) life, and it can even be discerned in the tables of contents of journals." He emphasizes, however, that he is expressing purely personal opinions and that "it is also essential to engage in problems not included in this 'list'" because "many remarkable discoveries and scientific accomplishments are unpredicted and unexpected."

31. These phrases are quoted from two volumes of conference proceedings that I happen to have at hand; equivalent phrases may be found in almost any similar publication in almost any field of science.

32. In a recent collection of personal reminiscences of "The Beginnings of Solid State Physics," *Proceedings of the Royal Society*, A CCCLXXI (1980), pp. 1–177, by various authors I found anecdotal accounts of about 50 episodes (in the lives of almost as many different scientists!) of "personal research planning." In more than 30 percent of these episodes, the decision to work on a particular problem was made because "it was the thesis topic assigned to me by my professor." Most of these episodes occurred 40 to 50 years ago, but this apprenticeship tradition dates back to the middle of the nineteenth century, and is still firmly entrenched in academic research in the natural sciences.

33. A further 25 per cent of the episodes referred to in the previous note came under this general heading.

34. I speak from personal experience. The correct answer is too cruel to be given; by analogy with Cornelius Vanderbilt's reply, when asked about the cost of running a yacht, it should be "If you have to ask that question, then you are not yet competent to undertake that investigation."

35. Kuhn, T.S., *The Structure of Scientific Revolutions* (Chicago: University of Chicago Press, 1962), p. 37.

36. Ibid., p. 82.

37. Kuhn (op. cit. pp. 18, 59 and 109), himself only draws attention to this feature. His supporters and critics (e.g., In Lakatos, I. and Musgrave, A., op. cit.) seem to ignore this aspect of science completely.

38. The clearest example would be the production of heat by friction, which was completely neglected, in theory and experiment, throughout the eighteenth century.

39. This point of view also resolves the ambivalence of serendipity in the Kuhnian scheme; see, e.g., Dean, C., "Are Serendipitous Discoveries a Part of Normal Science? The Case of the Pulsars," *The Sociological Review*, XXV (1977), pp. 73–86.

40. Hagstrom enunciates the relevant "norms of independence:" "First, the scientist is expected to be able to select research problems freely. Second, he is expected to be able to select freely the methods and techniques to be applied to them." Hagstrom, W.O., op. cit., p. 105.

41. "Importance" and "difficulty" are exactly the words that scientists themselves use in this

sense when talking about research plans; e.g., in a conversation with R.P. Feynman, reported to me verbatim by a friend. Although these terms are obviously correlated with "benefits" and "costs" respectively, I prefer not to follow the economic imagery too far towards the type of "coarse incremental cost-benefit model" postulated, for example, by Chubin, D.E. and Connolly, T., op. cit.

42. We thus deliberately evade the issue of what motivates scientists and whether they get their just deserts. This literature is reviewed by Fisch, R., in Spiegel-Rösing, I. and de Solla Price, D., (eds), *Science, Technology and Society* (London: Sage, 1977).

43. This schematization is similar in general principle to that of M. Nowakowska in her pioneering paper on "Measurable Aspects of the Concept of Scientific Career" in Knorr, K.D., Strasser, J. and Zilian, H.G., (eds.) *Determinants and Controls of Scientific Development* (Dordrecht: D. Reidel, 1975), pp. 295–322.

44. Schütz, A. and Luckman, T., op. cit., p. 92.

45. "A sense for 'the right kind of question' and for the character of its solution develops during the interaction between masters and apprentices and among the apprentices themselves as they pass judgement on the quality of scientific work, new and old, their own and that of others. It develops also as they speculate about the direction their field 'should take,' identity gaps in basic knowledge, and argue about which problems are 'ripe' for solution at the time and which are not. These matters are of evident and prime interest to scientists who intend to help shape the field in which they work; they might be of less interest to scientists who see themselves playing a more modest role." Zuckerman, H., *Scientific Elites: Nobel Laureates in the United States* (New York: The Free Press, 1977), p. 129.

46. This does not apply solely to scientists of limited ability with immodest ambitions. From personal observation I could refer to several cases of persons of high scientific ability who unwisely undertook research projects for which they had entirely misjudged their own intellectual or temperamental suitability. This is, of course, an unavoidable feature of academic science: it is a part of the price that must be paid for the benefits of academic autonomy.

47. Warren, K.S. (ed.), *Coping with the Biomedical Literature Explosion: A Qualitative Approach* (New York: Rockefeller Foundation, 1978).

48. Garfield, E., *Citation Indexing* (New York: Wiley, 1979).

49. Rescher, N. *Scientific Progress* (Oxford: Blackwell, 1978), p. 97 defines the λ-quality-level of research findings to be such that "when the total volume of (at least routine) findings is Q, the volume of findings of this category stands at Q^λ (for $0<\lambda\leq1$)." On this scale, "routine" findings are those for which $\lambda = 1$ whilst for really first-rate findings λ approaches 0. This parametric representation of the relative distribution of "important" papers in the scientific literature seems reasonably consistent with impressionistic assessments of this indefinable quality.

50. This often happens in mathematical research, where the solution of certain problems becomes an end in itself, simply because they have long resisted attack.

51. On this point I would not follow Nowakowska, M., op. cit., who considers cases where her parameter C is negative, i.e., where the "fear of failure" exceeds the "need of achievement." In such circumstances, the only rational decision is to give up research altogether—which is what many people actually do, although usually for much more complicated personal reasons.

52. What Kuhn, T.S., op. cit., describes as "normal science," i.e., research under the sway of a well-defined and essentially unquestioned paradigm—obviously provides many such apparently easy questions of marginal scientific importance. For want of a better term, we refer to these as "puzzles."

53. Szent-Gyorgyi, A., op. cit. Although his definition comes close to the customary denigration of "normal" science, it must be emphasized that the "convergent," "classical" mode of research is as intellectually demanding and as essential to science as the more "divergent" and "romantic" Dionysian style.

54. Haraszthy, A. and Szanto, L., in Andrews, F.M. (ed.), *Scientific Productivity* (Paris: UNESCO, 1979), p. 163, point out that although the average number of projects per capita in research organizations is about unity, the mean number of projects per research unit is about five, implying

that the members of research teams are normally working on several projects at once.

55. A very simple prosopographic analysis of the published work of about 100 recently deceased scientists (*Biographical Memoirs of Fellows of the Royal Society* (London: The Royal Society, 1974–79), vols. 20–25) suggested that about half of them had stayed entirely within the same subspeciality throughout their lives. That is to say, it was even money that a scientist whose Ph.D. thesis or first published paper had been on, say, marine worms, would die as the world's most eminent authority on marine worms.

56. Chubin, D.E. and Connolly, T., op. cit., call this "tinkering."

57. This point is well illustrated by a conversation reported in Judson, H.F., *The Eighth Day of Creation* (New York: Simon & Schuster, 1979), p. 588: "'...it seemed to me that people who claimed to be trying to isolate the repressor, and prove or disprove the theory, weren't really serious. Weren't really willing to take the kind of risks that were necessary.' What kind of risks were those? 'Well,' (Mark) Ptashne said, 'Psychic risks.' The risk of committing one's full effort to a difficult problem? 'Well yeah!... I mean the thing that was really tough about it was...nothing to show for it and always worrying...'"

58. Medawar, P.B., in *The Art of the Soluble* (London: Methuen, 1967), p. 7, pithily formulates this norm: "Good scientists study the most important problems they think they can solve."

59. Einstein spent the second half of his life in a fruitless attempt to solve one of the most important problems in physics—the relationship between gravitation and electromagnetism. But of course he was already so "well recognized" scientifically by then that he could afford to take the risk.

60. "...next to nothing is known in any systematically documented way about the effects of the Nobel prizes on problem choice in science." Zuckerman, H., op. cit., p. 246.

61. For example, by permitting speculative theories—often elaborately worked out—to proliferate uncritically.

62. About 25 per cent of the relatively recently deceased fellows of the Royal Society whose careers were analyzed seem to have followed some such policy. For such a scientist, an initial interest in marine worms might expand into significant contributions to various topics throughout the whole field of marine biology. Many of them evidently had "the kind of instinct which guides a scientist to the right problem at the right time before the rightness of either can be demonstrated...a gift of high order, never to be despised." Sayre, A., *Rosalind Franklin and DNA* (New York:Norton, 1975), p. 130.

63. See Chubin, D.E. and Connolly, T., op. cit.

64. This sort of change of course from one highly specialized field of research to quite a different one could be noted in the scientific careers of the remaining 25 per cent of my little sample of recently deceased fellows of the Royal Society. Thus, it might be that, at the age of 40, having elucidated some unusual biochemical phenomenon in one of his worms, our expert drops out of marine zoology and becomes a leading authority in the equally narrow field of neurological enzymology.

65. This migration is discussed in some detail for particular cases by Chubin, D.E., *Intellectual Mobility, Mentorship and Confluence: The Case of Reverse Transcriptase* (Atlanta, Georgia: Georgia Institute of Technology, 1978) and by Fleck, J., "Development and Establishment in Artificial Intelligence," Sociology of the Sciences Yearbook, VI (1980).

66. "The members of the first generation establishment [in the field of artificial intelligence] in fact were drawn from those who already had some standing or prestigious backing in another field and who could presumably, therefore, afford the risks of investing in a new approach, or the costs of moving field." Fleck, J., op. cit.: Allison, P.D., "Experimental Parapsychology as a Rejected Science," in Wallis, R., (ed.), *On the Margins of Science: The Social Construction of Rejected Knowledge* (Keele: University of Keele, 1979), p. 284, notes that "40% of his respondents said they had decided to become involved in parapsychology after already completing their education and having entered another field."

67. The research careers of the most influential members of the Australian scientific elite have followed this pattern. Bennett, D.J. and Glasner, P.E. "The Role of Scientific Establishments in the Promotion and Inhibition of Multidisciplinary Programs," *Sociology of the*

Sciences Yearbook, VI (1980).

68. Cases are described in Wallis, R., op. cit.

69. Ziman, J.M., *Puzzles, Problems and Enigmas* (Cambridge: Cambridge University Press, 1981), ch. 1.

70. A traditionally "unscientific" enigma taken up, of course, by Alfred Wegener in 1910, and eventually resolved, some 50 years later, by the theory of plate tectonics. See Hallam, A., *A Revolution in the Earth Sciences* (Oxford: Clarendon Press, 1973).

71. Ziman, J.M., "Some Pathologies of the Scientific Life," *Advancement of Science*, XXVII (1971), pp. 1–10.

72. The pay-off may indeed, be negative: "So strong was the feeling against continental drift [in the United States] until quite recently that in some institutions an open adherence to this doctrine would have put at serious risk the attainment of tenure by junior faculty members while their more secure senior colleagues would have been all but drummed out of their invisible college." Hallam, A., op. cit., p. 105.

73. "Every scientist should from time to time ask himself the following question: 'why has so-and-so contributed more to science than I have, although my understanding and mathematical skill are in no way inferior to his?' The answer is usually the same in every case: 'He has the courage to go through with unauthentic (i.e. novel) pieces of work, whereas I expend all my efforts on authentic ones.'" Migdal, A.B., "On the Psychology of Scientific Creativity," *Contemporary Physics*, XX (1979), pp. 121–148.

74. Ziman, J.M., "Conceptions of Science."

75. Ibid.

76. A deep analysis of the relationship between scientific progress and the exponentially growing effort needed to achieve it is made in Rescher, N., op. cit., pp. 54–78.

77. In Soviet science "power is measured first and foremost in terms of laboratory apparatus. The modern researcher needs apparatus almost as much as he does ideas. Indeed, for immediate purposes, a good new piece of equipment is even more important than a new idea since with the equipment you can do some kind of work in any case, but ideas are of little or no use on their own." Popovsky, M. op. cit., p. 62.

78. "Theoretical physics happens to be one of the few scientific disciplines which, together with mathematics, is ideally suited to development in a developing country. The reason is that no costly equipment is involved. It is inevitably one of the first sciences to be developed at the highest possible level; this was the case in Japan, in India, in Pakistan, in Brazil, in Lebanon, in Turkey, in Korea, in Argentina." Salam, Abdus, "The Isolation of the Scientist in Developing Countries," *Minerva*, IV (Summer 1966), p. 461.

79. For example, new apparatus for zone refining and crystal pulling opened up the whole field of solid state electronics by providing much larger, purer and more perfect crystals of semiconductor materials than were previously available. See Braun, E. and MacDonald, S., *Revolution in Miniature: The History and Impact of Semiconductor Electronics* (Cambridge: Cambridge University Press, 1978), p. 64.

80. A quick glance through the confidential citations of candidates for election to the Royal Society suggests that about 7 per cent are distinguished mainly for novel contributions to instrumental techniques rather than for direct contributions to scientific knowledge.

81. The fact that all the authors do not receive equal credit for a paper with multiple authors is another indication of the divergence of modern science from the traditional model of academic science. Meadows, A.J., *Communication in Science* (London: Butterworth, 1974), p. 198.

82. The apparatus with which Ernest Rutherford "split the atom" in 1919 might have been worth a month of his time, the apparatus required for a modern experiment in high-energy physics would cost many hundreds of years of the lives of the experimenters. See Ziman, J.M., *The Force of Knowledge* (Cambridge: Cambridge University Press, 1976)

83. Rescher quotes Planck's "principle of increasing effort:" "To be sure, with every advance [in science] the difficulty of the task is increased; ever larger demands are made on the achievements of researchers, and the need for a suitable diversion of labor becomes more pressing." Rescher, N., op. cit., p. 80.

84. Nevertheless the average productivity of scientists, in terms of papers published per author per year, has not changed greatly over several decades. See Ziman, J.M., "The Proliferation of the Scientific Literature," *Science*, CCVIII (25 April, 1980), pp. 369–371.
85. Dr. J. Malos tells me that the normal length of an experiment in high-energy physics is now about five years.
86. Dr. David Baltimore's experimental verification of the reverse transcriptase hypothesis took only two days: but he was only able to do this serendipitous research because he had taken some trouble, over a period of months, to lay hands, "unofficially" on the essential "apparatus—a large (and very valuable) stock of virus." Studer, K.F. and Chubin, D.E., *The Cancer Mission* (Beverly Hills: Sage, 1980), p. 137.
87. Lovelock was peculiarly fortunate in having exactly the right instrument under his personal control. Lovelock, J., op. cit.
88. By contrast, in 1951, Max Perutz spent only a weekend setting up X-ray diffraction apparatus to observe new phenomena predicted from Linus Pauling's model of the alpha helix. Judson, H.F., *The Search for Solutions* (New York: Holt, Rinehart, Winston, 1980), p. 136.
89. W.K.H. Panofsky illustrates this statement by detailed reference to four major instruments. Panofsky, W., op. cit., pp. 24–30.
90. "...an apparatus that costs a million marks often can be used only in a way that was thought out years ago by other persons, as they were developing that apparatus. To justify its operation and the outlay in personnel and money that are part of the operation, it is necessary to feed the apparatus with plenty of work. And no failure can be afforded. So easy jobs are selected—mostly operations that are already familiar to others in the scientific world: in other words nothing especially new." Maier-Leibnitz, H., *Report of the Deutsche Forschungsgemeinschaft* (Bonn: Deutsche Forschungsgemeinschaft, 1979). R. E. Marshak emphasizes that there is simply no point in attempting to start a university program of research in some of the "frontier areas" of science without the prospect of an adequate budget for equipment and supporting services. Marshak, R.E., op. cit., pp. 1,521–1,524.
91. In the more traditional humanistic and social science disciplines, it is still customary for doctoral candidates to propose the topics for their dissertations. Not only are such disciplines more individualistic than the natural sciences, they are also less dependent upon material apparatus.
92. Does this explain the extraordinary fact that the 14 students who took Ph.D.s under the supervision of C.G. Barkla at Edinburgh between 1924 and 1944 "seem to have had normal scientific careers," despite the fact that the phenomena they investigated were publicly—and quite correctly—regarded as illusory by almost all other physicists of the day? The implications of this pathological state of affairs are worth serious analysis. Wynne, B., in Wallis, R., op. cit., pp. 67–87.
93. Medawar, P.B., *Advice to a Young Scientist* (New York: Harper and Row, 1979), p. 14.
94. There is, of course, a wide variation in the terminology applied to the successive stages of advanced education in different countries. In Britain the Ph.D. degree is usually awarded three or four years after the three-year bachelor's degree—e.g., at the age of 24 or 25. In the United States—and probably in most other countries—six years of graduate study, up to the age of 27 or more, seem now to be the norm. See Pasachoff, J.M., "Six Years for a Ph.D.," *Nature*, CCXXXIII (17 September, 1971), p. 217.
95. Compare the "anomalous" role of the post-doctoral fellow in 1969 reported by Walsh, J. "Postdoctoral Education: Report Emphasizes Recognition Problem," *Science*, CLXVI (28 November, 1969), pp. 1,129–1,130 with its description as a "career pattern," 10 years later by Coggeshall, P.E., Norvell, J.C., Bogorad, L. and Bock, R.M., "Changing Postdoctoral Patterns for Biomedical Scientists," Science, CCII (3 November, 1978), pp. 487–493.
96. The situation in most scientific departments in British universities is probably much as it is now in the physics department at Bristol University, where almost all the academic staff below the age of 35 are on short-term appointments associated with specific research projects.
97. Walsh, J., "'Unfaculty' a Growing Factor in Research," *Science*, CCIV (20 April, 1979), p. 286.
98. Ziman, J.M., "Bounded Science: The Prospect of a Steady State," *Minerva*, XVI (Summer

1978), pp. 327–339.

99. In a recent issue of *Nature*, for example, there were 30 advertisements for research assistantships in academic institutions. In each case, the job description was along the following lines:—"Research assistant with Ph.D. in neurobiology with experience of electrophysiological and neuroanatomical techniques preferably including electron microscopy, required to study neuronal interaction in the cephalopod visual system." If all the technical terms in such a description are needed to differentiate significant categories of expertise then there must be hundreds of distinct scientific specialities which are not supposed to be interchangeable as components of a research team. In reality, there is usually considerable latitude in adjusting the research work to the capacities and interests of whoever is available to carry it out.

100. It is true that "open" research fellowships—e.g., "to enable outstanding young scientists...to devote their full time to original and independent research"—are offered by quite a number of institutions and agencies (U.K. Science Research Council, 1980). Without a careful investigation of the careers of scientists who have won such fellowships, often against fierce competition, it is impossible to say whether this policy has any significant effect in countering the general trend we are here discussing.

101. The consequences of this tradition in France are discussed by Cahn, R.W., "How French 'Postdocs' Are Left Out in the Cold," *Nature*, CCLXXIX (14 June, 1979), p. 972.

102. See Schwarz, J., "U.K. Researchers claim Rights over Short-term Contracts," *Nature*, CCLXXVI (21/28 December, 1978), p. 745.

103. See Schwarz, J., "Portrait of the Out-of-Work Scientist," *Nature*, CCLXXXII (1 November, 1979), pp. 7–8.

104. Brode, W.R., "Manpower in Science and Engineering: Based on a Saturation Model," *Science*, CLXXIII (16 July, 1971), pp. 206–213.

105. See Knorr, K.D., "Contextuality and Indexicality of Organizational Action: Toward a Transorganizational Theory of Organizations," *Social Science Information*, XVIII (1979), pp. 79–101.

106. See Orth, C.D., Bailey, J.C., and Wolek, F.W., *Administering Research and Development: The Behavior of Scientists and Engineers in Organizations* (London: Tavistock, 1965), pp. 107,364, 489, 532; and Cotgrove S. and Box, S., Science, *Industry and Society* (London: George Allen & Unwin, 1970), pp. 93–96.

107. Orth, C.D., et al., op. cit., p. 30; Cotgrove, S. and Box, S., op. cit., pp. 27–28.

108. Orth, C.D., et al., op. cit., pp. 147, 338. S. Marcson notes however, that "hedging," i.e., permitting a research scientist to spend a very small part of his time on a "risky" personal venture—may be no more than a management ploy to discourage the project without actually vetoing it. Marcson, S., *The Scientist in American Industry* (New York: Harper's, 1960), p. 112. This sort of psychological double bluffing is only one of the sources of indeterminacy of social action generated in organizational contexts, as discussed by Knorr, K.D., op. cit.

109. "'Bootlegging'—the diversion of small unaccounted amounts of time and resources to unofficial goals—is somewhat traditional in research institutions. Moreover, most research directors know that such work can be quite rewarding, perhaps because it is only the individual highly dedicated to a deviant or unorthodox idea who does it." Schmitt, R.W., "By Choice or by Chance, Human Values in Applied Science," unpublished paper. Cotgrove, S. and Box, S., op,. cit. p. 118, note that "bootlegging" is only one of several strategies by which research scientists can get to do the work that they personally prefer.

110. This policy still prevails for example in many governmental laboratories in India, often to the detriment of their more technological programs.

111. Ellis, N.D. "The Occupation of Science," *Technology and Society*, V (1969), pp. 33–41.

112. Hagstrom, W.O., op. cit., pp. 105–112.

113. Cotgrove, S. and Box, S., op. cit., p. 26.

114. K.D. Knorr points out that this notion is "by and for an agent at a particular time and place, and carried by local interpretations," and would make it the driving force of all research. Knorr, K.D., "Producing and Reproducing Knowledge: Descriptive or Constructive," *So-*

cial Science Information, XVI (1977), pp. 669–696. Although the whole trend of the present paper is towards recognizing the significance of such intermediate concepts in the calculus of personal motives, I would be more cautious in demythologizing the conventional scientific belief in the ultimate superiority of "objective" public knowledge.

115. E.C. Bullard remarks that "The preference of most scientists for working on what they wish to work on...[is a problem] that usually arises in establishments that are otherwise unsatisfactory. Most people's interests are not as firmly fixed as they believe..." Bullard, E.C., in Cockcroft, J. (ed.), *The Organization of Research Establishments* (Cambridge: Cambridge University Press, 1965), p. 265. A research organization that is obviously "satisfactory" in this respect is described by Orth, C.D., et al., op. cit., p. 532.

116. "The research staff member now begins to accept the laboratory's definition of broadened interests. What is more, the laboratory attempts to convince him that what he is doing is what he really wants to do. For his own ability to function effectively it is important that he internalizes an approximation of the view that he is working on what he really wants to. The change in definition of desirable research areas does not of itself lessen the individual's devotion to research." Marcson, S., op. cit., p. 66.

117. Lighthill, M.J., in Cockcroft, J. (ed.), op. cit., p. 32. Hammond, D.L., "Physicists in Industry," *Physics Today,* XXV (1972), pp. 42–45, emphasizes flexibility as a prime quality.

118. Bullard, E.C., op. cit., p. 266, refers to the difficulty of persuading a man to stop work on a project that is no longer important—with all the psychological factors that this entails. See also Davies, D., "Peculiar Problems of the British Civil Service," *Nature,* CCLV (22 May, 1975), pp. 293–296.

119. Marcson, S., op. cit., p. 142; Cotgrove, S. and Box, S., op. cit., p. 159.

120. Ravetz, J.R., *Scientific Knowledge and its Social Problems* (Oxford: Clarendon Press, 1971), pp. 31 ff.

121. It must not be thought, however, that all research organizations are now identical. This seems to have been an assumption of the UNESCO project reported in Andrews, F.M., op. cit. But the results reported by G.A. Cole in the same volume show that there are "striking differences in performance and influence patterns for research units in academic, government/cooperative, and industrial settings," with further distinctions, within the academic world, between "clusters comprised of the exact and natural sciences, the medical and social sciences, and the applied sciences and technology."

122. See Hagstrom, W.O., op. cit., pp. 112–124, regarding "free collaboration," especially among mathematicians.

123. Ziman, J.M., *Public Knowledge,* pp. 94–101.

124. The dialectical relationship between co-operation and competition between major research groups is discussed by Edge, D. and Mulkay, M.J., *Astronomy Transformed: The Emergence of Radio-Astronomy in Britain* (New York: Wiley, 1976). In so far as the research programs of these groups may be personally ascribed to the intentions of their leaders, these considerations are also relevant to our present theme.

125. Meadows, A.J., op. cit., pp. 199–206. The proportion of papers with plural authorship in the past—of the order of 20 per cent in the natural sciences—greatly overstates the degree of voluntary collaboration, since it includes many cases of unequal co-authorship by teacher and student.

126. See Morrison, D.R.O., "The Sociology of International Scientific Collaborations," in Armenteros, R., et al., *Physics from Friends* (Geneva: CERN, 1968).

127. Ibid.

128. Ben-David, J., *The Scientist's Role in Society* (Englewood Cliffs, N.J.: Prentice-Hall, 1971), pp. 130–132.

129. In this way, as J. Fleck points out, the foundation of the new universities of Sussex and Essex in the 1960s was a major factor in the development of research in artificial intelligence. Fleck, J., op. cit.

130. See, for example, Ziman, J.M., *Reliable Knowledge.*

131. The first comprehensive account of this process was probably by Kidd, C.V., American

Universities and Federal Research (Cambridge, Mass.: Harvard University Press, 1959). An elementary description is given by Ziman, J.M., *The Force of Knowledge.*

132. Just occasionally, however, a contrary decision may accidentally have very positive consequences. Among the 50 episodes reported in "The Beginnings of Solid State Physics," *Pro ceedings of the Royal Society*, 1980, there were four cases where what proved to be very fruitful new lines of research had to be taken up in place of other, apparently more favorable projects for which resources simply could not be obtained.

133. Merton, R.K., *The Sociology of Science* (Chicago: University of Chicago Press, 1973), p. 267.

134. See Cole, S, Rubin, I., and Cole, J.R., "Peer Review and the Support of Science," *Scientific American*, CCXXXVII (1977), pp. 34–41.

135. Muller, R.A., "Innovation and Scientific Funding," *Science*, CCIX (1980), pp. 880–883.

136. As in the case of Lovelock, J., op. cit.

137. Research on quantum electronics, which brought good fortune to C.H. Townes, did not apparently seem such a bright idea to four major industrial research laboratories (Westinghouse, Radio Corporation of America, General Electric and Bell), who must have been disagreeably surprised at their failure to anticipate the discovery of the laser only a few years after they had stopped research in this field. Townes, C.H., "Quantum Electronics and Surprise in Development of Technology," *Science*, CLIX (16 February, 1968), pp. 699–703.

138. This is the official terminology of the UK Science Research Council in its handbook of information on research grants (1979).

139. Ibid.

140. L. Cranberg quotes advertisements for senior faculty appointments requiring "Proven ability to generate grant support." He goes on to remark that "this closes the door to anyone whose research requires little or no grant support. Conversely, it tends to put a premium on research that is costly and requires, or has obtained, large support in the past...it tends to bar the innovative, the unconventional in favor of the familiar, the sure-fire, in the name of research!" Cranberg, L., "Grantsmanship in Advertising," *Physics Today*, XXXIII (1980), p. 15.

141. McCrone, J.D. and Hoppin, M.E., "Request for Proposals and Universities," *Science*, CLXXXIX (9 March, 1973), pp. 975–977.

142. See Crosland, M. "From Prizes to Grants in the Support of Scientific Research in France in the Nineteenth Century: The Montyon Legacy," *Minerva*, XVII (Autumn 1979), pp. 355–380.

143. Studer, K.F. and Chubin, D.E., op. cit., p. 95, plot this trend and quote L.D. Longo: "Contract research, which is largely for product delivery of procurement purposes, has the potential of undermining a scientist's commitment to patient, systematic and often frustrating discovery-oriented basic research." Longo, L.D., *Federation Proceedings*, XXXII (1973), pp. 2,078–2,085.

144. This applies in other professions. P. Elliott notes that "in spite of the apparent bureaucratization of [large] law firms, the so-called independent solo lawyers [in New York] were more constricted in several senses by their practice situation." Elliott, P., *The Sociology of the Professions* (London: Macmillan, 1972), p. 119

145. The rhetorical function of scientific papers is discussed in Ziman, J.M., *Public Knowledge* (Cambridge: Cambridge University Press, 1967) and *Reliable Knowledge.*

146. R.A. Muller was told by the National Science Foundation that "a proposal should be as 'polished' as a paper published in a major journal" and remarks that "Referees frequently expect all potential problems to be identified and their solutions outlined. Unfortunately, it is not an exaggeration to say that agencies expect a proposal to outline the anticipated discoveries." Muller, R.A., op. cit.

147 Garfinkel, H., *Studies in Ethnomethodology* (Englewood Cliffs, N.J.: Prentice-Hall, 1967), pp. 172–173.

148. This point is made in relation to science policy in general in Salomon, J.J., *Science and Politics* (London: Macmillan, 1973), p. 95 ff.

149. Garfinkel, H., op. cit., pp. 262–283.

150. Szent-Györgyi, A., op. cit. This has usually been the initial response of my own scientific colleagues and acquaintances to the theme of this paper.

151. Ziman, J.M., "Conceptions of Science."

152.This complementarity is more or less taken for granted by most scientists nowadays. For example, H.D. Daniel and R. Fisch questioned 3,000 German academic scientists about the motives and goals of research; in 75 per cent of the answers, both "internalist" and "externalist" factors were given equivalent weight. Daniel, H.D. and Fisch, R., (1980), private communication.

153.This was precisely the issue in the debate of the late 1930s. See McGucken, W., "On Freedom and Planning in Science: The Society for Freedom in Science 1940–1946," *Minerva*, XVI (Spring 1978), pp. 42–72.

154.W.T. Hanson Jr. describes the management philosophy needed to "design, develop and manufacture a completely new instant photographic film with its special camera, in just a few years," starting from basic chemical principles. From concept to practice this involved contributions from more than 2,000 scientists and engineers, from a wide range of disciplines. Hanson, W.R., Jr, "The Interdisciplinary Approach: A New Photographic Film," *Interdisciplinary Science Reviews*, IV (1979), pp. 290–297.

155.Marcson, S., op. cit. p. 115.

156.This is a grave issue of the politics of science. The physics community suffered a major crisis of morale after Hiroshima, when it was realized that it could be used collectively as an instrument for the most evil purposes without losing the peculiar "sweetness" of scientific discovery as an individual occupation. One may speculate that the fear of a similar trauma of "knowing sin" precipitated the recent debate among molecular biologists about the dangers of "genetic engineering." Until then, the possible instrumentality of their remarkable and beautiful science had seemed too hypothetical and remote for serious concern.

157.This attitude is revealed by the cliché metaphor: "In the present financial climate, we can hardly expect to get funds for..."

158.For example, K.F. Studer and D.E. Chubin point to the failure of biologists "with much responsibility, eminence, and clout within their respective research communities" to prevent the implementation of scientifically inept federal funding policies—chiefly because they took their stand on a "discovery" conception of the research process without allowing sufficiently for the overwhelming political instrumentality of the "Cancer Mission." Studer, K.F. and Chubin, D.E., op. cit. p. 226.

159.For some examples, see Greenberg, D.S., *The Politics of American Science* (Harmondsworth: Penguin, 1969).

160.See Weinberg, A.M., *Reflections on Big Science* (Cambridge, Mass.: MIT Press, 1967)

161.A clear example of this trend towards "positive instrumentalism" in the support of research was the decision to introduce the Rothschild "customer-contractor" principle into government science in Britain. See Gummett, Philip, *Scientists in Whitehall* (Manchester: Manchester University Press, 1980). It is relevant to our present theme that, even after this reform, most research sponsored by the Department of Health and Social Security "began with a proposal from researchers rather than customer initiative;" see Kogan, M., Korman, N. and Henkel, M., *Government's Commissioning of Research* (Uxbridge: Department of Government, Brunel University, 1980), p. 25.

162.This relationship of science to power is the central theme of Salomon, J.J., op. cit.

163.See Marcson, S., op. cit., and Studer, K.E. and Chubin, D.E., op. cit., p. 100. The latter point out that "through shifts in policy such as that embodied in the war on cancer [the National Institute of Health] can exert tremendous pressures on the selection of research topics. And scientists will pursue the opportunities which increased funding makes possible."

164.That is, it does not justify the policy described by Popovsky, M., op. cit., p. 182: "For the past half century, the chief principle [of the million scientists of the Soviet Union] has been that their work belongs to the state and that it is their duty to serve the state and the Soviet people."

165.Schmitt, R.W., op. cit.

166.Bennett, D.J. and Glasner, P.E., op. cit., point to breadth and interdisciplinary of interest and judgement as prime qualities of members of the Australian scientific establishment.

167.Campbell, L. and Garnett, W., *The Life of James Clerk Maxwell* (London: Macmillan, 1882).

Publish or Perish?

As they sip their mugs of instant coffee, scientists gossip about the follies of authors, editors, and referees. They deplore the abysmal quality of the papers that get accepted—and in the same breath curse the anonymous referee who has still not approved their own submission. They revile commercial publishers for encouraging the proliferation of new journals—and then agree to serve on the editorial board of just such a publication in their own favorite speciality. They commiserate with one another over the difficulty of tracking down an article in some very obscure publication—and then approve of the printing and distribution of the proceedings of the little conference they went to last week in Upper Volta. They complain bitterly of the cost of journal subscriptions—but will never allow their departmental librarian to cancel a serial which they have actually not consulted for about 17 years. In other words, they behave just as inconsistently about their communications system of their profession as they do in most other aspects of their lives.

Many of their complaints—which can be mirrored, of course, in other fields of scholarly endeavor—are justified. Citation counts and readership surveys confirm the deplorable skewness of the distribution of quality in the primary scientific literature. For every good paper, there are about nine others that are just worth looking at, and 90 that are not referred to again except by their proud authors and are apparently not read by anyone else either. The same applies to journals, of which a large proportion are bought by libraries, put on display, bound up in annual volumes—and not even noticed by the most assiduous browsers. As for "proliferation" (the pejorative terminology is obligatory when referring to this phenomenon), one has only to look at the length of shelf occupied by each successive year's volumes of an abstract journal to confirm that more science is being published now than ever before.

What extravagance! What superfluity! What redundancy! What a folly! Any sensible person must agree that the whole system of primary communication in science ought to be completely reformed along rational lines. A paper should only be accepted for publication after it has been judged to be really outstanding by acknowledged experts, and allotted to the single international journal specializing in the precise research area to which it contributes. Any journal that did not maintain high enough scientific standards should be struck off the list of authoritative scientific publications, and not allowed to be cited in reputable papers. A combined policy of quality control and rationalization of scope would soon suppress the symptoms of proliferation, even if the disease itself could not be altogether cured.

Alas for zealous rationality: this is very unlikely to happen, and if it does it will probably be for the wrong reasons. Until recently, in the happy era when research funds flowed like wine in a brothel, this was not an issue worth the breath of an argument. Apart from a few dyed-in-the-wool radicals, most scientists were reasonably satisfied with the system as a whole, even if they groused about its workings in detail. It was an historically evolved and socially molded institution, which they had learnt to manipulate as authors and readers, and which they had the resources to sustain as editors or librarians. If necessary, the disfunctions of proliferation could be fixed by advanced computer technology, in the form of systems for information retrieval such as citation indexing, KWIC (sorry, Key-Word-In-Context) indexing, selective dissemination of information, and other ingenious, if expensive devices. The endemic complaint has always been tardiness of publication. In an age when an image can be transmitted from the surface of Mars and reproduced world-wide within a couple of hours, it is a complete mystery why it has to take a couple of years—as long as in Newton's day—to get into print with a few thousand well chosen words on some equally profound scholarly matter.

Economic constraints put the pressure on for change. Should the communication system be thoroughly shaken up, to meet the supposed needs of a brand new, highly relevant research and development sector of the national economy, or will it be sufficient to lop off a few dead branches and introduce a few new gimmicks like the electronic journal? Judging by the very faint interest shown in all such matters by our scientific top brass, the latter policy will win by default, with the tacit approval of the lower ranks. If a crisis is upon us, it will have to appear much more alarming before it frightens the scientific and scholarly world to change its ways. In this, as in most things, belated adaptation to absolute necessity is its traditional response to a challenge.

But for would-be reformers and reactionary defenders alike, there must be some interest in why the present system has such obviously inefficient characteristics. This might be a suitable topic for an essay question at the end of a course of science studies, for it requires an understanding of all the nature of science itself, in all its personal, social, and cognitive aspects. The communication system of science is its core institution, where all these aspects converge and coalesce. There is *psychological* dimension to the question, as every scholar learns from bitter experience; the necessity of getting one's work published cannot be safely denied. There is a *sociological* dimension, since periodicals are the medium by which scientific communities are held together. There is obviously a *philosophical* dimension, since knowledge has to be made explicit when it is communicated. To complete the list of metascientific disciplines involved in such a question, the *history* of science would show the development of the communication system as a major factor in the origin and growth of modern science, and its incorporation into society as a whole.

The historical rate of growth was, of course, enormous. The very first lecture of the course should have been illustrated with Derek da Solla Price's famous graph of the, yes, "proliferation" of primary and secondary scientific journals over the centuries. It is an exponential curve, doubling regularly about every 15 years, with pauses for an occasional world war. It seems to have levelled off a bit lately in sympathy with the peaking of funding and employment for scientists, but has not yet gone into a decline. How is this continual expansion of the scientific literature to be explained? The simplest hypothesis—not, so far as I know, confirmed by a direct count of heads—is that it is proportional to the corresponding expansion in the number of active scientific authors. Taking account of the increase in the proportion of papers with several authors, scientists nowadays write about the same number of papers per head—on the average—as they have always done. Each periodical, as it tries to accommodate more and more papers, must either divide and subdivide into more manageable, more specialized sections, or it must see the papers it rejects being taken up by new journals eager to get into the market. It is just the biological process of *speciation*, that occurs when a population is growing and pressing to colonize a new territory.

The immediate "communication environment" of the individual scientist may not really have changed very much over several centuries. Every scientist, in every period, will be aware of the growth and speciation of the literature in which he or she is interested, and begin to complain of "the impossibility nowadays of keeping up with the proliferating literature of my subject." There have always been "too many scientific journals:"

but this complaint may be no more than a symptom of the competitive pressure of research on the individual psyche, obsessed with the thought that the very item of information one needs in order to make a truly original discovery might be hidden away in one of those journals one has not yet had time to read.

The cynical slogan "publish or perish" is said to dominate the psychological dimension of science. But surely there are more wayward factors than a calculated exchange of "contributions" for "recognition," by anthropological analogy with a potlatch ceremony. Why, for example, do some quite competent scientists pepper the periodicals with innumerable scraps of research, knowing perfectly well that they and their academic peers esteem quality far above quantity. Why, on the other hand, do some extraordinarily fertile scientists—Henry Cavendish is a famous case— publish only a fraction of all that they have discovered? Of course there is an enormous range in the ability of scientists, but is this really measured by indicators of productivity calculated in papers per annum. There is a personal variable here that is not at all irrelevant to the manner in which research results eventually become public property. The traditional system of scientific publication provides outlets for contributions of a wide range of quality. It is thus adapted to a certain breadth and balance in the distribution of research ability, career ambition and literary craftsmanship in the community it serves.

There is a school in the sociology of science that treats the primary journals as geological strata, to be examined for evidence of the evolution of cognitive species and genera. This school is hooked on the methodology of co-citation analysis. Scientific authors are regarded as linked if they cite one another, or are cited together by other authors. A linked cluster of authors is thus an objective definition of one of those "invisible colleges" of scientists supposedly working in the same problem area and sharing common intellectual interests. This methodology does indeed generate a preliminary map of the network of cognitive relationships within a speciality, although it must then be corrected for the informal influences that have actually shaped its development. It takes no account of journals as social institutions in their own right. There is a wealth of social action in the writing of papers, their selection for publication, their transformation into printed pages, and the manner in which they are made.

A learned journal has both practical and symbolic social significance. Try the experiment of starting a small scholarly society on any specialized subject. With a year or so, there will have to be a newsletter reporting group activities. This will soon be carrying reviews of relevant books, brief summaries of conferences, and comments on controversial issues within the field. In due course, members of the society will be sending in research papers, some of which will be so bad that the editor has to find some way of rejecting them without personal embarrassment. It can only be a matter of time before you find your hands full with a thriving scientific periodical, filling a specialized niche in the intellectual ecosphere, complete with a system of referees to maintain its quality. This is how many learned journals got started, and the same natural process can still be observed in many little corners of the academic world. Whether or not commercial publishers drive this process, or only exploit it for profit, a learned journal is a characteristic manifestation of communality in this world, and seems to have symbolic value far beyond its obvious practical functions.

At the very heart of the communication system of science and scholarship lies the practical problem of editorial selection for publication. The device of "peer review," as the Americans call it, is a brilliant social solution to this problem. This procedure is continually criticized, both by reference to hard cases of error and injustice and by all sorts of arguments of broad principle. Some of this criticism must be taken very seriously indeed—especially when it is backed up by very convincing evidence of ignorance or prejudice in the recommendations of referees. But any social institution that exists to mediate between the will of the individual and the priorities of the group is bound to be a focus for discontent. This task can be done in many ways, autocratically or democratically, bureaucratically or paternally, laboriously or lightly, anonymously or publicly, fairly or unfairly—but it is a mistake to believe the anarchist's claim that if only people were wiser and better it need not be done at all. Unfortunately, this is not an easy lesson to teach in an introductory course of science studies. The moment of real maturity for a scientist is when he or she is invited to act as a referee for somebody else's paper. Until then, the anonymous referee's reports on one's own submissions have seemed unreasonable attacks on ego's competence and confidence: now, in a dramatic reversal of roles, one must act as the representative of the community and guardian of its highest standards of rationality and integrity.

Our present-day system or selection of communications for open publication, with its delicate balance of rights and responsibilities between authors, editors and referees (in many cases, the same guys wearing different hats!) is more than a social practice that has evolved historically out of a traditional institution: it also embodies an epistemology. The philosophy of science has at last escaped from the logical chains of positivism into a much more open landscape, where anything is possible, even if not anything will go. The conjectures and refutations that are said to be the major constituents of the scientific method are also the characteristic constituents of the scientific papers in any primary scientific publication. If this "method" is, indeed, hypothetico-deductive, then how else could it be put into practice on a large scale except through a public social apparatus of this kind, where there is just enough separation of the hypothesizing, predicting, and verifying roles to keep the action going and keep it clean? The philosophical contribution to our course would thus be fully exercised by the proposed test question. Many other ways of organizing the publication of scientific information and argument could be devised, but if they altered significantly the balance between imagination and criticism, between the empirical and the conceptual, between individual originality and collective authority, they would generate a slightly different sort of knowledge, of a different degree of fundamental validity, and of a different standard of practical reliability.

I do not say that large changes in the communication system of science and scholarship will not, or should not happen. The nature and social functions of science are themselves changing rapidly, in response to the immense demands that society makes upon it, in almost every aspect. I merely would insist that there are deeper considerations about learned journals than whether there are too many of them, or they cost too much, or are of uneven quality, or take too long to produce—or even whether they have been made obsolete by technological progress.

Academic Science as a System of Markets

Market Models

It is the conventional wisdom of our age that the *market* model is superior to the *command* model. In straightforward politico-economic settings, where it is a matter, say, of getting consumer good into the shops, this is not now seriously to be disputed. But this wisdom is not compelling in more metaphorical applications. However instructive it may be to describe a particular social institution as being, say, "animated by market forces," this is often no more than an analogy of limited scope. The question to ask is whether, as an institution, it has all the essential features of a market *system* and may therefore be expected to exhibit similar dynamical characteristics—for good or ill.

The question needs to be answered carefully and urgently in relation to *research* systems. Whether the conventional wisdom is valid or not, it is certainly being applied wholeheartedly by some powerful operators. Nationally, in the UK, we seem to be somewhere near the middle of a decade of turbulent "restructuring," one of whose declared purposes is to give our academic science a much stronger market orientation (Ziman, 1989). In other countries, especially those of the former Communist world, there are compelling reasons for the radical replacement of planned and centralized national research organizations by much looser systems of quasi-autonomous institutions—i.e., for moving from a command model to a market model in one bold leap.

As so often in large policy matters, each debate centers on particular, local issues, without much reference to broader questions, whose answers are vaguely taken for granted by the various parties—often in contrary

senses. This is not to say that more consistent answers to these questions
would firmly settle the policy issues, but that an explication in very general terms might suggest a useful conceptual framework for all the actors
in such dramas. The following preliminary analysis is too schematic to
call upon the resources of sophisticated economic theory, but may suggest
further developments to experts in that discipline.

Vendors and Customers, Commodities and Currencies

What do we mean by a "market"? In the most primitive sense it is a place,
or a gathering, or an opportunity for buying and selling. In general, it is a
social institution for the systematic exchange of *commodities* for *currencies* between *vendors* and *customers*.

In the "perfect" form idealized by economists, numerous independent
vendors offer nearly interchangeable commodities to equally numerous
customers, with complete freedom on both sides to enter into mutually
satisfactory pairwise bargains, transacted in a stable common currency.
But vendors and customers are assumed to know all that is relevant about
the intrinsic value of the commodities, especially in relation to the prospects for their supply and demand. As everybody knows, these conditions
are very difficult to realize in practice, so that most commercial markets
are highly "imperfect" in a theoretical sense.

What needs to be emphasized here is that the concept of a market is not
solely economic in the narrow sense. It also includes the institutional
boundary conditions, the social conventions, the legal regulations, the politico-economic context, in which it is embedded. It requires, for example,
a procedure for establishing initial property rights and for enforcing the contracts by which commodities change hands. It also requires continual protection against extra-market forces, such as threats of personal violence, that
could coerce a vendor or customer into a highly disadvantageous bargain.

To put this point in another way: an essential characteristic of a "market" transaction is that it should be felt to be "fair" by both parties, in the
sense that they enter into it "voluntarily" and neither has been driven into
making the exchange by force majeure (e.g., impending bankruptcy). In
current political jargon, this is expressed in the demand for a "level playing field" in the industrial and commercial competition between nations.
We might say that one feature of a genuine market is a level agora in
which vendors *compete* equitably for trade.

These are some of the abstract characteristics of a real social institution that already ranges enormously in sophistication, from the open space in a small town where peasant farmers bring their pigs and potatoes for sale to the local butchers and greengrocers, to the global electronic network where symbols representing corporate stocks and shares are ceaselessly being processed in an effort to keep up with the changes in their equally symbolic ownership.

Our present study carries this abstraction further, into a more metaphorical realm What we shall find is that academic research in a country such as the UK does not conform to a simple market model. It actually seems to involve at least half a dozen interconnected "market systems," each working on a different level or in a different domain. In each system, one can identify typical market relationships, in that "commodities" are being exchanged for "currencies" by "vendors" and "customers." Some of these "commodities" and "currencies" are not at all like the goods and services, the cash and credits, of real commercial markets. Nevertheless, the exchange relationship generates social phenomena, such as competition, that are easy to understand in "economic" terms.

The National Institutional Market

The most general system of market relationships in UK academic science is what we might call the *institutional market* that has developed recently to modulate the funding of higher educational institutions. In effect, universities and other higher education institutions are now expected to make a living by selling educational and research services. Public attention is concentrated mainly on the educational side of this development, which is still totally confused by the disorganized retreat of the Universities Funding Council (UFC) from the imperial role previously assumed by the University Grants Committee (UGC). Our present concern is solely with the research side, which is also changing rapidly, although more quietly.

The death of Victor Rothschild is a reminder that, for two decades, all quasi-academic research establishments in the public sector have been officially cast in the role of research *contractors*, competing directly for part of their funding from governmental and industrial *customers*. This customer-contractor relationship (Rothschild, 1971), for all its deficiencies in practice (Kogan & Henkel, 1983), is now fully institutionalized and has been extended throughout the higher education system.

A regular market system is emerging, where the vendors are higher education institutions, quasi-academic, quasi-nongovernmental establishments and private sector consultants, and the customers are research councils, government departments, charitable foundations, commercial firms and (increasingly) international organizations such as the European Community. The commodity is essentially a technical service—that is, highly specialized scientific research—and the currency, directly or indirectly, is simply hard cash.

Since the vendor institutions are now very largely sustained (at least in their research activities) by the income they thus gain, the competition in this market is very fierce. A great deal of effort is put into preparing tempting proposals, negotiating the finer details of contracts and monitoring their performance. Even though much of this effort comes to nothing, an economic analysis would suggest that it is not necessarily all wasted, in that it has the effect of motivating effort and greatly improving the quality of the work that does, in fact, get funded. There is no need to rehearse these and other general arguments favoring market competition as a means of allocating resources and getting the best value for money.

Higher education institutions, as corporate bodies, are not apparently averse to the transition from the quasi-command system of funding that had developed under the UGC in the mid 1970s when the quinquennial grant cycles were disrupted by inflation. Their senior managers are finding that the uncertainties of the marketplace, where at least their successes and failures can be attributed to their *own* past decisions, are preferable to a situation where the major item in the institutional budget for each coming year—the block grant—was going to be decided at a distance by some *other* group on grounds that would seem to bear little relation to the health of the particular institution in question.

They have discovered, also, that an experienced and self-confident vendor in a reasonably stable market system has a much clearer rationale for making savings through improved efficiency, and for investing long term in outstanding staff, capital equipment and infra-structural facilities, than if he were still a subsidiary in a command structure. In other words—and this is all to the good—individual higher education institutions are once more being given the freedom and incentives to exercise the corporate autonomy that is their tradition and ancient pride.

Bumpy Ground in the Institutional Agora

I am not going to make any attempt to catalogue the many obvious things that could be said to temper this rather facile hurrah for a market approach to the funding of higher education institutions. It would be absurd to suppose, for example, that a market dealing in such intangible commodities as education and research could ever approximate to perfection in the economists' sense.

What I propose to do first is to discuss certain characteristic imperfections of the institutional market in research services, as they now present themselves in the UK. Again, this does not pretend to be comprehensive analysis, nor is it meant as a hard-nosed free-market critique of the management of academic science in the UK. On the contrary, it is only by understanding the origins and functions of these imperfections that the research system can be protected from destructive attacks based on an over-simplified analysis.

Taking the classical criteria of market efficiency at their face value, therefore, we observe that certain areas of the national "institutional agora" are very far from level. There are many excellent historical and circumstantial reasons why this should be so—for example, the overwhelming advantages to be gained from the intimate association of advanced research with higher education. The interesting point is that the bumpiness of the ground in some of these areas is not just due to arbitrary externalities, but to interactions with other, more symbolic, market systems operating within the science world. To make this point clearly, let us consider these "imperfections," very schematically, as they show up in practice. Then we can look in turn at each of these other market systems, noting its effects on the institutional market.

Monopsony
The national market for academic science at its more basic end is largely in the hands of a single customer—the Department of Education and Science—which not only controls and subdivides the total budget but also sets many of the contractual conditions for individual bargains.

Costing of Projects
Higher education institutions are not able to charge for individual research projects on an equitable basis to all their customers. This is a complicated exercise in accountancy, involving the "DR element" in UFC block grants to Universities (but not, incidentally, in the Polytechnics and Ccolleges Funding Council grants to other higher education institutions, the rules of the European Commission, etc. (ABRC, 1987; House of Lords, 1990). In effect, quite different conditions apply to research councils, charitable foundations, the EC and to private sector firms. The eventual cash income received by a higher education institution for undertaking a particular pro-

ject can vary by a factor of two or three, depending on the amount that it is permitted to charge for indirect costs, fixed overheads and academic staff time.

Performance Evaluation
Individual universities (but again, not other higher education institutions) receive a substantial proportion of the funds that they spend on research through the "JR" element in the UFC block grant (UFC, 1989). In effect, this procedure distorts detailed market mechanisms—especially customer evaluation of the commodities on offer—by subsidizing particular vendors on the basis of a very general ascription of past research capabilities.

Academic Staff Rigidities
Higher education institutions are not able to deploy flexibly the principal element in their research capabilities—academic staff time. This rigidity is not just due to academic tenure or to the obstacles to inter-institutional mobility. It is also inherent in the tradition of not accounting separately for the time that individuals spend on research and teaching, and in the barriers to role changes, such as interdisciplinary transfers, within institutions (Ziman, 1989).

Institutional Fragmentation
Considered as vendors of research services, higher education institutions often find great difficulty in acting as independent corporate bodies, each with an integrated structure and coherent policy. They are highly segmented administratively into departmental "cost centers" and research entities, which seek to go their own ways managerially, and to make their own independent living in the markets for resources. Institutional control over these sub-units is constrained by national coalitions organized around educational disciplines and research specialties.

Semi-employees and Sub-contractors
Higher education institutions' market strategies are confused by the activities of academic staff as commercial entrepreneurs in their own right. Thus, a contract by the higher education institution to provide a customer with a highly specialized research service may not be fully separable from a contract by an employee of the higher education institution to provide a similar service—even to the extent of using higher education institution infrastructural facilities—on his own account.

Intellectual Property
Higher education institutions find themselves making different arrange-

ments with different classes of customers concerning the ultimate rights in the intellectual property produced by the research they contract for. Although the actual sums at stake are not, on the average, very significant in the overall income of higher education institutions, this is important as an indication of the role of public sector customers for basic research: are they buying a global public good, an appropriable competitive advantage for a home-based firm, or a ticket for the higher education institution in a lottery with occasional large prizes?

Project Markets

A close scrutiny of the institutional market immediately reveals that institutions as such—whole universities—play a very limited role as active vendors of research services. Almost all the action takes place in numerous, highly differentiated, *project markets*, where *research entities*—that is, individual academics or relatively small, specialized research groups (Ziman, 1989)—formulate proposals for specific research projects and apply for grants or contract to undertake them.

Note, first of all, that it is quite usual for such a project to be proposed as a joint enterprise of researchers in cognate departments from several different higher education institutions, regardless of the fact that these are supposed to be in open competition in the institutional market as a whole. This is one of the causes (or symptoms?) of the managerial fragmentation of higher education institutions, which often tend to be treated as bankers, or distant holding companies, by trans-institutional coalitions of their staff members competing in the project markets of their respective disciplines.

The fact is that individual academics and leaders of research entities have a much more direct and active interest in maximizing their operations in the project market of their specialty than in the prosperity of their university or polytechnic. Indeed, they are strongly motivated to sell their services as cheaply as possible, resenting the "additional" indirect costs that need to be charged to the customer if the higher education institution is to remain solvent. This is why apparently arbitrary administrative variations in the incidence of these charges (ABRC, 1987) generate such discord and distortion in the institutional market.

The most serious feature of most project markets, however, is that they are near monopsonies. Most of the customers are systematically differentiated in their interests, and even make administrative arrangements so as never to compete directly for the same commodity. This is probably un-avoidable in the case of the research councils, where all the funds eventu-ally come through the same channel from the public purse, but there is also a tendency for charitable bodies to define specialized niches for themselves in the type of patronage they offer.

The result is well known. What a typical research entity has to offer is an expertly conceived project in a highly specialized scientific field. It is the responsibility of, say, a research council official to direct the proposal for such a project to the "appropriate" committee, in his own or some other organization. There it will be evaluated and selected in direct com-petition with all other proposals of a similar kind that are currently on of-fer. In effect, for the time being this committee may be almost the sole customer in this particular market, and is not under any competitive pres-sure from other customers to optimize its choices.

Some of the imperfections of monopsony in project markets can be re-duced administratively by making sure that customer groups such as re-search grant panels are widely representative of the potential "users" of the research results, that they are not dominated by an oligarchy of spe-cialist vendors, and that all proposals are subject to independent peer re-view. But the only "market" solution to this problem is to facilitate direct competition between customers in the same specialized market. This is the situation in the United States, where less centralized arrangements for funding a much larger volume of academic science make it worthwhile for research entities to submit essentially the same proposals simultane-ously to several independent funding bodies. In this context, the UK Treasure policy of "attributing" European Community grants received by UK scientists to the research budgets of UK government departments is ill advised (House of Lords, 1990). In the name of command rationality it creates a linkage between customer organizations which is detrimental to market optimality.

The conventions of "responsive mode" project funding do at least enti-tle research entities to have their proposals assessed on their intrinsic mer-its, and put the onus directly on customer committees to select wisely amongst them. The trend towards "directed mode" funding (ABRC, 1987) could have much more serious consequences for specialized project mar-kets. In effect, this means that the customers call for tenders to provide re-search services to meet stated specifications. The difficulty is that the most likely participants in the tendering process are precisely the research specialists who need to be consulted—often as an invited group—in drawing up these specifications, and who may thus gain insider advan-tages in the subsequent competition. In the end, this leads to a situation where research contractors become mere clients of the organizations which commission research projects from them—a decentralized "command" system.

Perhaps the only way to make the valuable process of program development consistent with market equity is to insist on a sharp break between the "cooperative" and the "competitive" phases. In the first phase, a great deal of useful work can be done by and for the researchers as a group, by way of surveying what is already known and formulating general questions capable of being tackled. But the call for "final project proposals" should be carefully phrased in general terms, it should be open to all applicants, regardless of their previous involvement in the process, and projects should then be selected, on their merits, by an independent panel of experts who have not been involved in the earlier phases of the process.

Internal Institutional Markets

As we have seen, the specialized project markets and the national institutional markets mesh loosely through the administrative and accountancy machinery within individual higher education institutions. In principle, this linkage should be quite positive both ways, since research entities are legally and financially subordinate groups, such a departments or "centers," whose academic staff are mostly employees of the institution. In practice for reasons already noted, this formal "command" system is usually very weak, and research entities enjoy (or suffer?) a great deal of independence in their scientific activities.

One response to this enduring characteristic of academic life is to deliberately encourage the managers of research entities to act as independent entrepreneurs *within their own institution*. Administrative mechanisms are devised that force them to compete directly with one another for institutional resources, such as technical services and funds for new research equipment.

This may be an effective way of managing a complex institution with a diversity of missions, but it stretches the metaphor to describe it as an internal market system. Who are the "vendors" and "customers," and what "commodities" are they exchanging? In quasi-economic language, it is much more like a franchise system, where research entities are independent entrepreneurs in the external marketplace but have to pay rents to their parent institution in return for technical services, senior personnel and access to venture capital. But there is obviously room here for a much more systematic exploration of the many novel ways in which large firms nowadays manage the financial and administrative links with their subsidiaries.

Academic Job Markets

One of the reasons for the internal fragmentation of higher education institutions is that their most effective academic staff are often competing as individuals in national or international job markets. Each is a vendor of a highly prized commodity—his or her own research promise and grant-earning capacity. Higher education institutions at home and abroad are ready customers for the highest-quality products in this market, seeking out what they judge to be the best and offering a very good price in terms of a personal stipend and research facilities.

This is not a novel analogy. In *The Academic Marketplace*, Theodore Caplow and Reece McGee (1961) described vividly how the academic profession in the USA hinges on the elaborate processes by which universities recruit their staff. Indeed, these processes have become so complex and institutionalized that other social metaphors suggest themselves—the "slave market" of post-docs at the annual subject convention, the "patronage" dispensed by the "godfather" of the discipline, the "head hunting" for "distinguished professors" to adorn the faculty, and so on.

In fact, this form of competition between higher education institutions dates back to at least the early 19th century, when it became customary for German universities to bid against one another for the most promising scholars, and where it was accepted that moves from university to university up the hierarchy of esteem were normal steps in a scholarly career (Ben-David, 1971). The same tradition of inter-institutional mobility became established in the UK, although never with such intensity as in the USA.

The driving force in this competition is, of course, the dependence of institutional performance on the quality of its academic staff. Research is extremely labor-intensive, yet research excellence is quite rare, and not always easy to spot. If a higher education institution is to be successful in obtaining research funds, it must have researchers able to hold their own in the project markets—and these are always in short supply. If, as was usual until quite recently, they are likely to gain early tenure, they have to be selected with all the care that a manufacturer gives to the purchase of a piece of capital equipment, such as a machine tool, which has got to perform well for another 30 years.

The academic job market thus plays a very important part in the institutional agora. Many significant features of contemporary academic life can be attributed to efforts to make this market more efficient—the abolition of tenure, increasing numbers of researchers on short-term contracts, separate accountancy of teaching and research, and the facilitation of entrepreneurial ventures by academic staff. Concern about the "brain drain," whether to the United States or to "Europe after 1992," indicates awareness that the UK is in a world market for academic talent.

What needs to be said, perhaps, is that some characteristics of a command system have developed in this area of academic life. Although UK universities are legally independent corporate bodies, they have long been subject to central financial controls, through the UGC, which inhibited competition for academic staff. Grades of seniority and pay scales were negotiated and standardized across the system, and tenure and pension rights accumulated in one institution were automatically transferred if one moved elsewhere. In some respects, academic employment in a British university constituted membership of a single elite corps, almost equivalent to the formal civil service status of academics in countries such as France. Conversely, it was a matter of principle that members of the UK scientific civil service doing quasi-academic research should feel almost as free as university researchers in their conditions of employment (Holdgate, 1980).

This is not the place to argue the pros and cons of command versus market modes of employment in academic science. A great deal depends on more detailed features of the system, such as facilities for active personnel management and performance appraisal, opportunities for role changes as between research, teaching and administration, incentives for inter-institutional and interspecialty mobility, strategic and technical autonomy in planning and undertaking research, reasonable continuity and stability of employment for individuals of proven competence, and so on (Ziman, 1989).

All that needs to be said is that the market mode need not be any more brutal and unfeeling than the command mode in its actual consequences. Those conditions of employment that are highly prized by academic scientists can be negotiated for in the job market—by collective bargaining if necessary—and established by contract. Conversely, if higher education institutions are to stand on their own feet as independent bodies, they can properly insist on those conditions they regard as essential for managerial purposes. In this corner of the agora, the long term balance of forces between "vendors" and "customers" is not inherently unfair, and it is not obvious that external interference by government regulation is needed to optimize the outcome. And as American experience shows, this outcome is not necessarily detrimental to academic values, to research excellence or to a decent style of life.

The Reputational Market

The position of a vendor—that is, of an individual academic scientist—in an academic job market is closely related to his or her status in the corresponding *reputational market*. This is a sociological metaphor proposed by Warren Hagstrom (1965) in his pioneering analysis of academic science, *The Scientific Community*.

The argument is that scientific research is driven by a non-contractual but highly formalized social process whereby scientists publicly present their "contributions to knowledge"—i.e., scientific papers reporting research results—in exchange for communal "recognition"—i.e., citations, employment, promotion, prizes, etc. In other words, it is a market where the vendors are individual researchers, the commodities are research results, the customers are "invisible colleges" of other researchers, and the currency is simply a public sign of personal esteem.

Although this style of functional sociology happens now to be out of fashion, the quasi-economic model is very instructive. It is no accident that Robert Merton (1942) named the traditional scientific norms so as to make up the acronym CUDOS: this is what the individual can expect to receive for obeying these norms in his or her social role. It is true that "recognition " is a very intangible currency, but nobody who has observed the ferocity with which scientists vie with one another for it will doubt its force as a psychological reality.

The difficulty that a classical economist might have with this whole concept is that the exchange process does not involve the negotiation of actual contracts between individual vendors and customers. Indeed, there is a strict taboo against any "trading" between those who seek esteem and those who award it. As Hagstrom pointed out, the proper analogy is not with monetary commerce or even barter, but with a traditional system of *potlatch*. At periodic intervals, the tribe indulges in an orgy of public gift-giving in which those who give away the most valuable goods receive the greatest honor. Even amongst scientists, the psycho-dynamics of such a system may be better understood through social anthropology than through economics.

The validity of this analogy is not, however, the essential point,. The job and reputational markets are very closely linked, and the former is not necessarily dominant. Many academic scientists would value, say, appointment or promotion to a full professorship less for its material returns than for its reputational implications. A highly cited paper is not only evidence of technically excellent research performance that will help win a more lucrative contract in the project market: it is a reputational asset that can be transformed into further symbols of public esteem such as membership of an academy or an honorary degree.

It would be wrong to suppose, moreover, that the reputational factor in academic science has been or will soon be nullified by the move to more collective modes of working. It is notorious that when a team of researchers makes a significant discovery, the acclaim is not shared out among its members, but goes almost exclusively to its leader. The unforced admiration of scientific colleagues for a good piece of work is being multiplied out of hand by the highly publicized award of lavish prizes and by increasing attention in the mass media. Who says that contemporary science no longer suffers from the "publish or perish" disease, or the "Nobel Prize syndrome?"

The most obvious sign of the continuing influence of the reputational market in academic science is the UGC/UFC policy of basing a major part of the block grant to each university on an assessment of its research performance (UFC, 1989). The indicators used in calculating this "Judgementally Related" element (the terminology is revealing) are essentially reputational. In principle (if very imperfectly in practice) they are just the same as would be used by, say, a sectional committee of the Royal Society in assessing a candidate for Fellowship. The senior academics who run the block grant system are so steeped in the criteria and traditions of the reputational market that they are trying to apply them directly to whole institutions.

There is certainly much to be said for making the reputational status of scientists and scientific institutions more explicit and transparent, right across the board (Ziman, 1989). Information on the past performance of research entities is an extremely important factor in project markets, and has significant managerial implications within higher education institutions. It is an axiom of economic theory that reliable information makes market competition more efficient. The expert judgement of peers is not necessarily unreliable, but it needs to be backed up with information on the actual contributions of researchers to the literature of their specialties.

The question is whether this information should be given direct operational status in the funding of academic science. One can understand the logic of a policy of giving public financial support to research organizations of proven scientific excellence, in the same spirit as one gives secure, well paid employment to individual scientists of correspondingly high quality. Indeed, support of elite institutions may be an effective way of supporting the outstanding individuals who set the pace in science. At first sight, there seems a good case for the current UFC practice of, so to speak, driving competition between universities in the national institutional agora by supplementing the intangible currency of the reputational market with large sums of hard cash.

But as soon as one puts it that way, one can see very serious objections to this development. The traditional reputational markets of academia are not run along commercial lines. Neither their "commodities" nor their "currencies" are quantifiable, and there are not formal schedules of prices. A potlatch system for the communal distribution of goods cannot be rationalized and made subject to legal contract. The "JR" formula, which puts an actual cash value on the research contributions of academic institutions, is bound to commercialize the whole system (Gibbons & Wittrock, 1985). Academics have grasped this point fully: who is to blame them if they tart up or adulterate the goods in their shop windows to attract these wealthy customers?

The other principal objection to the commercialization of the reputational market is that neither "contributions" not "recognition" are additive over institutions. They derive from and are attributable to the work of individuals or small groups, and vary enormously in their intrinsic value (Daniel, 1987). It is simply a fact of scientific activity that real progress is made through the contribution of a very small proportion of the competent researchers in a particular field. Under these circumstances, an average over a large group is extremely misleading (Platt, 1988), since it tells us nothing about most of the population except that they must be well below this average.

Here again, academics are well aware of the realities. They know that research reputation is hard won. It depends on exceptional individual effort and enterprise in the international marketplace of a scientific specialty (Ziman, 1987a). One of the important factors for success is originality in the choice of problems and methods—an originality that is not always demonstrable through projects that maximize institutional profits (Ziman, 1981).

Institutions as such have no standing in such markets. From the point of view of an academic researcher, the university is simply a managerial conglomerate, the holding company that provides a home base for the family firm. If a research entity builds up a reputation through which the university eventually gets a precisely calculable cash return, then the researchers think that the cash belongs to those who earned it, not to the institution as a whole. In other words, the "JR" funding formula is bound to aggravate the institutional fragmentation generated by competition in the specialized project markets, where reputational factors also play an important role.

The Market for Research Claims

In the reputational market, "recognition" is exchanged for "contributions." But what is the value of a "contribution to knowledge?" What is length of a piece of string? The typescript of a scientific paper submitted to a learned journal claims that the results that it reports are valid and significant. How is this claim tested, and perhaps transformed into an established item of scientific knowledge worthy of public recognition?

Research reports contribute nothing to science until they have been matched against one another in the world wide *research claims market* of the scientific community (Ziman, 1967). They are refereed before publication, subjected to expert questioning at seminars and conferences, reviewed critically in survey articles, cited favorably or unfavorably in other papers by other scientists, and tested by replication or experimental refutation. In effect, the literature of a research specialty is a public marketplace, where the pigs and potatoes produced by peasant farmers are prodded and weighed and shrewdly purchased by other peasants.

Evolutionary accounts of scientific and technological progress are built around this metaphor (Hull, 1988). As Darwin perceived, the invisible hand of selection for fitness is prodigal but precise. International competition for the validation of research claims is very harsh and very discriminating. A hundred flowers bloom for the one that is finally chosen for seed. One paper with a hundred favorable citations is worth infinitely more than a hundred papers with one citation each.

The research claims market is a worldwide social institution, beyond the control of any one national scientific community. It is more global, more formalized, and more objective than the reputational market which it resembles. It is a nice point, however, whether these two markets, which are obviously closely linked, should be considered separately. I would argue that they are different in principle because they operate in different dimensions. The sociologists of scientific knowledge would insist that they must be a single system, since there is no sure way of separating the epistemological from the sociological aspects of knowledge.

What we would all agree is that there is a connection in practice between the status of a research claim and the reputational interests of the claimant. The research claims market is not driven by some abstract striving for truth. It is driven by competition between individual "vendors" seeking public certification of their claims. The epistemological and sociological domains are linked dynamically by the social convention that a research finding is "owned" by the person who first reported and claimed it publicly. Without such a convention, the whole social institution that generates relatively reliable scientific knowledge would fall to pieces.

This is why, as every academic scientist or scholar knows, access to, and active participation in, all the relevant specialist submarkets of world science is so fundamental. It is not just a matter of maintaining efficiency in terms of competitive excellence in performance. The intellectual integrity of academic science in the UK depends on putting the findings of our researchers on public display internationally, and accepting fully the value put on them by the world community.

This fundamental principle obviously reinforces the individualism of the reputational marketplace. But it goes beyond that in flatly denying any role for universities or research institutes *as such* in the research claims market. No matter how much a university may hope to benefit eventually in the institutional agora from a valid scientific discovery by one of its staff, it must keep out of the validation process. As was clearly shown by the furor over the behavior of the University of Utah in the "cold fusion" affair, any attempt by an institution to assert "ownership" of a publicly communicated research finding is bound to discredit that finding. It is only because the "intellectual property rights" of academic researchers are tacit, personal and not negotiable that their public claims can be assessed reasonably objectively by the relevant research communities. This delicate mechanism would be destroyed if it were not firmly protected by convention from the corporate interests of large institutions.

Intellectual Property Market

The knowledge generated by research cannot only be traded metaphorically for "recognition": it can also be sold for cash. Basic science nowadays is seldom totally detached from its potential applications. Commercial firms are continually on the lookout for academic research findings that might be exploited technologically. From this point of view, all new scientific information is potentially "intellectual property," with a legal owner empowered to demand payment for its use.

The economics of scientific and technological information is extremely subtle. Knowledge of a novel research finding is an intangible commodity, whose value depends entirely on the context in which it is transferred from the vendor to the customer. Only a small proportion of the know-how that goes into a profitable technology can be protected by patents. Secret information can be bought and sold, but not publicly marketed. Nevertheless, there exists what might be called an *intellectual property market*, where, so to speak, "futures" in the exploitability of current research results can be sold on a normal commercial basis.

This market is, however, very imperfect in conventional economic terms. Quite apart from the well known problem of attributing and appropriating the return on scientific information which only has value as a public good, it is obviously highly speculative. It is also highly skewed, in that there is an immense variation in the amount that can ultimately be realized commercially out of a particular academic research finding, even when thoroughly protected by patents. In fact, 99% of what research scientists discover is absolutely worthless in commercial terms: the total value of the merchandise is large, but, like natural pearls in oysters, most of it is concentrated in only a tiny proportion of the goods on offer (Feller, 1990).

Nevertheless, from an institutional point of view, this market offers exciting prospects of windfall monetary profits out of its research activities. Institutional managements negotiate with funding bodies over the ownership of the rights to any intellectual property that may arise out of the research they undertake. This applies particularly to research contracts with private sector firms, which normally have a very active interest in the applicability of the work they commission. On the one hand, the intellectual property market is just an adjunct to the project market: on the other hand, it is an extremely important border area between the national institutional agora and the much greater marketplaces of commerce and industry.

From the viewpoint of the individual researcher, however, the *commercial* market in intellectual property competes uncomfortably with the *reputational* market, where the same research findings may have quite a different type of value. It is not even as if these alternative valuations were commensurable, since the commercial value may depend on keeping secret what has reputational value only if made public. The two markets may also have quite different conventions concerning the "ownership" of the intellectual property in question.

Thus, academics have become very concerned about the intellectual property rights conditions that institutional authorities and their research customers jointly impose on project contracts, because these conditions may restrict the kudos they feel they personally deserve for their contributions to knowledge. They are also being faced with a difficult career dilemma. On the one hand, there is a traditional academic career, where published research findings are accumulated, and eventually "capitalized" in the academic job market in the form of a prestigious, well paid post. On the other hand, there is now the possibility of an entrepreneurial career, where the same skills are used to earn a much higher (if less secure) income by generating, exploiting or selling research results. Academics and academic institutions are quite adept at combining and reconciling apparently contradictory roles, but this is a situation where satisfactory norms and rules have still to be established.

Market Mechanisms and Imperfections

This completes our survey of what the UK academic research system might look like through economic spectacles. What we have found is a very complicated set of quasi-market relationships operating in a number of different domains. Whether their different "commodities" and "currencies" are tangible or intangible, these markets are closely interactive. They are very far from perfect by the economist's criteria, but all are strongly competitive.

It is far beyond my purpose here to analyze the dynamics of this very complex system, or to suggest how its apparent imperfections might be reduced. Some of these imperfections are so serious that we should hesitate before advising our ex-communist colleagues simply to copy our system as it now stands. In any case, they might be wiser to take the lead from other countries, such as France, Germany or the United States, which have effective academic research systems working on rather different lines. Nevertheless, a few general points stand out.

The Competitive Tradition in Science
Academic science has always been highly competitive (Ben-David, 1971). Scientists have always striven for esteem and preferment in the reputational and academic job markets, and their findings have always been subject to searching selection in the international market of research claims. Selective pressure at the level of the individual researcher and his or her contributions to science is the key—the only key to excellent performance and valid results.

Commercializing the Academic Marketplace
The introduction of "market forces"—that is, commercial market relationships, with explicit contractual conditions and material rewards—has not necessarily strengthened this competitive tradition. This development was necessary, however (Ziman, 1987b, 1994), to rationalize the allocation of re-

sources to research projects, and to retreat from an increasingly centralized system of command over universities and other higher education institutions.

Harmonizing Institutional and Individual Interests

The conflict between the corporate interests of academic institutions and the personal interests of their academic research staff is not just the usual power struggle between employers and employees. It is a sign of the contrary forces generated by their commitment to different market systems, whose diverse values and trading principles have not been systematically harmonized. If this disharmony is to be reduced, the system will need to be restructured in detail at a number of points, including project funding procedures, internal institutional budgeting and management, conditions of employment, career prospects in teaching and/or research, and the assignment of intellectual property rights.

Entering and Leaving the Marketplace

Strenuous competition is invigorating for those who are established and reasonably successful. It is daunting for would-be entrants and demoralizing for the failures. A ruthlessly selective market system is not socially viable unless it has the means for limiting the wastage of human resources as people enter and leave the marketplace. This is where higher education institutions have an important role as multi-functioned corporate bodies, responsible for the training, apprenticeship and early stages of professional employment of academic researchers (Ziman, 1989), and for opening up fruitful careers in teaching, technical services, administration, etc., for those who have not been able to stand the heat of the research kitchen (Ziman, 1987a).

Competition by Quality, Not by Price

In basic science, quality of performance is so variable, and so decisive for successful achievement, that it must be the dominant criterion in every market transaction. Only the very best available projects are worth funding: only the very best available researchers are worth employing: only the very best research findings are worth believing. But the assessment of quality is always very difficult and uncertain. Information obtained systematically by peer review, staff appraisal, program evaluation, etc., is just as important in a market system as in a command system, but needs to be openly available to all potential "vendors" and "customers."

Satisfying National Priorities

The basic principle of all market systems is that they are driven by an "invisible hand" which optimizes the outcome of innumerable separate trans-actions. The problem is to steer the system as a whole to satisfy external criteria, without interfering seriously with its internal working. Although the "customers" for academic science—e.g., the research councils—have the responsibility of steering the research system to satisfy national priori-ties, this should be done through broad-brush budget allocations rather than by applying external criteria of "relevance" to individual research projects (Ziman, 1985).

References

ABRC (1987), *A Strategy for the Science Base* (Advisory Board for the Research Councils; HMSO, London).

Ben-David, Joseph (1971), *The Scientists's Role in Society* (Prentice-Hall, Englewood Cliffs, NJ).

Caplow, Theodore & McGee, Reece (1961), *The Academic Marketplace* (Basic Books, New York)

Daniel, Hans-Dieter (1987), "Research Performance Evaluation in German Universities" (pri-vate communication).

Feller, Irwin (1990), Universities as Engines of R&D-based Economic Growth: They Think They Can, *Research Policy*, 19, 335–48.

Gibbons, Michael & Wittrock Björn (Eds.) (1985), Science as a Commodity: Threats to the Open Community of Scholars (Longman, London).

Hagstrom, Warren O. (1965), The Scientific Community (Basic Books, New York).

Holdgate, Martin (1980), Review of the Scientific Civil Service (HMSO, London).

House of Lords (1990), A Community Framework for R&D: Report of the House of Lords Se-lect Committee on the European Communities (HMSO, London: HL Paper 66).

Hull, David L. (1988), *Science as a Process: An Evolutionary Account of the Social and Con-ceptual Development of Science* (Chicago U.P., Chicago, IL).

Kogan, Maurice & Henkel, Mary (1983), *Government and Research: The Rothschild Experi-ment in a Government Department* (London, Heinemann).

Merton, Robert K. (1942), Reprinted in *The Sociology of Science* (1973, U. of Chicago Press, Chicago, IL).

Platt, Jennifer (1988), Research Policy in British Higher Education and its Sociological As-sumptions, *Sociology*, 4, 513–29.

Rothschild, Lord (1971), *A Framework for Government R&D* (HMSO, London: CMND 4854).

UFC (1989), Circular Letter to Universities, September 1989 (Universities Funding Council, London).

Ziman, John (1967), *Public Knowledge* (Cambridge U.P., Cambridge).

Ziman, John (1981), What are the Options? Social Determinants of Personal Research Plans, *Minerva*, 19, 1–42.

Ziman, John (1985), Criteria for National Priorities in Research, *Rise and Fall of a Priority Field*, pp. 97–125 (European Science Foundation, Strasbourg).

Ziman, John (1987a), *Knowing Everything about Nothing: Specialization and Change in Re-search Careers* (Cambridge U.P., Cambridge).

Ziman, John (1987b), *Science in a "Steady State": The Research System in Transition* (Science Policy Support Group, London).

Ziman, John (1989), *Restructuring Academic Science: A New Framework for UK Policy* (Lon-don, Science Policy Support Group).

Ziman, John (1994) *Prometheus Bound: Science in a Dynamic 'Steady State'* (Cambridge U.P., Cambridge).

The Collectivization of Science

Within 40 years, J.D. Bernal's radical commentary on "the social function of science" has become the conventional wisdom of public science policy. It is now commonplace that science should be organized and financed on a large scale, and directed towards societal goals. This policy also includes basic research, which needs financial support for expensive apparatus. Science is thus being transformed from an individualistic community into a homogenous collective enterprise, which now covers all types of research from the academic to the technological.

The modern "R&D system" is undoubtedly beneficial to society, and to the advancement of knowledge. But collectivization has not only changed the societal function of science; it has also changed its internal sociology. Personal discretion in the choice of research problems is now severely limited, even in the university sector, because most projects are now funded by outside agencies. Tension between the individual norms of the academic tradition and managerial principles derived from the industrial tradition has made research an ambivalent profession. Should scientists be regarded as members of a transnational community devoted to "the search for truth," or are they simply typical employees of governmental and commercial organizations with very worldly aims? This ambivalence is evident in controversies over scientific freedom and responsibility, and in the ethical problems of military research.

Collectivization generates pressure for efficiency and public accountability. The R&D system is thus driven from the top towards more urgent and utilitarian programs. Is highly innovative research fostered adequately by the method of awarding project grants on the basis of peer review, which is often accused of being ponderous and unadventurous? It is suggested that this method of funding basic research should give way to the award of block grants to "institutes," e.g., whole university departments, on the basis of periodic evaluation of their past work. This reform would locate the risks and responsibilities of innovative research within individual institutes and would reduce the irrationalities and administrative rigidities of the present system; but it would need to be done rigorously, and with sincere concern for the dangers of institutional élitism and internal autocracy.

In 1939, John Desmond Bernal formulated high aspirations for science.[1] His utopian vision of scientific research as a major social force is now a commonplace. His thesis that the scientific method should be applied to every aspect of human existence is a generally accepted principle of our political and economic life. If we need war-like weapons, we do military research; if we hanker after peace, we foster peace research. If we fear for our health, we do medical research; if we are suspicious of the doctors, we sponsor a research project on iatrogenic disorders, or undertake a socio-economic investigation of the pharmaceutical industry. Science gives us, or will give us, the means for work and the means for play, the means for love and the means for hate, the means for power and the means for defiance of power, the means for communication and even the means for silence.

When I say that Bernal's aspirations for science have been realized, I certainly do not mean that it has given us all the benefits that he imagined. His prophetic masterpiece, *The Social Function of Science*, did not foresee the terrible war that was about to break out, and the resulting scientific intensification of the powers of destruction. He was altogether too optimistic about the capacity of science to solve the problématiques of food and population, wealth and employment, that seem always with us, clamoring for endless sympathy, charity and wisdom. The knowledge obtained by research is only one of the factors needed to change the way in which people do what they have to do, to the material world and to one another. Bernal himself was for a profound political transformation that would not only release the beneficial powers of science but would also, as he thought, put an end to the exploitation of the weak by the strong. This revolution has not taken place: we cannot know whether his hopes for science would have succeeded in the more benevolent society that he dreamed of.

"Yet," he wrote in 1964, looking back 25 years, "I failed to foresee how rapidly the tendencies I had observed were to bear fruit and to what extent the prophecy I made at that time was to be fulfilled and over-fulfilled."[2] What he meant was that science was now established as a major national institution. With the public realization of the practical value of science, it has been expanded and organized on an entirely new scale.

Every industrial company must have its research laboratory; if not, the shareholders will want to know the reason why. The machinery for military, agricultural and medical research has grown enormously in size and complexity. This machinery extends to the social services and education, where the results of officially sponsored research projects are allowed to have a significant influence on policy. Even in a highly contentious arena such as legal reform, or wage bargaining, the professors come down from their institutes of criminology or industrial relations to advise, or be consulted, or to judge, like Solomon, on the division of the spoils.

On the average, 1 or 2% of all our productive efforts are now being devoted to systematic scientific research and technological development. The distribution of these efforts is, of course, very uneven. We still argue fiercely about how much research money should be expended on this or that sector of life, or this or that problem. We keep tinkering with the machinery of research, and have wildly divergent opinions on what project is likely to succeed, or who would make the best job of running it. But we no longer need to be persuaded, as a nation, that organized science is one of the most likely ways of overcoming known difficulties, or of getting quite unexpected benefits, in whatever we are trying to do. This was what Bernal proclaimed, this is what most people now believe, and this is what we have now got.

This highly organized activity is the practical realization of the *instrumental* conception of science. There is nothing new about the idea that scientific knowledge can be put to practical use. The founders of this Society knew that very well, and made a great deal of it in their public relations handouts. But our contemporary style of "Research and Development'—"R&D," as I shall call it, to emphasize the inseparability of its constituents—clearly derives from the late nineteenth century. Applied science was then established in a variety of small institutions, such as the research laboratories of the German chemical industry, the agricultural research stations of the U.S. land-grant colleges, and numerous government bureaux of astronomy, meteorology, geodesy, geology and metrology. By the 1930s, many of these had developed into large and elaborate organizations, such as the Bell Telephone Laboratories, the Institut Pasteur, or the Royal Aircraft Establishment at Farnborough. Few R&D laboratories are actually as large and successful as these great institutions, but these were the archetypes on which all modern R&D organizations are modelled.

Farnborough is a particularly apt model, since the outstanding effectiveness of applied science in the Second World War was the main reason why it is now established on such a very large scale. But there is really not that much difference between military and civil R&D in the way that they are organized and conducted. The detailed management of applied science has always varied much less from country to country, than, say the administration of the criminal law or of the local school system. I shall not hesitate to draw from the experience of other countries, since I am here concerned with a worldwide phenomenon, observable in every nation with an advanced scientific sector, from the United States and Russia to Israel, India and Argentina. The instrumental function of science is such a new development that there has simply been no time for the evolution of different styles of organization to match the different cultures into which it has been introduced.

But the enlargement of applied science in scope, scale and sophistication is only half the story. We must not overlook the parallel process of changes that overtook so-called "pure" science during the same period. The expansion of academic research in the past 40 years owes much to the experience of World War II. It is also closely connected with the expansion of higher education as such. But there has also developed an insatiable need for expensive apparatus and other material resources. Before the War, these resources could be provided out of the normal budget of a university laboratory, eked out by occasional grants from charitable foundations, industrial concerns or governmental agencies. Bernal rightly complained of the state of genteel penury of academic science in Britain at that time,[3] but this was the same all over the world.

Nowadays it is taken for granted that basic science should be supported on a very large scale, primarily by the state. Many hundreds of millions of pounds are provided for laboratory equipment and supplies, for technical services, for computing facilities, for research students and for fully qualified post-doctoral assistants. Immense sums are also spent on major research instruments, such as particle accelerators, space satellites and ocean-going ships, whose prime use is for the advancement of knowledge. Never mind, for the moment, where this money comes from, how it is distributed or what it is supposed to buy. The main point is that the typical modern scientist doing "pure" research, whether as a university teacher or as a full-time employee of a research institution, is sustained by an organization of assistants, technicians and professional colleagues numbering several hundred people. The annual budget of such an organization will be several times the aggregate of the personal salaries of the research workers who share its facilities. It is no longer practicable to undertake worth-while research in the natural sciences unless one belongs to a departmental laboratory, or a research institute or some similar establishment. In other words, most "pure" scientists are just as organized as their "applied" friends and acquaintances.

Those of us who have lived and worked as scientists in the midst of this continuous quantitative change may be a little hesitant to acknowledge its qualitative effects. Let us look at science now, and forget, for the moment, how it was before, and how it got the way that it is. The stodgy documents prepared by government people for other government people refer to something called "the R&D enterprise," or "the national research system."[4] This enterprise incorporates all the activities that were previously termed "pure science," "applied science" and "technological development." It is a system combining various administrative sectors: universities and polytechnics, research councils, government departments, private industry and commerce, and so on, depending upon the country being surveyed. But it is a single system. Almost everything that is said about "science policy" conveys the same presumption: modern science is essentially a unitary undertaking, whose internal segmentation need not be taken into account in a primary analysis of policy.[5]

This macropolitical presumption is also valid microsociologically. At the level of everyday "laboratory life," the practice of science is remarkably homogeneous.[6] Let me suggest a simple thought experiment. Imagine that you are blindfolded, and driven around for several hours, and then released in a typical laboratory. You identify various pieces of apparatus—an electron microscope, an ultracentrifuge, racks of test-tubes, incubators, biochemical reagents, and so on. You observe people in white coats—those are the technicians, aren't they—and some very argumentative people in rollneck sweaters—presumably the theoreticians. They make chemical and biological manipulations, they consult books and journals, they enter notes on index cards, and they commune for hours with computer terminals. They also discuss, endlessly, the work that they are doing. The question is: how long will it take you to decide where you are? At this level in the R&D system, how would you distinguish between, say, the Laboratory of Molecular Biology at Cambridge, doing absolutely "pure" fundamental basic science, and the laboratory of some whizzkid company trying to break into the tough commercial world of biotechnology? If you happen to be a professional scientist, you can probably guess the answer to this question from such subtle clues as the quality of the apparatus, or the personal interactions between senior and junior researchers. But if you are not very familiar with the research environment, you may have to wait for several days before somebody refers to the paper that they must hurry to get published—perhaps to win another Nobel Prize—or, alternatively, to the adverse result that must be kept secret so as not to affect the standing of the company's shares on the Stock Exchange.

The organization chart of the R&D system states the objectives of each major laboratory, institute or establishment. These objectives range along the traditional scale of "utility," from the "purest" to the most "applied." But the degree of utility demanded of a major R&D institution does not determine the nature of its intellectual and technical resources, nor the way that they are put to work.[7] Wherever it lies along this scale, it will employ people with the same basic education and technical training, it will use the same sort of advanced scientific apparatus, it will need access to the same body of information, it will rely upon the same theories and it will be faced by the same areas of ignorance.

The characteristic parameter in modern laboratory life is the time scale on which results are desired. The utility of the answer to a question obviously affects the urgency with which it must be answered. In the later stages of technological development, this time scale is very short indeed; "next week" may be just too long to wait for an answer to the question why a new jet engine is not up to its designed performance. But in the exploration of remote and irrelevant realms of being, such as cosmology, there is no real hurry at all; so far as most of us are concerned, we can wait for ever for the key to the Universe.

But the correlation between utility and urgency is not as strong as we tend to believe. Particle physics is of no known utility, whereas plasma research has an enormous potential payoff in fusion power. I doubt, however, whether there is less urgency in experiments at CERN than there is in the experimental work at JET The progress of research in any large R&D organization is related to a complex mixture of time scales, as plans and projects are nested inside each other.[8] We may be able to wait for ever for the key to the Universe, but the satellite is to be launched in June 1986, and we must get our cosmic ray detector working within 18 months, and that means we only have 6 weeks to find out how to quench the luminescence in the acrylic photoconverter sheet, and so on. The scientifically exciting objective of the project as a whole may be an important source of inspiration for those taking part in it, but it does not affect the amount of perspiration that will be needed if this objective is to be achieved.

The function of science policy is, of course, to *plan* the research enterprise. Strictly speaking, as Bernal's opponents argued in the 1940s,[9] it is logically impossible to plan an activity like research, whose outcome is, by definition, still unknown. But this argument neglects the intentionality of the notion of research.[10] A scientific investigation is not a spontaneous walkabout motivated by casual curiosity: it is a deliberate undertaking, contrived to bring Nature under systematic observation, interpretation or control. Apparatus must be purchased, assistants hired, and experiments very carefully designed and performed, before one can get anywhere near an unforeseen, or inexplicable, result. Scientific research is a highly rational and orderly activity which *must* be planned to the point of defining a specific question and procuring the means by which this question might be answered. The uncertainty lies in what should be done after a result has been obtained: what investigation should *next* be undertaken, and how should *it* be planned?

The real issue in the planning of science is: what mechanism is used to prescribe the questions to be tackled? The outstanding characteristic of the modern R&D system is that this is seldom left entirely to the discretion of the individual scientist. Those who worked in "applied" science and technology always took this for granted. The instructions always came from outside the research organization. "We"—that is, the company, the Navy, the farmers, the patients, or whoever else was paying the piper—"require a solution to *this* problem; go thou and solve it." In practice, a great deal of discretion had to be allowed on how the problem should be approached, but this remains the underlying logic of the situation. The more urgent and utilitarian objectives of R&D can only be reached by a bureaucratic organizational structure where specific research goals are imposed from above.

The distinguishing feature of "pure" science, on the other hand, was that the individual scientist was free to follow his or her own nose in the choice of research problems. Nowadays, that freedom is severely constrained. The resources for research in a particular field are only made available for projects that are specifically approved by the grant-awarding bodies in that field. You may be a competent scientist with a permanent post in a university or polytechnic. You may formulate a scientific problem that you consider interesting and soluble. Nonetheless, you may never have the means of doing research on this problem, because it does not commend itself to some committee, or board, or panel of a research council, or foundation, or government department. The non-urgent, open-ended, wing of the national research enterprise is not organized bureaucratically, but it is no longer permitted to advance erratically in response to the personal research plans of a horde of individuals. The crucial decision in the research process—*this* is the question to be studied by *this* method—now rests in the hands of organized groups of people, rather than in the hands of independent scientists.

Jerome Ravetz[11] described this as the *industrialization of production* in science. It seems to me that this term overemphasizes the instrumental intention of the process, as if all science were now directed towards strictly utilitarian ends. As I shall argue later, there is an inevitable tendency in that direction, but I do not think that the alternative conception of science, as unprompted discovery,[12] is yet out of date. I should prefer to say that science has been *collectivized*, thus indicating that the traditional *individualism* of the academic mode of research has been decisively and irreversibly curbed.

This term is not merely a reference to the growth of team research and "big science," nor to the expansion of industrial research and of the scientific civil service. It refers to a profound transformation of the social relations of science, comparable in significance with the rise of industrial research itself, in the nineteenth century.[13] In the house of the Royal Society, beneath the inscrutable gaze of our celebrated predecessors, it would be easy to fall into a sentiment of regret at the passing of an era to which they gave so much. But I see no point in lamenting a historical transition that cannot be avoided. As Bernal put it[14] "Science is as little able as industry to function on an individualist basis." Society demands the results of research that is organized to meet its multifarious needs, and scientists must have research facilities that society is only willing to provide on a collective basis. This is the social contract for the contemporary R&D system, and almost everybody agrees that it is a good bargain. It is beneficial to society, and to the advancement of knowledge; neither science nor society can go back to the pre-War condition that Bernal deplored.

Nonetheless, the transition of a new social function for science has not gone as smoothly as he expected.[15] A number of structural issues still dominate the science policy agenda: how should research goals be established?; how should resources be matched to goals?; by what criteria should one choose between research projects?—and so on. The Merrison Report[16] on the Dual Support system depicts a disastrous crisis of means and morale in the university sector of the "national research enterprise." Science policy is turning out to be rather more difficult than science itself. Some of these difficulties may be arising because we are not sufficiently aware of the influence of science policy on the way that research is done. The collectivization of science has not only changed the *external* social function of science; it has also altered the *internal* sociology of the world of research. Two main effects can be observed. There has been a change in the professional and social role of the individual scientific worker, and there has been a change in the distribution of risks and responsibilities in innovative research. The remainder of this lecture will be devoted to these two topics.

First let us consider a little formal sociology. In 1942, Robert Merton[17] characterized science as the activity of a community governed by a special set of norms. These norms, which may be cryptically labelled *Communalism, Universalism, Disinterestedness, Originality* and *Skepticism,* constitute the scientific ethos. Thus, for example, the norm of communalism tells us that the results of research belong to the whole community, and should be published at once, without fee or favor. By a strange coincidence, the initial letters of these terms spell CUDOS, which is a reminder that the real reward for scientific achievement should be professional esteem.

Sociologists of science are deeply divided about the theoretical validity of this sort of scheme.[18] Nonetheless, it does characterize the *academic* tradition in science, especially in "pure" research. What is seldom emphasized is that most organized forms of "applied" science have developed on quite a different sociological pattern.[19] The industrial tradition in science is primarily bureaucratic, and is governed by principles that are almost antithetical to the Mertonian norms. Industrial research is directly commissioned by the management of a firm, it is performed by local "in-house" experts, under the authority of the management, and its results are considered to be company property that may have to be kept secret. In other words, it is *Proprietary, Local, Authoritarian, Commissioned* and *Expert*; by no coincidence, the initial letters of these terms spell PLACE, to remind us what the researcher mainly gets out of working in this branch of science.

The modern R&D system is the heir to both the academic and the industrial traditions of research. But it has not yet developed a consistent internal sociology to match its technical homogeneity or the overall coherence of its goals. Research programs and institutional forms from the two traditions are intermingled, but they have not yet been merged.[20]

As a result of this, people find themselves in very ambivalent situations, torn between the norms that they have personally internalized and the duties that they are being called upon to perform. This is the situation for many scientists directly employed in the establishments of Research Councils.[21] Although formally organized along Civil Service lines, their patterns of work and career paths were actually much closer to the academic than to the industrial tradition. But now, as a result of the Rothschild Report and other external pressures, they are being ordered to seek out and undertake commissions for research on problems that they have not chosen for themselves, and which seem to them to be of little intrinsic scientific interest. These orders may possibly be for the good of the nation, but to the scientists asked to carry them out they seem to contravene the norms of disinterestedness and originality which they have previously striven so hard to follow. People do, in fact, manage to adapt remarkably well to this inversion of their scheme of professional values, and can even get new life from this challenge and the change. Nonetheless, their immediate plight draws attention to one of the major dilemmas in the management of R&D. Is it possible to reconcile the personal norms and organization principles derived from these two traditions in a single way of life, with a single career structure to the top, or should they be openly differentiated by providing a "dual ladder" for promotion on the basis of "individual merit" outside the normal managerial hierarchy?[22] Let me emphasize that the collectivization of science need not be a one-way process in which scientists who were previously free to seek CUDOS are being put in their bureaucratic PLACE. In many larger industrial research organizations, successful achievements in R&D can be recognized by promotion to the rank of "Senior Scientist" or "Research Fellow," with minimal direct administrative responsibilities, and encouragement to do science in the traditional individualist style. This seems good sense, except that it may induce less talented scientists to do research that can easily be published rather than what the organization really needs. The Scientific Civil Service and the Research Councils had better think hard about the way in which "individual merit promotions" are handled and the purposes that this scheme should serve.

In the university sector, also, collectivization is corroding the central pillar of the career structure: the principle of academic tenure. The Mertonian scheme is based upon the unstated presumption that the individual scientist is actually free to observe the scientific norms. The social foundation for the outright individualism of the academic ethos is security of employment. One could at least make a modest living as a university teacher, even if one failed to win a great deal of CUDOS for one's research results. But as the ultimate responsibility for the performance of research is shifted onto the grant-awarding bodies, academic tenure comes to be regarded as an irrelevance, or even an extravagance. Many senior scientific policy-makers[23] now seem to take the view that it is more efficient to tie the research job to the project, and are apparently incensed at the thought of wasting public funds on scientists whose research proposals do not earn such support. We are now in a situation where older academics can risk redundancy just because they are not successful in research, and where many younger research workers go on for years on a succession of short-term appointments without the prospect of getting tenure. This policy is logically consistent with a narrowly managerial view of the R&D enterprise, but fails to take account of the fact that time, patience and human understanding will be needed to shift the internal sociology of academic life from the individualist to the collectivist mode. It also seems to ignore the fact that a policy that deters people from obedience to the norms of disinterestedness, originality and skepticism may eventually prove damaging to the advancement of natural knowledge. I wish I could be sure that all our senior scientists understand this issue, and do not hesitate to warn their political overlords against this path of folly.

The internal inconsistency of the R&D system gives scientists a very ambivalent role in modern society. This comes out very clearly in every discussion of their social responsibilities and human rights.[24] On the one hand, they are heirs to the academic tradition, which sets them a little apart from other people. According to this tradition, they are members of a universal community, pursuing knowledge "for its own sake," in the sure belief that this will eventually turn out for the good of all humanity. They may therefore lay claim to transnational rights of communication and travel, and need take no blame for the unfortunate ways in which some of their discoveries are actually used. On the other hand, they can mostly be identified as the employees of particular governments or corporations, encouraged to work on clearly defined projects whose likely consequences have been carefully discussed. From this point of view, their rights and responsibilities are the same as those of other employees of such organizations and other citizens—civil servants, military personnel, teachers, salesmen or bus drivers—and come under the normal rules of ethics and of law. As Bernal put it "The scientist as citizen is not in the first place a scientist, only in the second."[25]

This ambivalence is exploited by both sides in controversies over scientific freedom and responsibility. Take the case of Andrei Sakharov.[26] We in the West expressed our personal solidarity with him, and insisted that he should be given special preference, just because he was a great scientist. The Soviet authorities, on the other hand, argued that he deserved to be punished as a traitor, just because he was no longer willing to give his talents uncritically to the service of the State. The ambiguities of the situation could not be resolved by reference to the internal values, norms and conventions of the scientific world. These lines of argument will never meet until the two great traditions of collectivized science are harmoniously combined and merged. That is why it is essential to call upon general legal and moral codes, such as the International Code of Human Rights, if we are to protect other scientists from oppression, or take them to task for their crimes. Sakharov himself recognized this necessity, and asked for no more than his due under that Code.

Sakharov's case really touches upon a larger issue than the denial of elementary political and civil rights to a particular citizen of a particular State. It arises from the subordination of a great part of all scientific activity to the devilish business of war. This is a characteristic feature of every national R&D system. We could scarcely expect it to be otherwise. Preparation for war is one of the main concerns of every collectivity that calls itself a nation: collectivized science is inevitably mobilized in support of these preparation. Many of us, without being pacifists, used to believe that we could at least contrive to line up our own personal research enterprise on the side of the peace-loving angels. This belief has become very hard to sustain. Few scientists nowadays can afford to be fastidious about where the money is coming from to support their research; if one's own Ministry of Defence is the only agency interested in funding a basic research project, it is difficult to produce convincing arguments against accepting their support. The traditional appeal to the spirit of international cooperation and understanding in the search for truth has also begun to ring a little hollow. International cooperation in science is no longer the personal fad of cosmopolitan intellectuals: it has been collectivized, and brought under political control by governments and intergovernmental organizations. In this sphere, also, science cannot be considered an independent activity or institution, disconnected from all the turmoil and anguish of the affairs of the nation or of the nations.

The other main effect of collectivization on the internal structures of science is to change the distribution of responsibilities and risks in highly innovative research. This change comes about almost inevitably, as a policy apparatus is imposed on scientific work. Collectivization cannot be separated from rationalization, in the broadest sense of the word. Organizational criteria of efficiency[27] and accountability[28] are automatically formulated, to monitor and control the way in which policy is put into practice. Whatever characteristics happen to be considered good indicators of performance, these criteria tend to be applied more and more strictly, at lower and lower levels in the decision-making process.[29] I cannot recall whether this is one of Parkinson's Laws, but it arises irresistibly out of the logic of collectivism. We all know that administrative zeal expands to use all the resources that it can lay its hands on, including all the time that it can get out of the members of advisory committees.

The dominant criteria of efficiency and accountability in a national research system must surely be utilitarian. In one way or another, the budget for science has to be drawn up by the Government, and approved by Parliament. Whether presented in large chunks, or in disconnected fragments, it has to be justified in the same political and economic language as the budget for motorways or for prisons. That is to say, it must give the impression that it is directed towards the usual objectives of democratic politics: the work, wealth and happiness of taxpayers, and things like that.

There is nothing wrong with this. Heaven help the nation where such utilitarian considerations are not absolutely paramount in the decisions of central government. But if there is any pressure on science policy from above, it is bound to inject similar considerations into all decisions lower down the system. Thus, as collectivization proceeds, and opens up new channels of influence from the top, there is an irresistible tendency to phrase the objectives of research projects in utilitarian terms.[30] The members of a research review committee may have a high opinion of a project for its purely scientific interest, but they are pushed into trying to justify it formally for its social merit, even though it appears to have only a very uncertain pay-off in a very distant future.[31]

Inside the R&D system, we are all now aware of much higher levels of internal stress and strain. This is usually attributed to something called "the present climate" in our political and economic affairs, which is causing a severe drought in the funds for research. We do indeed have to learn to live with the idea that the postwar years have been a very lush growing season for science, which could not have continued for ever.[32] I do not myself think that our present discontents are merely seasonal—a winter with a joyful spring to follow. In any case, the public demand for "useful" results is of longer term than an economic cycle or a swing of the political pendulum and penetrates more deeply into the details of the scientific enterprise, making scientists themselves much more self-conscious about the utility of their work. This is a world-wide phenomenon, as much a matter for concern among scientists in Europe and the United States as in Britain.[33]

Science policy-makers and analysts agree that the shift of emphasis from the "basic" to the "technological" end of the utility scale has been overdone, and that some areas of science are beginning to live off their seed corn.[34] My argument here goes further. I do not believe that these stresses and strains will be reduced by government officials and top scientists making New Year resolutions to devote a higher proportion of their R&D budgets to basic research. The troubled feeling inside science is not really due to the accent of utility as such: it is generated by the attempt to impress greater *urgency* on the performance of research, in all sectors and at all levels. As I have already remarked, a sense of urgency is correlated with utility, but is not synonymous with it. It is also produced by organizational insistence on greater efficiency and accountability. The Rothschild Report, for example, may or may not have had much effect on the real amount of applicable research done by the Research Councils and government departments, but it has undoubtedly shortened the time scale on which "useful results are being asked for and "accounts" have to be given of intended work and of results achieved.[35]

The collectivization of science is designed to reduce certain undesirable features of the academic style of research:[36] Inefficient use of resources, duplication of projects, swings of fashion, aggressive and divisive competition, as well as irrelevance to social needs. But these individualist vices cannot be eliminated without repressing some corresponding virtues, such as curiosity, independence of mind, flexibility in exploiting opportunity, and personal willingness to take intellectual risks. The extreme individualism of academic science is not mere irresponsibility or anarchy: it is governed by behavioral norms which derive from its peculiar social structure[37] and which determine the nature and quality of the knowledge that they generate. This may not be the best, or the only, social arrangement for getting such knowledge,[38] but it is the way that the scientific method has been practiced since the seventeenth century.[39] Change the sociology of science, and you are bound to change its epistemology.

It is very hard to guess how much this change of method will actually affect the contents of science in the future. We accept, of course, that a greater emphasis on utility will reduce the rate of progress in those branches of science, such as high energy physics and cosmology, that appeal enormously to researchers as individuals but that seem to have no practical applications. But there seems to be no reason why the main stream of basic and strategic research should not flow serenely on, sustained by its promise of unspecified benefits for everyday life or commercial profit or military power. There is no obvious threat to the general reliability of scientific information and understanding. The sort of collectivization that I am talking about here has nothing in common with the Stalinist usage of this term.[40] Intellectual pathologies like Lysenkoism can only take hold when the whole body politic is in a pathological condition, and that would be cause for much graver concern than the mere perversion of scientific truth.

The real worry is that we may be producing a climate in which the tender growth tips of knowledge shrivel up before they have had a chance to establish themselves as worthy of further investigation.[41] In a highly collectivized and rationalized system, how long could a wild conjecture—for example Wegener's absurd theory of drifting continents—be allowed to linger on, against all the authoritative opinion of its times? Would a modern Wegener, or one of his misguided disciples, get funded for a research project associated with such a ridiculous notion? It is not merely imaginative theorizing that would be curbed. As everybody knows, serendipity plays a vital part in scientific progress. *Surprise* has always been one of the names of the scientific game.[42] But a surprising phenomenon makes no contribution to knowledge, except as an opportunity to be exploited: we must be able, at times, to drop everything in order to follow up some peculiar observation or idiotic thought that happens to come our way.[43]

Every scientist knows the value of speculative, defiant, contentious, serendipitous research. The problem is: how do we arrange things so that (to quote the Merrison Report) "new ideas, which may seem at first not only odd, but positively perverse and outlandish, can be nurtured in the tender initial period"?[44] Of course, 99% of such ideas are not better than a load of old codswallop: but science would be very much the poorer without the 1% that turn out to have some sense in them after all.

Academic traditionalists argue that the only way to keep the spirit of discovery alive in science is to return to the outright individualism of "pure" research done out entirely "for its own sake." But who would give them the apparatus that they would need, just to follow their own devices in this way? In any case, the same spirit is essential in all forms of applied science and technology.[45] Even in a very down-to-earth industrial research laboratory, one must be able at times to respond promptly to a surprising discovery, or try out some crazy invention, without upsetting the major program for which the laboratory is supposed to exist.

Scientists who work in an urgent utilitarian atmosphere of R&D know how difficult this is. They often complain that the management will not take any notice of their brilliant and novel suggestions. Research managers, for their part, are sensitive to this criticism, but their discretion is limited by organizational objectives and imperatives. They often exercise a good deal of private wisdom in protecting and fostering promising new ideas that lie outside the official R&D program,[46] but there is no public wisdom on how to respond flexibly to the unexpected within a normal bureaucratic organization.

But the imperatives of utility and urgency do not apply to basic or strategic research in the university sector. How can there be any difficulty in getting support for original, innovative research? Is not that the whole purpose of the system of making separate grants to individual academic scientists, so as to leave them free to have bright ideas and yet give them the means to follow up the brightest of these ideas with a fully equipped investigation? Are we not getting the best of both worlds, individualist and collectivist, by this brilliant administrative device?

Well, we have been collectivizing academic science by this device for 30 or 40 years, and have adapted to it as a way of life. It has obvious advantages in getting research funded in a big way, without interfering with the formal autonomy of universities and their staffs. None the less, the so-called "peer review system" is coming under increasing criticism. In Britain this criticism centers on the breakdown of the dual support system for university research. The Merrison Committee was undoubtedly right to urge that a much higher proportion of the science vote should flow through the books of the University Grants Committee, to provide a solid base of permanent staff and facilities for research in university science departments. As I have already remarked, the traditional intellectual virtues of openness and independence of mind are founded upon the social practices of openness and independence of academic office. But I think that we should also cast a cool eye on the procedures by which the Research Councils and other grant-awarding bodies actually support R&D in the universities. In particular, does the peer review system really provide a fertile climate for highly innovative research?

Most of the public criticism of the peer review system comes from the United States, and focuses on the grossly excessive amounts of time and effort that scientists have to spend on writing their own research proposals and reviewing the proposals of others.[47] This is a scandalous waste of a limited resource, although it is hard to see how it can be held in check without a fundamental change of organizational goals and constraints. The collectivist demands of efficiency and accountability are bound to have this effect in a highly dispersed "system" where the organs of policy and performance are deliberately kept at arm's length from one another. Project proposals flow inwards, and grant decisions flow outwards, with almost no interactive dialogue between the two streams. Fortunately, science is not a railroad; if it were, then what a way to run it!

This lack of administrative reflexivity makes the system very ponderous and slow to take advantage of serendipitous opportunities. An unexpected discovery suggests a new avenue of research that ought to be explored. This may not call for very large resources, and is not obviously urgent by external standards of judgement: is there any point in applying for a modest grant for this project, knowing that one will have to wait for many months before the review panel can meet to approve it, and that there is no certain prospect of getting it even then? It is not enough to say that grant-awarding bodies ought to be giving lots of small, short-term grants.[48] For the "quick and dirty" type of investigation, the Italian beggar's request is appropriate: "poco, maladetto, ma subito" (a little bit, with a curse, but at once)! The basic principles of the whole peer review system—collective responsibility for research funding decoupled from individual responsibilities for research initiatives—practically forbid a policy of rapid response and flexible opportunism in research performance.

Is the system unadventurous because funding decisions rely on the views of representative groups of the scientific "peers" of applicants? One would naturally suspect so, but this worldly wise suspicion does not seem to have been proved by actual research. What has been proved is that experts are usually wildly at variance with one another in their assessments of research proposals.[49] This is not really very surprising. The chief characteristic of the opinions of well informed scientists on current scientific questions—that is, on just those questions that might be worth researching—is wide diversity. This diversity is entirely natural and healthy: the fact that it shows up in the way that they rate research proposals merely demonstrates how honest and conscientious scientists are in giving those opinions. But this does not mean that the most orthodox proposals will get the highest marks when the ratings of a number of reviewers have been added up or when a panel or reviewers has gone through the psycho-dynamic drama of making a collective decision. There is really no saying what considerations will eventually hold sway in a committee trying to decide a matter on which individual opinions are so very subjective and diverse. Consult your own experience on this, ladies and gentlemen, and write a paper on it if you think you can prove a general principle.

The nub of the matter is that the peer review panel is being asked to do an impossible job. On the face of it, an expert assessment of the "timeliness and promise" of a research project is just as reasonable as an assessment of the originality and competence of a doctoral dissertation or of a scientific paper. But they are really quiet different propositions. The Ph.D. examiner or journal referee is working in the context of *justification*, where there are universal standards of factual accuracy and logical consistency.[50] The project grant reviewer is being pushed into the context of *discovery*, where there is nothing but personal intuition, experience and skill to be pitted against the uncertainties and contradictions that one hopes to resolve. The effort to articulate such tacit considerations into a persuasive public document leads to one of the higher forms of nonsense.[51] A research proposal may be very instructive on a number of relevant points, such as the knowledge that the proposer has of the subject and the facilities available for the research, but the more rational it tries to appear, the closer it comes to pretending to answer the absurd question "what result are you going to get out of this investigation?"[52] This is the question that is most nonsensical when applied to the most innovative, exploratory, outlandish research. That is the main danger of the peer review system to the progress of science.

Rustum Roy has given a convincing account of the weakness of peer review, particularly in the form now practiced in the United States.[53] Many of these weaknesses are probably unavoidable in any system that tries to combine individualist and collectivist principles. There is no way of resolving the eternal dialectic between deviance and orthodoxy, or between private interest and public good. But the present system is trapped in a non-dialectical confusion between prospective and retrospective judgements, between risk-taking and responsibility, between deciding what to do next and accounting for what has already taken place.

The only way to resolve this confusion is to move from pre- to post-project support for basic and strategic science.[54] There is nothing novel in providing support for research on the basis of scientific achievement rather than of research intention. This principle obviously underlies the traditional practice of appointing people to permanent academic posts on the basis of their published work. Success in science is an enduring characteristic of individual scientists. It is not a characteristic that endures for ever, but it can be estimated as accurately, and relied upon for as long, as any prior assessment of the timeliness and promise of a research project.

Any new scheme for providing prizes to individual scientists for their research achievements would, of course, simply encourage the cult of personality. But the persistence of success is a characteristic of institutions as well as of individuals. As I have already emphasized, the resource of apparatus and technical staff that are needed for research cannot be provided separately to individuals: they must be shared by dozens, or even hundreds of scientists, in a distinctive "institute," such as a university department or research council establishment. It is entirely appropriate, therefore, that the funds needed for the facilities of the institute should depend upon the scientific contribution of the institute as a whole.

There is nothing really novel about this way of funding quasi-academic research. The permanent establishments of the Research Councils used to receive block grants, guaranteed for relatively long terms in advance, but subject to periodic review on the basis of their actual achievements. The excellent scientific record of many of these establishments speaks for the practicability of the "institute grant" method in general. From a British point of view, the only radical element in the proposal is to treat each university department as a single "institute," to be funded for research by a single grant, instead of receiving support for a variety of individual projects.

This means, of course, that each institute would be free to allocate funds internally to research projects, without reference to the main funding authority. According to at least one careful study of a system of this kind, which has been operating for some years in the United States, this has no significant effect on the quality of the research that is actually done.[55] And yet it cuts out most of the unproductive activity of generating elaborate, and slightly bogus, project proposals and of making elaborate, and equally bogus, assessments of their merit. From our present point of view, it also permits much more flexibility in responding to serendipitous opportunities, or exploring unorthodox hypotheses as they occur, without all the palaver of making a formal application to the grant-awarding body.

Now that I have made a specific proposal for the reform of a cherished system, I am at the mercy of every experienced scientist. There is no way of defending such a radical proposal from all the objections that they will undoubtedly be able to think up against it. Indeed, I should have grave objections to any such scheme myself, unless it is carefully designed to satisfy three conditions: it must be critical, it must be open, and it must be egalitarian.

When I insist that it must be critical, I mean that the periodic review of the performance of each institute should not be mere formality. Rustum Roy[56] would like to save a great deal of trouble and expense by calculating the annual grant according to a strict formula. This would be compounded by a 3 year rolling average of objective productivity indicators, such as the number of papers published, the number of masters and doctors degrees awarded, and the amount of research directly commissioned by government departments and by industry. I guess that this mechanistic approach would not be politically or administratively acceptable, although it might give as good as answer as any more elaborate procedure of assessment. Certainly, in any such retrospective evaluation of the actual scientific performance of a research institute it would be instructive to use the scientometric methods pioneered by John Irvine and Ben Martin.[57] But most of us would also expect the decision to be based upon a direct visitation and detailed report by a body of experts: a systematic and thorough peer review in the fullest sense.

This sort of critical review of research performance is an imperative of collectivization. We cannot go back to the old University Grants Committee system of making block grants to universities for their research work, without any external checks of the objectives or quality of that research. In any case, institute grants would be made within the framework of general science policy. They would be given for research in particular fields, often in relation to a particular range of social needs, and would be tapered down and terminated if the work strayed too far outside these broad requirements. Would this degree of retrospective accountability inhibit innovative or unorthodox research? My own feeling is that the risk of undertaking it, or encouraging it in an institute, in the intervals between periodic reviews, would be acceptable to members of that institute, because they would all have a stake in its possible success.

It is much more difficult to keep the system open and dynamic. One of the benefits claimed for the conventional peer review system is that it does give relatively unknown scientists, in out of the way institutions, a chance of winning support for their research. The most serious objection to a comprehensive system of institute grants is that it could become systematically elitist. The only way to get into basic research would be to get

356INTOTHESOCIALDIMENSION

inside the charmed circle of institutes as a student and try desperately to stay on. This is not a new problem in science. An effective mechanism would have to be set up to discover, and encourage, scientific talent as it matured outside the elite institutions, and to allow new institutes, often in entirely new fields, to be nucleated and grow to viable size. This is why the central review procedure must be ruthless, so that obsolete or decadent institutes are killed off in good time.[58]

Would an institute grant system damage the egalitarian relationships between the tenured staff in a large university department? We tend to think that any process of collectivization will formalize and strengthen the authority of one individual over others. Is this necessarily so? There is no reason in principle why an institute devoted to non-urgent research should be organized hierarchically. This brings me back, once more, to Bernal, who understood, and personally exemplified, the role of leadership in any scientific enterprise. He also understood the value of colleagueship. In the *Social Function of Science* he insisted that a research laboratory must be "a voluntary association of scientists for an object they all have in common,"[59] and discussed very realistically the conditions for cooperation, collaboration and internal democracy. Many scientists with unhappy experiences of being managed by, or trying to manage, other scientists, may think this is a counsel of perfection, but Terry Shinn has been studying the internal organization of various types of research unit in France, and finds that the Kibbutz is a more effective model for a highly innovative research institutes than the Commando.[60] Somewhere in this direction, we shall have to find our way to a new internal sociology of science, combining individualist and collectivist strands in the pursuit of knowledge.

Finally, I have an uneasy feeling that the grandeur of Desmond Bernal's vision of science has tempted me into extravagant generalizations that go far beyond the evident facts. I have undoubtedly tried to cover too many issues, and not succeeded in dealing properly with any of them. But within this ragbag of topics there is a single theme: the change in the essential character of the scientific enterprise in the past few decades. The term that I have applied to this change, "collectivization," is ambivalent. On the one hand, it refers to the way in which the goals of research have been brought under the control of the general "collective," that is the governmental and other organs of society at large. This instrumental conception of science has been the main concern of science policy, in theory and in practice. On the other hand, the word "collectivization" may be interpreted as "acting contrary to individualism." This seems the best word to describe a profound transformation that is taking place in the relationships between one scientist and another, as they go about their work of problem solving and discovery. This change is more subtle than the triumph of bureaucratic utilitarianism, and cannot always be shaped by the crude device of administrative and financial policy. None the less, we need to think and talk about it from time to time, to clarify our own notions of what the scientific life might be like in the future, in the hope that it may be just as good as it has been for many of us in the past.

References

1. Bernal, J.D. 1939 *The Social Function of Science*. London: Routledge.
2. Bernal, J.D. 1964 After twenty-five years. *In the Science of Science: Society in the Technological Age* (ed. M. Goldsmith & A. Mackay), pp. 201–228. London: Souvenir Press.
3. For example: Bernal, J.D. 1949 *The Freedom of Necessity*, p. 127. London: Routledge and Kegan Paul.
4. For example, in innumerable publications of O.E.C.D., Unesco and similar intergovernmental organizations.
5. This presumption is seldom made as explicit as by Salomon, J.J. 1973 *Science and Politics*. London: Macmillan.
6. It would be instructive to verify this informal observation by detailed ethnographic research, along the lines of Latour, B. & Woolgar, S. 1979 *Laboratory Life: The Social Construction of Scientific Facts*. London: Sage.
7. "It is a mistake to think that all socially relevant research is short-term in nature." Gibbons, M. 1982 The funding of research. In *The Future of Research* (ed. G. Oldham), pp. 48–82. Guildford: Society for Research in Higher Education.
8. Ziman, J.M. 1981 What are the options? Social determinants of personal research plans. *Minerva* 19, 1–42.
9. See, e.g., McGucken, W. 1978 On freedom and planning in science: the Society for Freedom in Science 1940–46. *Minerva* **16**, 42–72.
10. As note 8.
11. Ravetz, J.R. 1971 *Scientific Knowledge and Its Social Problems*, p. 44. Oxford: Clarendon Press.
12. Ziman, J.M. 1983 Conceptions of science. In *Realizing Social Science Knowledge* (ed. B. Holzner, K.D. Knorr & H. Strasser), pp. 179–196.Vienna: Physica-Verlag.
13. Ben-David, J. 1974 *The Scientist's Role in Society*. Englewood Cliff, New Jersey: Prentice-Hall.
14. Bernal, J.D. 1949 loc. cit. (note 3), p. 271.
15. Freeman, C. 1980 Bernal and the "social function of science." Draft chapter for book on Bernal edited by F. Aprahamian.
16. Report of a joint working party on the support of university scientific research 1982 Cmnd 8567. London: H.M.S.O.
17. This is reprinted, with a number of other papers on the same subject, in Merton, R.K. 1973 *The Sociology of Science*. Chicago University Press.
18. For example, Mulkay, M.J. 1978 *Science and the Sociology of Knowledge*. London: Allen & Unwin.
19. Bernal, J.D. 1939 (note 1), p. 387 shows his awareness of this aspect of the scientific life.
20. Much of the empirical evidence cited against the Mertonian scheme is really more relevant to this point; cf. Mitroff, I.I. 1974 *The Subjective Side of Science*. Amsterdam: Elsevier.
21. This is apparent from unpublished interview material collected by Mr. R.J. Beverton and myself, in connection with research on specialization and change in scientific careers.

22. Griffiths, D. 1981 Job evaluation, technical expertise and dual ladders in research and development. *Personnel review* **10**, 14–17.
23. Witness the failure of such bodies as the Committee of Vice-Chancellors and Principals to speak out for academic tenure in public.
24. Ziman, J.M. 1978 Solidarity within the republic of science. *Minerva* **16**, 4–19. Ziman, J.M. 1982 Basic principles of social responsibility of scientists. In *Scientists, The Arms Race and Disarmament* (ed. J. Rotbalt), pp. 161–178. London: Taylor & Francis.
25. Bernal, J.D. 1964 ibid. (note 2), p. 227. He always insisted that the scientist should "be recognized as a common worker with the fortune and ability to deal with new instead of established things." (Bernal 1939, p. 314). It must be admitted, however, that he was himself an exceedingly uncommon worker, and was prone to assign a correspondingly uncommon role to science and to scientists in society at large. On this issue, see Werskey, G. 1978 *The Visible College*. London: Allen Lane.
26. Ziman, J.M. 1981 The dark side of science. *Index on Censorship* **10**, 29–30.
27. As note 15.
28. Bernal, J.D. 1964 ibid. (note 2), p. 222. For a formal exposition of this theme see a report of the U.S. National Commission on Research 1980, Accountability: restoring the quality of the partnership. *Science*, **207**, 1177–1182.
29. Irvine J. & Martin B.R. What direction for basic scientific research? 1982 in Science and Technology Policy in the 1980s and Beyond (ed. Gibbons, M., Grimmett, P., and Udgaonkar, B.M.) pp. 67–98, London: Longman.
30. This tendency has been noted in Studies of the working of research review committees of the U.S. National Institutes of Health: see Weiss, C.H. & Bucuvalas, M.J. 1980 *Social Science Research and Decision Making*, pp. 213–229. New York: Columbia University Press.
31. Weinberg, A.M. 1963 Criteria for scientific choice. *Minerva* **1**, 159–71.
32. Ziman, J.M. 1978 Bounded science. *Minerva* 16, 327–339.
33. Blume, S.S. 1982 A frameword for analysis. In *The Future of Research* (ed. G. Oldham), pp. 5–47. Guildford: Society for Research in Higher Education.
34. Striking evidence of the long-term utility of basic research is given by Comroe, J.H. & Dripps, R.D. 1976 Scientific basis for the support of biomedical science. *Science*, **192**, 105–111. The concern of high-level policy-makers on this point is expressed by Press, F. 1982 Rethinking Science Policy, *Science* **218**, 28–30.
35. Gummett, P. 1980 *Scientists in Whitehall*, pp. 202–206. Manchester: Manchester University Press. This tendency is also clearly observable in the interview material referred to in note 21.
36. References to the "inefficiency" of research are to be found in many places in Bernal's writings, e.g. Bernal 1939 (note 1), pp. 94–125 and later.
37. Hjern, B. & Porter, D.O. 1983 Implementation structures: a new unit of administrative analysis. In *Realizing Social Science Knowledge* (ed. B. Holzner, K.D. Knorr & H. Strasser), pp. 265–280. Vienna: Physica-Verlag.
38. The case for individualism as an optimal social structure of science was argued eloquently over many years by Michael Polanyi: see, e.g. Baker, J.R. 1978 Michael Polanyi's contributions to the cause of freedom in science. *Minerva* 16, 382–396.
39. Ziman, J.M. 1967 *Public Knowledge*. Cambridge University Press. Ziman, J.M. 1978 *Reliable Knowledge*. Cambridge University Press.
40. Bernal's public position on this matter is sympathetically analyzed by Werskey 1978 (see note 25).
41 Disturbing evidence on this point is given my Muller, R.A. 1980 Innovation and Scientific Funding. *Science*, **209**, 880–883.
42. Causey, R.L. 1968 The importance of being surprised in scientific research. *Agric. Sci. Rev.* **6**, 27–31. Townes, C.H. 1968 Quantum electronics, and surprise in development of technology. *Science*, **159**, 699–703.
43. Ziman, J.M. 1981 loc. cit. (note 26).
44. See note 16.

45. Bernal was always very alert to the dangers of producing a "rigid and unprogressive" research organization; see, e.g., Bernal 1949 (note 3), p. 132.
46. This point, also, has come out in the research interviews referred to in note 21.
47. Leopold, A. C. 1979 The burden of competitive grants. *Science*, **203**. Leopold estimates that 6600 persons' research time is spent in writing and reviewing proposals.
48. Data on this point are given by Kalberer, J.T., 1981 Grant length and budget stability at the National Institutes of Health. *Science*, **211**, pp. 675–680.
49. Cole, S., Cole, J.R. & Simon, G.A. 1981 Chance and consensus in peer review. *Science* **214**, 881–886. It is instructive to follow the controversy generated by this research report, in *Science*, and in other journals.
50. Ziman, J.M. 1967 (note 39), 1978 (note 39).
51. Ziman, J.M. 1981 (note 8).
52. Peres, A. 1982 *Physics Today* (July), p. 15. Peres ironically proffers a standard referee's report, recommending that any brief, vague research proposal with imprecise goals in an active field of research should be supported, whereas any detailed proposal with precise goals and implications should not be funded!
53. Roy, R. 1979 Proposals, peer review and research results. *Science*, **204**, 1153–1156. Roy, R. 1982 Peer review of proposals—rationale, practice and performance. *Bull. Sci. Tech. Soc.* **2**, 405–422.
54. Sognnaes, R.F. 1974 Post-project research grants. *Science*, **184**, 940. There seems to be a move in this direction in France; see Walgate, R. 1982 *Nature*, **296**, 289.
55. Ling, J.G. & Hand, M.A. 1980 Federal funding in materials research. *Science*, **209**, 1203–1207.
56. Roy, R. 1981 An alternative funding mechanism. *Science*, **211**, 1377.
57. Irvine, J. & Martin, B.R. 1983 Assessing basic research: the case of the Isaac Newton telescope. *Social Stud. Sci.* **13**, 49–86.
58. Irvine, J. & Martin, B.R. What direction for basic scientific research. (To be published.)
59. Bernal, J.D 1939 loc. cit (note 1), p. 263.
60. Shinn, T. 1982 Scientific disciplines and organizational specificity. The social and cognitive configuration of laboratory activities. In *Scientific Establishments and Hierarchies* (ed. N. Elias, H. Martins & R. Whitley), pp. 239–264. Dordrecht: Reidel.

The Individual in a Collectivized Profession

The Cult of the Individual

Let me begin with a sentence that any sociologist of science might record in an interview:—"We are all so busy at the Laboratory these days that I never seem to get a chance to do any of *my own work*." The professor or lecturer who says this is a full time employee of a university or polytechnic. He or she holds an appointment with tenure until late middle age, and is paid quite good wages to provide certain specific services—primarily teaching, but also research.

In what sense can such a person be said to be doing "their own work?" They are supposed to be fully occupied in serving the institution that employs them. During ordinary working hours, they go into the laboratory or study provided by this institution, use apparatus owned by the institution, expend materials paid for by the institution, instruct assistants and technicians on the institutional payroll, have their papers typed by institutional secretaries—yet they evidently feel that they are entitled to call it "their own work!"

What this phrase indicates, of course, is the ideological *individualism* of the traditional *academic* mode of research. The social practices and norms of this mode of science are legitimated by reference to the role model of the independent entrepreneur. Progress is attributed to the work of specific individuals, whose achievements are personally recognized and rewarded. Scientists are thus strongly motivated to make their results public, and to compete with one another to be first with a discovery. They themselves see competitive individualism as the force that maintains the essential tension between originality and criticism.

But there have to be cooperative elements in the scientific way of life, to hold this force in check. The functional sociology of science is primarily concerned with the rules and conventions that balance its competitiveness. In principle, the "academic ethos" delineated by R.K. Merton is maintained by a self-governing scientific community, which not only rewards scientific achievements, but also punishes infringements of its social norms. In practice, as many sociologists of science have pointed out, the mechanisms of social control in science are not very effective, and the actual behavior of scientists is often more individualistic than the normative ideal. Scientists seldom exhibit all those qualities of Communalism, Universalism, Disinterestedness, Originality, or Skepticism that they ought to, if they are to gain any CUDOS for their work!

The *academic* mode of science is thus designed around the stereotype of the scientist as a highly individualistic, inner-directed person, deeply engaged in the exercise of an extremely difficult and highly specialized craft, and seeking a reputation for excellence in the performance of that craft. This was the role institutionalized in the German universities in the early 19th Century, when science came to be recognized as a *profession*. Even if scientists are not all born that way, that is what they mostly become. They see themselves and their peers as following *reputational* careers (Whitley 1984) within the loosely articulated *implementation structures* (Hjern & Porter 1983) of the academic world, and act out their parts accordingly.

Organizational Careers

In reality, academic scientists are not independent entrepreneurs, gaining a living by selling their professional services directly to the public. Unlike certain professions, such as dentistry or law, they are very seldom "self-employed." As I have already pointed out, they are usually full-time functionaries of educational, governmental or commercial organizations, where they are expected to play out an *organizational* career.

The individualist ideology emphasizes the reputational aspects of the scientific profession, and plays down its organizational aspects. That is why scientists are licensed to talk of doing "their own work," as if this had nothing to do with the jobs for which they are paid. Organizational employment or preferment is supposed to come as "recognition" for reputational success, and is so protected by "tenure" that it cannot be lost by the most disastrous performance thereafter. The academic ideology likewise discounts the "Matthew Effect" (Merton 1973)—the means by which organizational office can be used, inversely, to boost a reputational career.

Despite its fictions and reticence, academic science has proved a remarkably stable social institution. For a century and a half, it has been accepted as the standard way of getting people to undertake research on every aspect of the natural and social worlds, primarily, as they would say "for its own sake." Of course there are many subtle variants on the same general theme (Whitley 1984). Informal "reputational communities" and formal academic organizations differ in structural detail from discipline to discipline, from country to country, and from sector to sector within each country. The professional career of an Italian professor of physics, for example, may follow a very different pattern from the career of his or her opposite number in Britain—or, indeed, from a colleague in the very same university studying and teaching sociology. One of the major tasks for the sociology of science in the next few years is to record and interpret these differences, in order to understand what is truly characteristic of science as a profession.

Industrial Science

The sociology of science, being itself an academic enterprise, has tended to focus on the academic mode of research, and to overlook the growing importance of *industrial* science. This, again, is a very loosely defined term, covering a very wide range of activity. Its most typical organizational manifestation is the "R&D Laboratory" of a large industrial firm, where many hundreds of "QSE's"—Qualified Scientists and Engineers— are employed to invent new products and processes, and to bring them to marketable form. Organizations of this kind began to be set up about a century ago, and are now, of course, major institutions in manufacturing industry and in many branches of government.

In the academic ideology, all the work that is done in these institutions is referred to as "Applied Science" or "Technology," and sharply distinguished from "Pure," "Basic" or "Fundamental" science. But this terminology is primarily ideological: it affirms that the search for knowledge "for its own sake" is morally superior to endeavors with more pragmatic goals. From the point of view of science as a profession, the most important distinction between "academic" and "industrial" science is in their internal social structure.

In the industrial mode of research, a scientist only has an "organizational" career. A reputation gained by public contributions to scientific knowledge may be of some advantage in such a career, but does not weigh heavily by comparison with contributions to the success of the organization according to its own lights—problem solving, trouble shooting, technical creativity, project leadership, etc. Far from doing "their own work," industrial scientists are expected to do the firm's work, and are rewarded accordingly.

As a consequence, the academic norms encapsulated in the CUDOS formula simply do not apply to industrial science. Indeed, one could say that these norms are actually inverted—that Communalism and Universalism, for example, are precisely what is forbidden to the scientist working for Ciba-Geigy, or Lockheed, or the national Atomic Energy Commission. But the sociology of such organizations does not depend on the "norms" internalized by their members. From the point of view of the individual, an organizational career in industrial science is governed by principles that might be characterized by a set of adjectives such as Proprietary, Local, Authoritarian, Commissioned, and Expert. I will not try here to explain the precise significance of these terms—but it is no accident that they add up to the acronym PLACE.

This dominance of PLACE in the industrial mode of research has a number of sociological and epistemological implications which deserve further exploration. It is significant, for example that many governmental and quasi-governmental research organizations with very practical objectives, as in agriculture and medicine, have traditionally given priority to the reputational careers of their scientific staffs, even though they are apparently organized as standard civil-service bureaucracies. On the other hand, it should not be thought that industrial science lacks genuine "scientificity" because it does not respect the CUDOS ethos. There are evidently other ways of ensuring that originality and criticism are both brought into play in the research process. Sociological studies of R&D organizations have shown, moreover, that scientists brought up in the "academic style are able to adapt quite quickly to stronger organizational constraints, and do not hanker after "reputational" careers (Cotgrove & Box 1970).

The Collectivization of Science

Until, say, the end of the Second World War, the academic and industrial modes of science tended to keep somewhat apart and to develop independently at a moderate rate. Since then, however, the pace of change has quickened, especially in industrial science. There is no need to present formal evidence of the enormous scale of scientific and technological activity in contemporary civil and military life. R&D budgets have now become standard items in national and international political, economic and diplomatic affairs. But for industrial science this has been more a change of scale and influence than of internal structure.

For academic science, on the other hand, the last thirty years have a been a period of profound structural change. To appreciate the extent of this change, let us accompany our professor, or senior scientific officer, or docent, or maître de recherche into his or her work place. We see immediately that the modern scientific laboratory is not designed for solitary pursuits. It is crammed with elaborate and expensive apparatus, serviced by gangs of technicians, and managed by a hierarchy of bureaucrats. In some cases, each piece of apparatus is so large that a single experiment is being undertaken by a whole team of researchers—a hundred scientists working closely together, over a period of many years. In other cases, our guide will tell us that their use of a major research facility had to be requested months in advance, and justified to a committee, before the investigation could begin. All around us we shall see standard instruments such as electron microscopes and high resolution mass spectrographs, and standard laboratory facilities such as animal houses and computers, which are evidently so expensive, and are needed in such variety, that they have to be brought together in large buildings and shared by many researchers. Even the postdoctoral fellow, or student, or technician, to whom we are introduced, and whose devoted labor will later be gratefully acknowledged, is not there simply as a personal assistant. He or she is supported by a grant to do a particular piece of research, and that grant had to be applied for to a funding agency, and graciously approved by a peer review committee, before the work could be set going.

The *collectivization* of academic science (Ziman 1983, 1984) has been going on now for something like half a century. The trend is irresistible. The term "Big Science" is usually applied to fields such as high energy physics where there is simply no way of getting the results we think we need except by pooling all the scientific resources of a whole continent in a few enormously expensive instruments. But all science is "big" these days. The facilities needed for research are becoming so diverse and sophisticated that they can only be operated and used efficiently in an institutional setting involving hundreds of people. Scientific work has thus become a "collective" activity, in that many people have to collaborate, directly or indirectly, to produce any particular research results.

These facilities are also so costly that they exceed the financial resources of universities, and charitable foundations. The expense of providing them and running them could only be met by very large corporations, or, as is now the normal case, by national governments. And the scale on which the governments have responded to this demand is truly remarkable. The budgets for "discovery-oriented" research are now so large that they also appear on national and international political and economic agendas.

But governments and commercial companies are already deeply involved in R&D with strictly utilitarian objectives. Immense sums go into research and development in industry, agriculture, medicine, military technology, and so on in the expectation of profits, benefits, or victory in war. All scientific work carried out in the "industrial" mode is inevitably "collectivized" from the outset. Its objectives, and even its detailed plans, are not decided by the scientists who carry it out, but are laid down by various non-scientific "collectives," from the state downwards.

In these circumstances, it is difficult for the patrons of "pure" science not to feel that they should have some say in the way their money is spent. They may be quite disinterested in the immediate outcome, but they want to be quite sure that their benevolence has been devoted to a worthy cause. So they do not hand it out to individual scientists to spend as they think fit; they set up a system of panels and boards and committees to scrutinize research proposals, and choose those which seem most promising. Whether or not the members of these committees are scientists, they are functioning as a "collective" in making such decisions.

Let me make it quite clear that I am not deploring this rapid transformation of the social relations of the "academic" mode of science. It seems an inevitable consequence of the increasing sophistication of all scientific techniques, and of the increasing application of science in all other human techniques.

Some people suggest that if only we were clever enough we could find all sorts of beautiful science that could be done with elementary apparatus. Well, if they are clever enough, let us welcome their ingenuity, but I do not believe that we can now go back to sealing wax and string—or, for that matter, vacuum tubes and slide rules.

Other people are fearful that all science is becoming "industrialized" or "militarized," and that all scientists will soon be enrolled in bureaucratic hierarchies, simply doing the bidding of the powers that be. There are certainly disturbing trends in that direction, but the predominance of the organizational factors, as in the conventional "industrial" mode of research, is not a necessary consequence of collectivization as such. Many quasi-governmental scientific organizations have contrived to preserve a "quasi-academic" atmosphere, where scientists can still follow "reputational" careers within a large administrative frame. The extensive use of "peer review" procedures has shown that the collective goals and the division of effort in research can be decided within quite a large group by egalitarian, consultative procedures, without imposing a heavy managerial hand.

The Individual in Collectivized Science

The tide of collectivization is bringing change to every aspect of science. In particular, it is changing the place of the individual scientist in the enterprise as a whole. In any study of science as a profession, it is essential to take account of the widening gulf between the individualist ethos of science and its collectivist practice. The academic mode of science maintains a delicate balance between the extreme individualism of scientists as would-be discoverers, and their commitment to the common cause of scientific progress. This balance is likely to be disturbed by a process which is changing the role of the individual scientist in the actual work of making and validating discoveries. The academic ethos itself then comes into question by those who are called on to live by it but cannot see themselves as receiving its promised rewards.

This is a very complex subject, where it is easy to appear simply as the defender of an established tradition. It is certainly a subject where every assertion must be qualified by reference to numerous exceptions. For example, there are fields such as pure mathematics where the place of the individual researcher is as secure as ever. There are major scientific organizations, in some countries, which have protected the individualism of their members whilst keeping up high standards of instrumental technique in their research. The "reputational" structure of the scientific community also appears quite sound. The social machinery for recognizing and rewarding outstanding scientific achievement is still in full working order, and the persons whom it honors are probably as worthy of such recognition as they ever were.

Nevertheless, profound changes are taking place in the professional role of the ordinary scientist in an academic or quasi-academic institution. Some of these changes are quite concrete. Scientists nowadays have less autonomy in the choice of their research problems and less opportunity to show that they can solve such problems on their own. They can no longer make independent careers for themselves simply by competing for recognition as specialists in particular fields of knowledge, but must demonstrate their expertise as members of research teams covering several disciplines. They must work within a more organized social framework, where they must either subordinate themselves to managers or else become managers themselves.

These changes in the material and social conditions of scientific life are amply documented, and are not in dispute. But they need to be considered in relation to the academic mode of research, which is still widely regarded as the best way of motivating scientists and ensuring scientific progress. Is it possible, perhaps, that this mode of research is no longer compatible with the realities of scientific work? Is it being gradually superseded by a new ethos and a new set of social relations? These are the questions, lying across the indefinable boundary between sociology and social psychology, which much be addressed in any serious discussion of contemporary science as a profession.

Autonomy in Problem Choice

Research is not aimless tourism through the world of nature: it is purposive action (Ziman 1981). It begins with the recognition of a question to be answered, or a problem to be solved. How is electricity conducted through pure silicon? Is there any way of reducing the dispersion of velocities in a packet of protons being accelerated in a synchrotron? This question may originally have been formulated very vaguely, but it is what eventually gives meaning to any scientific investigation. The very first step—the recognition and formulation of a problem which might yield to determined study—is usually the step that really counts. Anyone could have thought of possible evidence for or against continental drift: it was the genius of Alfred Wegener to see that this was a genuine problem, worthy of sustained attack.

The first principle of academic science is, therefore, *autonomy in problem choice* (Hagstrom 1965). Strangely enough, apart from the work of Gieryn (1978, 1979), there has been little systematic study of personal patterns of problem choice in science. Yet this is the key factor in making a successful reputational career. It is a commonplace of academic life that the professional standing of a scientist is assessed on the assumption that he or she undertook each particular investigation as a free agent, and is to be credited with having seen that there was a soluble problem, as well as with having solved it.

Needless to say, this assumption is very far from reality. Scientists never were as free as all that. They have always been severely limited by their training and experience, by the resources available to them, by the authority of their professors, and by the formal conditions of their appointment. They are not to be blamed for going on for years in the same narrow field, using the skills they have acquired, and the apparatus they have accumulated, to tackle the little problems that are thought to be soluble by the leading authorities in their subject. This is what Thomas Kuhn called "normal science," not to disparage it but to indicate that this is how most research is really done. In fact, part of the skill in choosing a research problem is to be aware of ones own likely limitations in the task of solving it (Reuter et al. 1978).

Nevertheless, even within such limits, the ability to choose worthwhile problems is easily recognized as a major component of scientific talent. Academic science not only rewards achievement: it also offers the opportunity for future achievement to those who have shown "promise" in just this way. The length of a scientist's list of publications is not the only criterion of professional competence (Cole & Cole 1973). For promotion to the higher strata of the profession, much more account is taken of the *quality* of these contributions, judged as much by the imagination prompting the questions as by the technical competence with which an answer is sought.

Collectivization puts further severe constraints on problem choice. At the worst, it takes this choice entirely out of the hands of the individual. For example, in the last few years, many research council establishments in Britain have had to take on research projects of a fairly routine character commissioned by government departments. Scientists involved in such projects fear that this will seriously prejudice their chances of promotion, which are still largely based on the assessment of their published papers by external academic authorities. They may have carried out this research with immense skill and devotion, but it was not "their own work" in the traditional academic sense (Ziman, 1987).

Even in academia, where many scientists are still free to formulate their own research projects, this autonomy is seriously limited to the need to win the approval of a panel of reviewers. The academic ethos allows some credit for the *formulation* of an original question, but the real prize has always gone to the scientist who is prepared to take on this question and demonstrate in action that it can be answered. Autonomy in "problem choice" extends beyond freedom to point out the possibilities of an unconventional research project: it must include the opportunity to stake ones own professional career on getting somewhere with it, regardless of the opinions of ones scientific colleagues or competitors. This is exemplified by the recognition eventually given to some of the pioneers of molecular biology, such as Max Perutz, who labored for twenty years to determine the structure of a protein by X-ray crystallography. The experts insisted that this project was hopeless, but he persevered to outstanding success.

The "peer review" system is designed to protect the professional independence of academic scientists, but it fails on this vital point. The full range of individual talent in research cannot be demonstrated within a system where every research project has to be "proposed" in formal terms—almost up the point of stating what the results will be—and approved by a collectivity of other scientists before it can be undertaken (Ziman 1983). Personal success in winning grants is certainly indicative of professional talent of a sort, but not necessarily of the sort that is beneficial to the advancement of knowledge.

Serendipity and Curiosity

The phenomenon of "serendipity" is an unexplicated anomaly in the philosophy and sociology of science. The historical record demonstrates conclusively that science owes a great deal to apparently chance discoveries, but these are either attributed to the hidden influence of some social or intellectual trend of which the actors are unaware, or else treated as singular personal events which cannot be fitted into a general pattern.

Scientists themselves, naturally, offer individualist explanations of such events (Barber & Fox 1958). They say, in effect, that such discoveries come to people whose minds are prepared to receive them—which is a truism—and that scientists ought to be prepared to make such discoveries—which is a platitude. In my opinion, there is not much more that can be said about this from a purely psychological point of view. Some people are very much more alert to an anomaly, more observant of an unusual detail, more curious about the cause of an unexpected event, than the bulk of humanity, and such people often make splendid scientists. But this may well be a deep trait of personality, either inborn or imprinted by early childhood experience, and although it can be disciplined and exercised by an imaginative scientific education, it is very doubtful whether it can be generated by formal training.

The important point about a serendipitous discovery is not, so to speak, how it was originally conceived, but how it managed to be implanted and nourished until it could be brought to birth. The intimation of a discovery—the notion that there might be something worth looking into—is fruitless unless the thought can be followed up and confirmed by systematic research. Jocelyn Bell and Anthony Hewish would not have discovered pulsars if they had not been in a position to turn their attentions to the strange signals that they were receiving, and convince themselves that these were really coming from outer space. A serendipitous discovery owes as much to the working conditions in which it is made as it does to the worker who makes it. If those conditions do not permit a deviation from the original plan of the research, then the anomalous event or unusual observation cannot be explored and the opportunity is lost (Ziman, 1981).

Serendipity plays a very important part in science as a profession. It is a notorious feature of the academic mode of research that it does not distinguish between good luck and good work in those whom it rewards. It is the quality of the discovery that counts, not the quality of the discoverer. This may be unfair, but it does encourage individual scientists to take full advantage of any serendipitous opportunities that may occur to them. In any case, freedom to exploit the opportunities presented by fortune is the essential condition for showing that good luck in science is not just a matter of luck! The power of the outstanding individual to impose his or her will on highly uncertain material is one of the key myths of the scientific vocation.

The general effect of collectivization is to reduce this freedom. It is hard enough to decide for oneself to deviate from a carefully planned research project in order to follow up some little point that does not seem strictly relevant to the main problem. This psychological barrier is strongly reinforced if the project is one that has been approved in detail by a grant-awarding agency, or by higher management in the research organization. The deviation has to be justified to others, before the idea behind it has any real substance. It may be necessary, for example, to apply for a specific grant simply to undertake further investigation of a mere hunch—a project that is often difficult to justify to a critical panel of expert reviewers.

The discovery of pulsars again illustrates this point. Bell and Hewish had to do months of work before they could even persuade themselves that they had hit upon something of real scientific interest. Would a typical peer review panel have given them a grant to do this work? Would a typical academic scientist, in Hewish's place, have gone to all the trouble to apply for such a grant, when the most likely explanation was that Jocelyn was accidentally picking up the signals from somebody's electric clock? The likelihood is that this particular discovery would not have been made under these conditions.

No doubt the same discovery would eventually have been made by somebody else, as a result of more systematic observations within a more explicit organizational frame. From a strictly epistemological point of view, this would not make any difference: the social context in which a discovery claim is made is held to be irrelevant, provided that it is satisfactorily validated. The same attitude seems to be taken in the sociology of knowledge (Branigan 1981), where the interpretation of such an event as a "discovery" is regarded as a construction put upon it subsequently by the scientific community. But for the sociology of science as a profession, the precise manner in which serendipitous discoveries actually occur—or, in many cases, are inhibited by organizational influences—can be a very sensitive indicator of structural and functional change.

Team Research

Every graduate student quickly learns the basic principle of academic life: "The only scientific work that you can count is your own work that you publish in your own name"—more crudely "Publish or Perish." What does it say in the job advertisements: "Applicants should submit a curriculum vitae, including a full list of publications." How are promotions decided: external referees are invited to comment on the publications list of the candidates. Who gets elected to a National Academy or wins a Nobel Prize: the author of the paper originally claiming an important scientific discovery.

This is the mechanism that makes academic science tick. Some sociologists of science (Hagstrom 1965) have suggested that it is essentially a gift exchange ritual of the kind studied by social anthropologists. To win "recognition" in the scientific community, scientists go around giving away, not pigs, not stone axes, but "contributions" to the scientific literature. This comparison does not take account of other, more consciously rational aspects of the process by which scientific knowledge is produced, but it does not exaggerate the importance to the individual scientist of getting into print (Meadows 1974).

The academic mode of research equates personal scientific *achievement* with public scientific *authorship*. A list of publications is deemed to be *objective* evidence of scientific competence, not tainted by the personality or social standing of its author. This principle is reasonable enough when most of the papers are by a single author—but what should be done when the scientist in question is only one out of the many names listed as the "authors" of a scientific publication?

Academic science has always had its difficulties with multi-authored papers. For example, there is still no settled convention on the order in which the names of the authors should appear, since the first name on the paper gets the significant "reputational" advantage of appearing in the citation index. The problem of the senior professor who insists on having his name on all the papers of his junior colleagues is notorious. For this reason, academic science has always had to give a little weight to information obtained through informal, confidential channels of communication, in order to assess the actual contributions of individual scientists to the published literature.

But the whole system is breaking down in fields such as experimental high energy physics, where the heading of a single paper might list the names of a hundred professional scientists, from graduate students to professors, not to mention another hundred technical workers whose assistance is "acknowledged" in a footnote at the end. This situation is becoming more and more common. It is an inevitable consequence of the process of collectivization, which absolutely compels scientists to collaborate. Nowadays, in many fields, research can only be done in large teams, involving many different tasks, and many different skills. The paper reporting the results of such research cannot be broken down into a number of separate items, each attributable to a single member of the team; indeed, if it really was a team effort, then it is only fair that the names of all members of the team should appear as its author.

As a consequence, the published work of a mature scientist of the highest calibre—a candidate for a professorial chair, for example—may be contained in a dozen or so papers in each of which he or she has not more than a fiftieth share of nominal authorship. This is quite insufficient information on which to base an objective assessment of a person's scientific ability. This assessment then comes to depend mainly on informal, semi-public knowledge—that is, hearsay—concerning the candidate's "reputation" in the field, or on unattributable, undisclosed information from more senior colleagues concerning the part that he or she has actually played in various projects, such as the contribution of key ideas, determination and energy in pushing through to a conclusion, capacity for working with others, and so on.

This is not to suggest that the opinions of other scientists about the personal scientific achievements or scientific abilities of a particular colleague are necessarily less reliable than the evidence that can be got from reading the candidate's original papers. In any case, this evidence seldom speaks for itself, and has always had to be expertly interpreted and evaluated by those same other scientists. It is obvious, nevertheless, that appointments or prizes awarded without reference to a public archive of certified scientific authorship will be much more sensitive to organizational influences than is allowed for in the academic ethos. This is clearly a significant change in the sociology of the scientific profession, since it means, in effect, that the traditional distinction between the "reputational" and "organizational" aspects of a scientific career can no longer be sustained.

From the Research Bench to the Office Desk

The collectivisation of academic science is a more complex and subtle process than mere "industrialization" or "bureaucratization." These terms do not adequately cover the peculiar practices and protocols that have evolved around the peer review system, the "request for proposals" procedure adopted by some grant-awarding agencies, the quasi-egalitarian structure of an international "collaboration" in elementary particle physics, or, indeed, the wide variation of managerial structures to be found in academic research laboratories (Shinn 1982).

Nevertheless, a collectivized activity has to be collectively organized. At certain points in the system, people, apparatus, and other resources have to be brought together on a large scale, and directed towards fruitful problems. In modern science this task has become so heavy that it can no longer be performed by a few devoted secretaries and professional administrators under the benevolent eye of a senior scientist who is still active in research. The role of the scientific *leader* has been enlarged and transformed into that of the research *manager*. The work of many of their scientific colleagues has similarly been transformed into the *administration* of various aspects of the scientific enterprise.

This movement from the research bench to the office desk is not envisaged in the academic ethos. Of course, there is a manifest contradiction in this ethos between the norm of "universalism" and the recognition of individual achievement. The notorious stratification of authority within the scientific community has always been a source of personal anguish (Weber 1918). The power of the senior professor over the research assistant was petty, but it was sustained by patronage in jobs and promotion, and often exercised in a spirit of extreme individualism. Sociologically speaking, however, there was no such thing as an "organizational" career in science, except as an adjunct to a "reputational" career. Any move into a seat of institutional power in academia had to be justified as a form of "recognition," awarded to individuals on the basis of their achievements in research, with little regard to their competence in human affairs. By the same tokens, a scientist who moved into a relatively subordinate position

within the administrative apparatus of a university or research institution
was regarded as having failed in research, however demanding and useful
this post might be. The amateurish way in which most scientific institutions were run could even be considered a perverse source of pride, indicating adherence to the principle of academic individualism and professional autonomy.

The organizational needs of collectivized science can no longer be met
in this way. The people who run our universities, our research establishments, our industrial laboratories, and our big research teams have to
be chosen primarily for their managerial abilities rather than for their scientific prowess, and they are supported by an extensive apparatus of expert administrators with scientific experience. A distinct profession of "research management" is growing up, as a stratum of influence and
authority over scientific work. The leading members of this profession are
not necessarily members of the traditional academic elite—in Britain, for
example, they are not all drawn from, nor do they always become, Fellows of the Royal Society—but they clearly hold the reins of financial and
organizational power in the scientific community.

Again, this must be accepted as just another aspect of the general process of collectivization, but it subtly changes the professional standards
and goals of scientists away from excellence in the *performance* of research towards excellence in the *organization* of research. This is not to
say that scientists are all ceasing to be "inner directed," and are becoming
"organization" men and women. Even within "industrial" science there is
no lack of opportunity to show individualist traits in search of PLACE.
Indeed, modern scientific life exemplifies a stage in the transition from a
paternalistic or oligarchic structure to a more formal hierarchical structure,
where the competitive individualism enjoined by a charismatic ideology is
not quenched, but is transferred to new objectives and spheres of influence.

Many scientists in academic and quasi-academic institutions are thus
faced with a dilemma: should they strive to acquire "reputational" credit
by concentrating on work that will add to their list of published papers, or
should they deliberately seek administrative and managerial responsibilities in the hope of winning promotion in that direction? This is a serious
issue in many research organizations, ranging from universities to industrial laboratories, where there is an uneasy balance between the traditional
"academic" criteria for career advancement and the obvious need to give
the power to people with managerial skills (Ziman, 1987).

Specialists and Generalists

The academic mode of science owes a great deal of its effectiveness to the way in which the labor of research is divided and subdivided into innumerable "specialties" (Ziman 1987). This is not a planned process: it occurs through the complex interaction of uncoordinated influences and individual choices by which scientists enter specific disciplines, find themselves committed to specific research programs, and end up as *specialists* in particular problem areas.

The conventional scorn for the scientific specialist who is said to "know more and more about less and less, until he knows everything about nothing" misses an extremely important sociological point about science as a profession. It is not so much that the academic mode of research encourages and sustains accuracy in detail, which is one of the features that makes science as a whole so effective. It is that the extreme specialization of their interests and functions gives personal status to each scientific worker as an individual, and permits the wide distribution of CUDOS over a large community. Each "invisible college" or other specialty grouping becomes, so to speak, a village, where even a person of modest competence is somebody. In this respect, academic science has the feeling of being a "cottage industry" rather than a "factory" for the manufacture of knowledge.

The division of labor in a "collectivized" research project has to be more explicit. The members of a multidisciplinary research team must be specialists, indeed, but usually in the techniques of their respective disciplines rather than in the problems. It is difficult to arrange a project where one's expertise in a closely-defined "subject" will be exercised to the best advantage, and it may be necessary to learn a good deal about the work of one's colleagues if one is to collaborate harmoniously with them. In other words, the tendency of collectivization is to force scientists into more "generalist" roles, both technical and administrative. Instead of devoting a lifetime to the investigation of one very narrow field, they have to become professional "problem solvers" and "trouble shooters" over a considerable area of science and technology (Ziman, 1987).

This is probably all to the good. There are pathological aspects of the traditional academic career, such as intellectual staleness, and "undue persistence" in the pursuit of outmoded problems (Chubin & Connolly 1982). Scientists are well advised to diversify their interests and skills, to avoid these traps. Both science and society need people with a broader understanding of science, and a broader vision of what it might do. But the academic mode of research has no mechanism for encouraging such diversification, and the academic ideology does not allot any CUDOS to the "generalist." There has long been unease about this amongst thoughtful scientists: it is a worthy topic for the sociology of the scientific profession.

Ideology and Vocation

The overt competitive individualism of scientists is a professional characteristic which seems inconsistent with their ethos. Nobody could say, for example, that Jim Watson's account of how he and Francis Crick won their Nobel Prize was in accordance with the norm of "disinterestedness" (not to say "humility")—yet nobody now doubts that this was a genuine slice of scientific life and work, where the stereotype of the "honest seeker after truth" barely concealed the reality of the unscrupulous egotist in pursuit of fame.

But from a psychological point of view we are all incurable egotists and individualists. Looking at it more sociologically, we would see the stereotype of the "honest seeker after truth" simply as the role model demanded by the academic mode of science, and note that this social institution seems to go on working as a whole, despite the fact that its members often deviate widely from their assigned roles. The CUDOS ethos can then be interpreted as a means of keeping such deviations under control: in other words, it functions as an "internal ideology" of the scientific community.

The thrust of this paper is that this ideology is being demystified, and is losing its rhetorical grip. This is very obvious in its "external" aspects, where it no longer protects the scientific profession from the intervention of the state, and other societal forces. It is no longer plausible to insist that every scientist has a perfectly disinterested, perfectly rational, "scientific attitude" by which the problems of the world could be solved. The best we can do is to admit that we are very fallible and very much on the make as individuals, but that our combined efforts can produce a great deal of quite reliable information of very considerable utility. We appear in the national economic statistics as "QSE's"—Qualified Scientists and Engineers"—and are relieved to hear that our professional trade union has won us an increase of 5%, together with an extra week's annual vacation, in the current round of pay talks. "Research is a job," we say, "like any other."

But these and other effects of collectivization are weakening the academic ethos as an internal ideology of the scientific profession. For all its pious fictions, this did instill in young scientists a certain sense of reverence for their work, representing it to them as a morally worthy vocation. As my current research on specialization and change in scientific careers is showing (Ziman, 1987), this mystique is certainly not dead. Nevertheless, it is clearly under attack. One can see this, for example, in the way that many of the older generation of scientists have participated in the popular movements against nuclear weapons: they had always felt that science is an essentially benevolent undertaking, and they have devoted their lives to an activity that can be harnessed to such obscene purposes.

This disenchantment is a slow process, for individuals and for communities. The process of collectivization, which I have described so simplistically, is also a slow process, and is going on at different rates, and in different directions, in different countries and disciplines. A deeply entrenched social institution with a charismatic tradition can go on for centuries resisting formal change—especially if, as always in Italy, there is informal adaptation by individuals to external necessities (Statera 1984). Nevertheless, I believe that the nature of science really has changed in the past forty years, as a social institution and as a profession.

The primary effect of the traditional scientific ethos has always been to reinforce the individualism of the academic mode of research. As that mode declines under the influence of collectivization, a different ethos will inevitably take its place—an ethos that puts much more emphasis on the participation of each individual in the activities and objectives of organized groups. The obvious candidate for such a collectivized internal ideology of science is what I have called PLACE. But we can all think of reasons why that mode of organization, which is entirely appropriate for the instrumental aspect of science, could not cope with the epistemological challenge of the "discovery" process in non-utilitarian fields. I therefore share the reluctance of some of my colleagues (Lemaine 1984; Broesterhuizen & Rip 1984) to believe that the "industrial" mode of research will simply take over all science in a bureaucratic wave.

References

Barber, B. & Fox, R.C. *Am. J. of Sociol.*, **64**, 128–36 (1958)
Branigan, A., *The Social Basis of Scientific Discoveries* (Cambridge: University Press: 1981)
Broesterhuizen, E. & Rip, A., *EASST Newsletter* 3 (3) (1984)
Chubin, D.E. & Connolly, T., "Research trails and science policies," pp. 293–311 in Elias, Martins & Whitley (1982)
Cole, J.R. & Cole S., *Social Stratification in Science* (Chicago: University of Chicago Press: 1973)
Cotgrove, S. & Box, S., *Science, Industry and Society* (London: George Allen & Unwin: 1970)

Elias, N., Martins, H. & Whitley, R., (eds) *Scientific Establishments and Hierarchies* (Dordrecht: Reidel: 1982)

Gieryn, T.F. *Sociological Inquiry* **48**, 96–115 (1978)

Gieryn, T.F. "Patterns in the selection of problems for scientific research: American astronomers 1970–5" (unpublished Ph.D. dissertation: Columbia University: 1979)

Hagstrom, W.O., *The Scientific Community* (New York: Basic Books: 1965)

Hjern, B. & Porter, D.O., "Implementation structures: a new unit of administrative analysis," pp. 265–280 in Holzner, Knorr & Strasser (1983)

Holzner, B., Knorr, K.D. & Strasser, H. (eds) *Realizing Social Science Knowledge* (Vienna: Physica-verlag: 1983)

Lemaine, G., *EASST Newsletter* 3 (2) (1984)

Meadows, A.J., *Communication in Science* (London: Butterworth: 1974)

Merton, R.K., *The Sociology of Science* (Chicago: University of Chicago Press: 1973)

Reuter, H., Tripier, F., Hubert, F. & Lahon, D., *Le Travail de Recherche dans l'Universite* (Rouen: Centre de Documentation et de Recherche en Sciences Sociales: 1978)

Shinn, T., "Scientific disciplines and organized specificity: the social and cognitive configurations of laboratory organizations," pp. 293–312 in Elias, Martins & Whitley (1982)

Statera, G., *EASST Newsletter* 3 (2), (1984)

Weber, M., "Science as a Vocation", (1918) reprinted in Gerth, H.H., Mills, C.W. (eds) *From Max Weber* (London: Ronteledge & Kegan Paul: 1948) pp. 129–156.

Whitley, R., *The Intellectual and Social Organization of the Sciences* (Oxford: Clarendon Press: 1984)

Ziman, J.M., *Minerva* 19, 1–42, (1981)

Ziman, J.M., *Proc. Roy. Soc.* B219, 1–19 (1983)

Ziman, J.M., *An Introduction to Science Studies* (Cambridge: Cambridge University Press: 1984)

Ziman, J.M., *Knowing Everything about Nothing: Specialization and Change in Scientific Careers* (Cambridge: Cambridge University Press: 1987)

On Being a Physicist

The Country Called Physics

When I am asked nowadays to define myself academically, I sometimes say that I am a "lapsed physicist." Amongst non-physicists, this usually produces a nervous laugh: but to physicists it may not seem quite such a joke and they may feel uneasy at being addressed by a self-confessed "defector" from their community. They know that "being a physicist" is not just a way of earning a living; it is a deeply rooted mode of personal existence. Even now, after working for several years in an entirely different academic Department, engaged in entirely different scholarly pursuits, I still feel great nostalgia for the time when I was active in physics research, and know that this self-identity cannot be thrown off like an old overcoat just because I now want to parade around in a new set of intellectual clothes. Even if some of you think of me as some sort of traitor, at least I still appreciate the virtues and delights of the land in which I lived happily for so many years.

The geographical image is apt. A major scholarly discipline such as physics is rather like a great country, inhabited by large numbers of people sharing a common language and a common culture. Few of us—Albert Einstein was a notable exception—could be said to have been *born* physicists, in the way that we might have been born Spanish, or Belgian, or Polish. We all remember how long it took to learn to "think physically"—how many years of puzzling out the meaning of the Cornu spiral, or trying to solve ghastly problems about discs of resistivity r rotating with a constant angular velocity w about an axis making an angle A with a magnetic field H, or determining the mechanical equivalent of heat by rubbing two sticks together in some ingenious way? But in the end, it became second nature for us to talk, and think, and act as physicists, so that now we are impatient with our students for making such heavy weather of these perfectly obvious things, just as adults become very impatient with their children for not knowing yet how to behave as good Spaniards, or Belgians, or Poles.

What sort of a country is physics? What is the nature of this culture which we have taken so much trouble to acquire? There is more to being a physicist than understanding the Second Law of Thermodynamics, or knowing how to calculate the resistivity of copper, or being able to set up a laser. Physicists differ from other people—even from other scientists—in their approach to the world. The visible culture of physics—its elaborate instruments, its superabundant data, its intricate theories and its awe-inspiring technical capabilities—springs from a particular attitude of mind which all physicists have to learn, and which they all share. There is something about the notion of "being a physicist" which goes beyond an account of the contents of physics as a science, and a chronicle of its contributions to technology.

What I have to say on this topic can only be very personal and subjective. Indeed, the great majority of physicists would probably say that it is not the sort of topic on which anything useful can be said at all, because it is so vague and ill-defined. But it is sometimes quite instructive to muse on such matters without pretending that one's opinions can be made precise, and formulated quantitatively, and represented mathematically, and tested experimentally, and thereby falsified, or corroborated, in the correct scientific manner.

Physics as a Subject

The land of physics surely covers a much wider territory than the subjects that are usually taught in university courses under that heading. Such courses only touch upon a large proportion of the topics and themes that physicists actually work on in their careers. Atmospheric physicists, say, or medical physicists, could explain the direct connections between their work and their original training in the basic principles of physics, but this work would have spread into fields of research and realms of nature very far from those considered in the undergraduate lecture room or teaching laboratory. On the other hand, many graduates in physics eventually move into other professional domains, such as engineering, or technical management, where they would be the first to admit that they were no longer "physicists" in the sense that we are considering. This is not a process to deplore or discourage. It is surely beneficial to physics as an academic discipline to be able to point to a wide range of employment opportunities for physics graduates. This vocational mobility also spread some of the elements of the culture of physics into other intellectual and practical domains, where they may make such distinctive contributions as the wartime development of operations research. By the same tokens, we must surely welcome migrants form other disciplines into the land of physics, and benefit from their experience and insights. The frontiers between adjacent departments or faculties on the official academic maps are not to be seen on the ground, and have about as little meaning as the frontiers between the colonies into which Africa was divided in the Nineteenth Century.

For this reason, any attempt to draw demarcation lines between physics and other disciplines is bound to be arbitrary. It makes more sense to concentrate, instead, on the "heart land" of the subject, and to consider whether the only way to "be a physicist" in the fullest sense is to be doing research on one of its traditional themes. Many people simply define all these themes as "the discovery of the ultimate laws of nature," and use this as their criterion for deciding whether a particular activity is or is not "physics." It is certainly true that the investigation of such mysteries as the primary constitution of matter, the interactions between these constituents, and their role in the overall structure of the universe, remains a central concern of physics and of physicists. But the claim that this investigation is the only "real" physics (all the rest being, presumably, in Rutherford's scornful phrase "stamp collecting") is a pretension that would not be accepted by about three out of four physicists. None of us can doubt the excitement of the process of unravelling these deeper and deeper secrets, or the extraordinary beauty of the concepts in which the answers to some of these questions are now presented. But to say that "being a physicist" consists simply in taking part in that vast undertaking, or having the wit to appreciate what has been discovered by the particle physicists, the astrophysicists, and their attendant theoreticians would turn this essay into yet another review of recent progress in these fields, which would be far beyond anything I could personally undertake.

Maps of Reality

The essence of physics is that it is the science in which every endeavor is made to represent nature in mathematical terms. This may sound like a (lapsed) theoretician's way of looking at the subject, but I think it can be justified as a definition of the whole discipline. The primary objective of all experimental and observational research in physics is to produce numerical data which can eventually be fitted together into mathematical formulae. The other major scientific disciplines are characterized primarily by their subject matter. Geology, for example, can be defined as the study of the solid substance of the Earth, and biology as the study of living organisms. In the corresponding branches of physics—that is, geophysics and biophysics—research is similarly concentrated on specific objects and directed to the discovery of relevant types of mechanism, or structure, or explanation. But such research is not called "physics" simply because observations are made with "physical" apparatus or interpreted by reference to "physical" laws: the essential point is that quantitative information is being sought by these means, with the intention of formulating mathematically precise theoretical concepts about this particular aspect of nature. To "be a physicist" is to exercise such methods and concepts in a scientific or technological context.

Now it has to be admitted—indeed, 90% of people freely admit it, without embarrassment—that mathematics is not a common mode of human thought. In natural languages, most concepts are vague, laden with values, and even self-contradictory. In the everyday life-world, precise prediction of the outcome of events is impossible or undesirable, and we struggle to keep our footing in the stream of events and action in which we are immersed. By systematic exercise of our powers of observation and pattern discrimination, by social communication and mutual criticism, by solving various practical problems by sheer trial and error, humanity has built up its knowledge of the world, and its skills in overcoming many obstacles to comfort and life. A very large part of this theoretical knowledge and practical skill is now embodied in very accurate numerical data, meticulously draughted diagrams, rigorous symbolic formulae, and geometrically precise models—that is, in essentially mathematical terms. But most of the time, most people get on very well without having to make more formal calculations than, say, working out from a map how long it will take to make a journey, reading the bus timetable for a suitable departure, and adding up the change for the fare.

I have deliberately chosen this trivial example of the use of mathematical reasoning in everyday life, because what physics does is to generate metaphorical "maps" of a great many other aspects of the world and what physicists can do, above all, is to "read" these "maps." This is obviously a very abstract way of talking about an immense body of factual and theoretical information, but it seems to me the most appropriate metaphor for the knowledge generated by a rigorous science such as physics. It is not so concrete and mechanical as the term "model" which the philosophers of science have taken over from physics and allowed to be applied to almost any conceptual scheme, however loose and implausible. I personally find the "map" analogy more instructive. Let me try and explain it in more detail.

The first thing that has to be done in establishing a body of scientific knowledge is to define all the terms very precisely. As every student discovers, all sorts of loose ways of talking have to be tightened up, and ambiguities removed by careful redefinition of all the terms. Thus, for example, the first step in all the biological sciences is to discriminate between various natural species, to describe them precisely, and name and classify them systematically. But physics goes beyond verbal or pictorial descriptions: it only begins seriously when the relationships between various features of the natural world can be measured quantitatively, and the data summed up in algebraic formulae or in geometrical figures. The history of physics began with Archimedes' formulae for densities and Ptolemy's epicycles and came to maturity with Kepler's ellipses and Boyle's Law. Since that time, the prime task of every physicist has been to represent what he discovers about the world in the unambiguous language of number and geometrical form, and to use all his genius of experimentation and theoretical argument to extend and connect these representations as widely and precisely as possible. In other words, he is engaged in "exploring," "surveying," and "mapping" a particular aspect of the world about him. The theoretical formulae in which the final results of physics research are embodied are just like maps in that they summarize very succinctly the inferred relationships between a vast quantity of past observations. But these formulae can also be "read" for their detailed predictions of the outcome of future observations, just as a map can be "read" for its predictions of what will be encountered on a journey that we are about to take for the first time. Whether or not these equations also constitute a "model" of the real world in a more concrete sense is largely a matter of ones personal philosophy, and need not affect their use in physics.

Learning to Think Physically

"Map reading"—that is, relating a symbolic structure to the "reality" it represents—is a very common human activity. Every day, lots of people take out their road maps, locate their present position, work out their route, decide that they will have to turn at a certain intersection, and identify it by a landmark such as church or a post office shown by its conventional symbol on the map. Most people can learn to do this sort of thing reasonably well, although it is certainly a skill that improves with practice. For some professions it is developed to a very high order of expertise. An experienced engineer or architect can "read" a blueprint as if it were a photograph of an actual machine or building, and distinguish subtle features of construction or design which would scarcely be obvious if one had the object itself before ones eyes. That is to say, they can manipulate a symbolic "map" in their own minds, and discover in it characteristics of the real world that had not previously been noted. This is the sort of skill that physicists have to acquire, in relation to their sorts of "maps," in order to "think physically" about natural phenomena and human artifacts.

In most practical professions, and in most of the sciences, the "maps" are fairly close to "pictures." Although they are abstract and symbolic, they still convey the spatial topology of the object or landscape that they represent. One can imagine oneself actually standing at some point in the diagram, and checking that the staircase does not obscure the window or that the head of the bolt holding the pinion in position will not interfere with the motion of the lever. A stereochemical "model" of the molecule can be built and manipulated under our hands; the sequence of codons on a gene simply follows the order in which they would be encountered if one could walk along an even larger model of a DNA molecule.

Many of the "maps" we have to learn to use in physics are similarly spatial, even though the features they emphasize may not be directly visible. Geometrical optics, for example, has this characteristic, even though the ray paths to be traced out on the diagram are mere abstractions whose properties have to be inferred by reference to trigonometric formulae. The conceptual frameworks of classical dynamics and electromagnetism are also set up in the everyday world of three dimensional space, although the equations of Newton and Maxwell have many unexpected properties, and there are many non-intuitive steps of argument to understand and accept before one can say that one truly thinks with them. In fact, the engineer who has had to grapple for years with the disturbing effects of gyroscopic forces, or has learnt to harness the linkages between the fields in an electromagnetic wave, may be better at manipulating such systems in imagination than the average physicist who has not specialized in that sort of thinking.

The real problem of getting to "think physically" comes when the abstractions to be manipulated have no counterparts in life-world structures. This is very evident in the central theme of education in physics—the concept of energy. Although the word "energy" is quite widely used in relation to many different familiar aspects of life—health, food, exercise, warmth, fuel, etc. —its meaning in physics is highly abstracted and not at all consistent with any one of these everyday usages. It requires a real effort of imagination to grasp that this same entity is present in a flying bullet, a car battery, a warm bath, and a strong magnetic field. School children can easily learn how to manipulate the symbols for energy, heat, work, etc., in the abstract thermodynamic domain—for that is just algebra, or calculus—but recent educational research has shown that they find it extraordinarily difficult to set up these equations in the first place. Given a familiar everyday situation, such as car running out of fuel, or a child eating an ice cream, they cannot easily translate it into the language of energy transformation, and explain what is happening in thermodynamic terms.

This is not just an educational problem. The hardest part in "thinking physically" is "mapping" or "modeling" the situation in the domain of everyday things into the simplified, abstracted, formalized concepts of the domain of physics. Of course every physicist needs to have a sufficient grasp of mathematics to appreciate the significance of the symbolic relations in this domain, and to have some idea of how they can give rise to new results which can be interpreted back into the real world. But sheer mathematical ability—that is, skill at manipulations within a particular formalism or a conceptual structure—is not usually the narrowest bottle neck along the path to a solution of a problem in the physical sciences. In fact, as is well known among physicists, the best of theoretical physicists are not necessarily superb mathematicians, whilst very good mathematicians often fail as theoretical physicists because they are trapped in the formal domain and never relate their work to the world that it is designed to mirror.

This is not the place for a debate on whether the world is essentially "mathematical" in its underlying nature, or on whether mathematics itself already exists to be discovered, or is merely constructed as we go along. Many people outside physics, including many philosophers of science, are so impressed by the beautiful abstract maps of "reality" generated by physics that they tend to think of them as being produced by mental activity alone, and greatly underestimate the labor that has gone into exploring the "reality" they are designed to represent. They fail to appreciate the practicality of physical thinking, and its essential linkages with the world of things and events. The physicists' notion of "being a physicist" would encompass an even balance between formal manipulations in the "physical" domain and manipulations of the corresponding objects in the material world. It would encompass the ability to set up an abstract representation of a material system, and the ability to design and construct real objects—that is, experimental apparatus that would perform according to the abstract formulae. Within the community of physics, the role of the experimenter and instrument maker is just as important as that of the theoretician. The philosophers and sociologists of science have been misled by the division of labor between those who work mainly with material apparatus, in the real world, and those whose specialized work is with symbolic apparatus in the mathematical world. We should explain to them that theoretical physicists do not constitute some sort of upper class in the physics community, but are themselves slaves to the facts uncovered by the genius of the experimentalists.

In making this point, I have in mind the career and talents of the late Dr. C.R. Burch, who was a colleague of Cecil Powell at Bristol for more than 30 years. Bill Burch was, one might say, primarily an inventor, for his work eventually produced useful devices—a reflecting microscope, a machine for separating tin from low-grade ores, a very hot flame from ordinary domestic coal gas—rather than new physical laws or data. But he was also a first class physicist, because he knew how to take a physical system and interpret its behavior, in the abstract domain of geometrical optics, or hydrodynamics, or thermodynamics. There he would worry away at it until he had got out of it the results he needed and could realize them in material form. Many of the most creative engineers have this same capacity, but my impression is that their imaginative domain is closer to the life-world, and less concerned with purely symbolic entities, than the world of physics.

Hierarchies of Abstraction

The fascination of physics is its power to generate symbolic systems of greater and greater abstraction that can still be related to the real world. The concept of energy, leading into the formalisms of thermodynamics, is only the first (but by no means the easiest to grasp) of the succession of structures that are encountered on the way to "becoming a physicist." When the student comes to quantum mechanics, or relativity, a whole new layer of abstraction has to be mastered. Of course, the first stage in learning quantum theory can be made to look like a simple extension of thinking in the wave language of optics. It is when one has to accept that the waves themselves do not have the permanence and "reality" which had been given to them in optics that the difficulties arise. Philosophers of science pounce on the paradoxes associated with the wave-particle duality, and make a big thing of them epistemologically. We, as practicing physicists, may feel that these paradoxes can be kept out of the clockwork by a systematic probabilistic interpretation, but we certainly appreciate the initial difficulties of learning to think that way. It is a nice question of pedagogy whether one should acquire a vivid sense of the "reality" of wave functions, in order to manipulate them mentally—for example, to get the feel of bonding and antibonding orbitals in molecules and crystalline solids—or whether one should try from the beginning to grasp the much more abstract formalism of states and operators, which is more powerful and general, but which is very far away from our life-world concepts. And then again, many theoretical physicists would argue that the "Heisenberg-Dirac" domain does not provide sufficient insight concerning the most fundamental entities of physical reality, and that we should master the language of propagators in the "Feynman" domain in order to think clearly about quarks, say, or neutrinos.

But I would be a little cautious about the claim that in order to "be a physicist" nowadays one must master all these abstract domains, right up to the top (or is it down to the bottom?). This claim echoes the corresponding claim, which I have already tried to rebut, that the essence of physics is to discover the "fundamental" laws of nature. It may be, one day, that we shall have a complete theoretical scheme in which it will be possible to express the problem of, say, weather prediction, in terms of quantum chromodynamic interaction bundles (or whatever it will have become), and actually to operate mentally on such terms to calculate whether it will rain in Bangor on Tuesday week. But that day is far beyond Tuesday week, and even our most powerful computers have to learn to "think physically" at the much more mundane level of temperature, and pressure, and wind velocity. The essence of "being a physicist" is to be able to manipulate the concepts and relationships in the abstract domain that is most relevant, in scale and symbolic structure, to the aspect of the real world that happens to be under study. For the chemical physicist, this may be the domain of statistical mechanics, where the appropriate entities are state functions in the phase space of innumerable dimensions. For the cosmologist, it may be the four-dimensional space-time continuum, with its warped metrics and black hole singularities. For the solid state physicist, the appropriate abstract domain might be electron momentum space, with its Fermi surfaces and Brillouin zones. For physicists studying liquid crystals, the hard thinking has to be done in real space, but with subtle quasi-geometical entities such as disclinations. Every specialized field of physics develops its own special "maps," where the relations between its characteristic objects and phenomena can be represented in the clearest

form. To "be a physicist" is to have mastery of one or more of these domains, not only in being able, say, to answer examination questions on it,
or explain some of its typical phenomena by reference to a standard theoretical "map," but also in being able to work creatively in that domain,
moving easily back and forth between the abstract world of theory and the
real world of empirical facts.

The Unity of Physics

The plurality and diversity of the specialized "maps" and "models" used
by physicists is of great interest, both philosophically and practically. A
philosopher might ask, for example, whether this might be interpreted as a
genuine plurality and diversity in the nature of physical reality? Theories
that seem to say different things about the same subject matter—Newtonian dynamics and relativity, say, or the chemical and quantum-mechani-
cal representations of an atom—are said by some philosophers to be essentially "incommensurable." If, indeed, these maps are so different that
they cannot be logically related to one another, then it becomes difficult to
believe in a single underlying "physical reality" just waiting to be discovered. In other words, all our physical concepts and theoretical models are
no more than conventions that happen to work well in describing the facts
currently available.

Whatever the case in principle for such skepticism, this is certainly not
the attitude with which physicists normally undertake research. Like any
sane person, they believe firmly in the existence of an external world and
normally have very little doubt that this is faithfully represented by
"physical reality." Nevertheless thoughtful physicists are well aware that
our current conceptions of this reality have evolved historically, and cannot be absolutely "true" as they now stand. There is always an uneasy
awareness of the corrigibility of just those facts and theories on which one
would most like to rely.

My own feeling is that one of the arts of "being a physicist" is to accept the ambivalence of this situation, and to waver irresolutely between
realism and conventionalism. When working within a particular field—
space navigation, say, the physicist takes an entirely realistic attitude to
the appropriate theoretical scheme—in this case, of course, Newtonian
dynamics. But he or she would be quite ready to admit that this scheme
could only be regarded as a convention in a larger frame, where relativistic effects were significant, and would throw it away altogether if it were

envisaged that the space vehicle might be entering the gravitational field of a black hole. Sometimes, as with the shell model and liquid-drop model of a nucleus, the alternative maps look frankly contradictory, and their domains of validity must be carefully discriminated. But the "incommensurability" about which the philosophers seem so concerned is often no more serious in practice than the fact that the bus map and the subway map of a city may look entirely different, because they emphasize different features of the same basic landscape.

Does this mean that to "be a physicist" one has to positively assert the existence of a single underlying reality to which all such discussions tend to turn? Must there be a single, coherent, unified, "physical world" which we cannot yet represent unambiguously as a whole, but which will eventually be mapped in its entirely and in all detail? That may be our unstated working hypothesis, but such "metaphysical" considerations play little part in physics research, even at the most extreme levels of intellectual abstraction or of space-time extent.

Nevertheless, the theoretical unification of physics is strongly favored by one of the basic principles of scientific research. This principle lays it down that whenever the fields represented by two different "maps" overlap, every effort should be made to resolve any contradictions that appear in the regions where the maps overlay one another. Thus, for example, the problem of reconciling relativity theory with quantum mechanics has now become a practical necessity. Without such a reconciliation of concepts and formalisms, physicists will not be able to solve problems about certain types of physical system, such as black holes, where both these systems of thought have to be applied simultaneously. We need to be sure of the piecewise continuity of physical science, even though we are in no position to insist that the whole system is unitary and unambiguous in its deepest principles.

Finalization and Collectivization

The *practical* utility of physics is vividly demonstrated by the immense projects in which many different branches of physics are combined to achieve a single scientific or technological goal. There is no need here for me to wax eloquent about the particle accelerators, the electronic computers, the interplanetary space probes, and the intercontinental ballistic missile systems, which humanity has inherited from its physics community. It is impossible nowadays to "be a physicist" without being involved in the construction or use of such incredible products of our science.

Such projects, whether in basic science or in technological development, are only made possible by the theoretical maturity of the branches
of physics they involve and combine. Physics is the archetype of a "finalized" science—that is, one where there is a well-established "paradigm"
around which most of what is known can be rationally ordered. In such a
science, it is possible to direct research projects to the achievement of
specific technical objectives, rather than simply working away in the hope
of discovering something that might turn out to be useful or interesting.
Thus, for example, modern microelectronic components are not simply
"invented": they are conceived in principle, on the basis of our knowledge
of the basic physics of semiconductors, and designed from the beginning
to perform certain specific functions. A great deal of research and development may still be needed to get the idea to work in practice, but this research
itself will be guided by a rationale that goes beyond crude trial and error.

Because so many of the different branches of physics have reached a
similar phase of maturity, it is possible to combine them into such large
and complex projects as the construction and operation of a particle accelerator—with all its associated instrumentation—or, on the darker side of
physics, a cruise missile or surveillance satellite system. An extraordinary
example of a "finalized" project in physics is, of course, the endeavor to
generate electric power by nuclear fusion in a magnetically confined
plasma. This is not only a very large and expensive project: it also brings
together knowledge and specialized skills from all over physics—nuclear
physics, plasma physics, electrodynamics, chemical physics, the physics
of materials, atomic physics and optics. There is no sure proof that this
project will achieve its objective, yet the nations are prepared to spend
billions on it, over a period of something like half a century. This confidence in such an unproven concept is compelling evidence of the intellectual coherence of physics and the confidence that physicists have in the
technical unity of their subject. It is a striking illustration of the way in
which one of the most remote and abstract of human activities may eventually show itself to be one of the most practical.

The difficulty is, of course, that research projects to link specialty domains, or to explore new areas of nature require the cooperation of physicists with different specialized skills. There was a time—how long ago it
now seems—when physics was an individual vocation. For more than a
century, of course, it has been a professional activity, in that it could only
be carried out in a laboratory providing technical support services to a
number of researchers. But each project could still be the work of a single
physicist, reporting the results as the single author of a paper, or reporting

on the development of an invention to the stage where a properly engi-neered prototype might be called for. Even when a large group of re-searchers gathered around a particularly effective leader of research, such as Enrico Fermi or Ernest Rutherford, they would undertake their research projects and gain personal credit from them as independent individuals.

Nowadays, physics is very largely a *collective* enterprise, where most of the work is done in closely coordinated teams. There are still areas where the individual style is appropriate and feasible, usually under the umbrella of a well-founded institution and often under the personal influ-ence of a senior scientist. But to be a typical physicist these days is to give ones specialized skills and knowledge to a project that is larger and more complex than one could manage on ones own. This is particularly striking in the physics of technological development, where the combined efforts of hundreds of scientists may be needed to see a project to completion. But the same phenomenon is to be observed in the most basic areas of pure physics, such as high energy physics and astrophysics.

The way in which these collective enterprises are organized and man-aged is obviously of vital concern to physicists. One of the striking fea-tures of team research in high energy physics is that it is not highly organ-ized in a bureaucratic manner, but is carried out by "collaborations" drawn from many institutions and countries. It speaks volumes for the continued sense of the unity of physics as a discipline that such loosely structured groups are able to coordinate their efforts over a very wide range of specialized skills to perform experiments of the highest subtlety and technical virtuosity. It is difficult to think of any other human activity where such precise coordination of effort is achieved by such voluntary cooperation. The successes of "big physics" in the last twenty years or so are usually attributed to inspiring leadership, of the kind that Cecil Powell himself demonstrated in the years immediately after the War. But all such projects also owe a great deal to the enthusiasm, and team spirit of the hundreds of physicists involved.

Unfortunately, there is nothing comparable to these voluntary "collabo-rations" in other branches of physics. Generally speaking, the majority of present day physicists are employed by the large laboratories and research establishments of enormous organizations such as industrial corporations or government ministries, and must conform to the administrative con-ventions of such organizations. The projects on which they work are laid down for them by the management, and the results of their research can-not be claimed as their own, to make public and to earn personal esteem. Within these constraints, they may have a great deal of freedom to plan and execute their work, and to enjoy the traditional gratifications of physics as a creative craft. We all know that advanced research in physics is fascinating for its sheer technical challenge, whatever the problem on which we are working. Nevertheless, it has to be recognized that one has little chance of "being a physicist" nowadays except as a member of a large, essentially bureaucratic organization.

The multiplicity, diversity, and sophistication of the various "maps" devised and used by physicists has the unfortunate effect of fragmenting our discipline into many "specialties," which seem at times quite foreign to one another. This can have a devastating effect on a scientific career. There was a time when an academic physicist could safely settle down to research on a particular phenomenon, such as atomic scattering, or to the exploitation of a particular technique, such as cryogenics, for a lifetime. That is no longer a safe way to "be a physicist." A physicist who has spent twenty years specializing in one field of research may suddenly be called on to move into another field, as a consequence of administrative reorganization, financial retrenchment, changes of government policy, loss of employment, or other mischance. Of course the specialty of every active scientist is slowly changing, simply through scientific and technological progress. But adaptation to this natural process of intellectual evolution or drift is not at all equivalent to an enforced migration to a remote province of physics. It is true that a basic education in physics to first degree level can give a good feeling for the characteristic principles and formal "language" of all the physical sciences, from theoretical mechanics to chemical engineering and from observational astronomy to molecular biophysics. But one cannot discount the further years of study and experience that have to be put in before one can get to the frontier of knowledge in any one of the several dozen sub-disciplines into which even physics itself is now divided. This knowledge is not to be got from books of reference: it is the tacit understanding of concepts and data that underlies the choice of appropriate experiments and theoretical initiatives in a particular field of research. Physicists have proved themselves to be surprisingly adaptable and flexible in entering other fields of science and applying their skills on new problems, but I do not think such moves are easy, even from one domain of physics research to another.

On Not Being Totally a Physicist

Nobody can go through the experience of becoming and being a physicist without a significant effect on their personality. In extreme cases, the effect is to produce people who live so entirely in the abstract world of physics that they seem unable to see the everyday world except in terms of physical concepts. One knows the type. All their conversation is not only loaded with jargon and concerned almost entirely with technical matters, large and small: all professional workers—lawyers, doctors, engineers, car salesmen—are the same in that respect. But even when the dedicated physicist has to give an opinion on, say, politics, or economics, or poetry (if he can recognize the existence of such a thing) or even love, he or she will express their thoughts in the same language. They will talk of the feedback loops between politicians and their supporters, they will want to know what the resultant of the forces is in the eternal triangle of passion, they will count the number of words in the lines of a sonnet, and will no doubt assure you that the blood of the martyrs would not look red at all under the light of a sodium lamp.

That is, of course, a bit of a caricature: but the temptations of "physicalism" are not easily resisted. It is often very instructive to quantify the variable in the situation, to set up a model of their interactions, to predict the consequences of certain actions and even to carry out experiments to verify those predictions. It is neither surprising nor deplorable that many physics graduates go in for accountancy, where just such attitudes may prove extremely effective. But there is a serious danger that in becoming a physicist one may unwittingly cut oneself off from the experience, insights and values of other people.

Most physicists do, of course, perform their duties, and enjoy their roles, as parents, spouses, taxpayers, football addicts, house-owners, tourists, etc., etc. Many of them eventually become responsible managers of major enterprises, in commerce and in government. My impression is that they often perform such responsible duties quite competently, perhaps because of their talent for, or upbringing in, orderly, practical thought and action.

Nevertheless, we do have to recognize that if one gives oneself too completely to a particular domain of being one may find it difficult to appreciate the strength and validity of alternative points of view. Philosophical "physicalism"—the tendency to scorn other scholarly endeavors, such as psychology, or history, because they lack the primary characteristics of physics and must therefore be misguided, or "soft"—is a relatively harmless malady which is usually cured by the experience that real life is not so simple after all. But ethical physicalism—the tendency to overvalue the pursuit of knowledge and the generations of technical devices as ends in themselves—can be a mortal sin against humanity itself. Thus, for example, the person who justifies his part in the development of some devilish weapon on the grounds that it is "all good physics," is morally sick,

and should be repudiated by the physics community. That is not to say
that there may be convincing reasons for taking part in military R&D as a
physicist; but these reasons derive from considerations such as patriotic
duty, or the need to make a living, which lie right outside the domains of
physics and its representations of the physical world.

And that brings me back to Cecil Powell. He was, as all of us who
knew him will recall with pleasure, a warm-hearted man, full of fun and
the spirit of life. And at a time when the technical resources of physics, to
which he had contributed so notably, suddenly became threatening to the
survival of humanity, he did not shirk his ethical responsibilities. Whether
or not one happens to agree with the particular actions and policies that he
fostered in the later years of his life—especially his outstanding role as
President of Pugwash—one must celebrate his magnificent example of
what it is required of one as a physicist in the modern era.

Now, alas, Cecil is no longer with us, so that his influence can be ex-
erted only by historical example and personal memory. But the times are
even more severely out of joint than ever, and the need for such examples
is yet more desperate. To my mind, the person who has shown us the most
powerful example of what is truly involved in "being a physicist" was Andrei Sakharov. I fully appreciate that the particular political and social
policies for which Sakharov put himself into such personal jeopardy are
very far from those which Cecil Powell advocated in his lifetime, and
which he might perhaps advocate now. But the impulse which drove Powell away from the merely technical aspects of physics to its ethical, social
and political implications also drove Sakharov, and to much greater personal cost. He moved slowly but steadily away from the dark side of
physics, where it is subordinated to nationalism and military force, towards the light of human values and rights. He thus came to realize that
there is a fundamental inconsistency between being totally a physicist and
being truly human. This is the uneasy thought with which I feel obliged to
end this essay.

Acknowledgments

The publisher and the author wish to express their grateful acknowledgment to the following publications and publishers for permission to reprint previously published articles.

WHAT EXACTLY DID HE DISCOVER? first appeared in the 3–16 May 1984 issue of the *London Review of Books*, pages 8–9, as a review of the four books discussed in this chapter.

PORTRAIT OF A DISAPPOINTED SCIENTIFIC SOUL was a review first published in the *New University Quarterly*, Spring 1980, pages 271–274.

LANDAU AND HIS SCHOOL, a review in the *London Review of Books,* was published in their 18 December 1980–21 January 1981 issue, page 10.

SEPARATION: ON THE REFUSENIK MARK AZBEL, also a review in the *London Review of Books*, appeared in the issue dated 4–17 August 1983, pages 15-16.

THE DARK SIDE OF SCIENCE, subtitled "A British scientist's response to the example of Andrei Sakharov," was an article first appearing in the May 1981 issue of *Index on Censorship*, pages 29–30 .

RESOLVING LITTLE LOCAL DIFFICULTIES was a book review published July 25,1986, page 15, in the *Times Higher Education Supplement*.

SERENDIPITY appeared as an Opinion column in Volume 32, 1981 of *Physics Bulletin*.

IRREVERSIBILITY was originally published as a book review in the *London Review of Books*, 18–31 March 1982, page 24.

A BILLION YEARS A WEEK, a book review, appeared in the *London Review of Books*, 19 September 1985, pages 20–21.

PROCESSING WORDS AND THOUGHTS, retitled for this collection, was the "Shelf Life" column in the September 1988 issue of the *Times Higher Education Supplement*.

THE EYES HAVE IT is an excerpt taken from the author's book *Reliable Knowledge: An Exploration of the Grounds for Belief in Science*, copyrighted 1978 by Cambridge University Press. This excerpt appeared as an article in *The Sciences* in January 1979, pages 13–15, 18.

SCIENCE IN AT LEAST THREE DIMENSIONS is a chapter the author wrote for *Theory of Knowledge and Science Policy*, pages 394–410, edited by M. de Meg and published by The University of Ghent in 1979.

PUSHING BACK FRONTIERS—OR REDRAWING MAPS! was originally a chapter the author wrote for *The Identification of Progress in Learning*, pages 1–12, edited by T. Hagerstrand and published by The Cambridge University Press in 1985.

WHAT IS YOUR SPECIALTY? was originally a lecture in San Antonio, Texas dated 16 November 1986.

A NATURAL PHILOSOPHER, originally the "Shelf Life" column of the 26 April 1991 issue of the *Times Higher Education Supplement*. was retitled for this collection.

OUT OF THE PARLOR, INTO THE LABORATORY first appeared as a book review in the 23 March 1984 issue of the *Times Higher Education Supplement*.

DEVASTATING THE DISSEMBLERS first saw print as a book review for the *Times Literary Supplement* of December 25, 1981.

FUDGING THE FACTS was originally a book review for the *Times Literary Supplement* of September 9, 1983, page 955.

RECONSTRUCTING THE REALITY OF SCIENTIFIC GROWTH was first printed as a book review in Volume XVII, 1979, of *Minerva*, pages 321–327.

WHAT SHALL WE LOOK INTO NOW? first was published as a book review in the *London Review of Books*, May 21, 1987, page 16.

EXPANSIONISTS AND RESTRICTIONISTS, a book review for the *Times Literary Supplement* of May 21, 1982, appeared on page 561 of that issue.

SCIENTISTS—AND OTHER PEOPLE was a paper delivered in December 1982 at MIT.

THE SOCIAL RESPONSIBILITY OF SCIENTISTS: BASIC PRINCIPLES first appeared as Chapter 9 "Basic Principles" (pages 161–178) of *Scientists the Arms Race and Disarmament*, edited by J. Rotblat, and published by Taylor & Francis in 1982.

SOCIAL RESPONSIBILITY IN VICTORIAN SCIENCE was an essay included in *Science. Technology & Society in the Time of Alfred Nobel*, pages 21–43, edited by C.G. Bernhard, E. Crawford, and P. Sorbohm, and published by Pergamon in 1982.

RIGHTS AND RESPONSIBILITIES IN RESEARCH, an article originally entitled "Rights and Responsibilities in Research—Dichotomy or Dynamic Balance," first appeared in the journal *Society and Science*, Volume 3 Number 4, November 1980, pages 109–116.

SCIENCE EDUCATION FOR THE REAL WORLD was first printed as an article in *New Scientist*, 16 October 1980, pages 169–170.

CONCEPTIONS OF SCIENCE made its first appearance as an essay in *Realizing Social Science Knowledge*, copyrighted 1983 by PhysicaVerlag (pages 179–196), and edited by N. Elias, H. Martins, and R. Whitley.

WHAT ARE THE OPTIONS? was first published in 1981 in *Minerva*, Volume XIX, pages 1–42.

PUBLISH OR PERISH? appeared as a column in the *Times Higher Education Supplement*, August 10, 1982, on page 13.

ACADEMIC SCIENCE AS A SYSTEM OF MARKETS was first seen in the Volume 45, Number 1, Winter 1991 issue of *Higher Education Quarterly*.

THE COLLECTIVIZATION OF SCIENCE was the Bernal Lecture, 1983 (delivered April 14 of that year) and printed in the *Proceedings of the Royal Society of London, Series B* (B 219, pages 1–19).

THE INDIVIDUAL IN A COLLECTIVIZED PROFESSION was the inaugural address at the conference on "La Professionalita Scientifica" in Rome, March 20–23, 1985.

ON BEING A PHYSICIST was the Cecil Powell Memorial Lecture,which was to have been delivered at the conference of the European Physical Society, in Prague, 1984.

Index

About the Author

John Ziman, Emeritus Professor of Physics of the University of Bristol, is a noted authority on the social relations of science and technology. For more than thirty years, he has devoted himself to enlightening his peers and lay audiences alike with his insights into social aspects of science and technology through popular books, articles, and involvement in numerous organizations, both at home and abroad.

Brought up in New Zealand, he received his MSc. from Victoria University College (Wellington, New Zealand) in 1946. He then studied at Oxford, completing a doctorate in theoretical physics in 1952. In 1954, he moved to Cambridge University as a lecturer at the Cavendish Laboratory, and later as a Fellow of Kings's College. In 1964, he was appointed to a professorial chair at Bristol where he was Director of the H. H. Wills Physics Laboratory from 1976 to 1981. Following voluntary early retirement from Bristol in 1982, he was Visiting Professor at the Department of Social and Economic Studies at Imperial College in London. From 1986 to 1991, he headed the Science Policy Support Group.

His work in the theory of the electrical and magnetic properties of solid and liquid metals led to his election to the Royal Society in 1967. His extensive list of appointments include serving as Joint Editor of the *Cambridge Review*, Member of the Scientific Council of the International Centre for Theoretical Physics, Chairman of the Council for Science and Society, and Chairman of the European Association for the Study of Science and Technology. In 1985, he received an honorary DSc. from Victoria University.

Dr. Ziman, author of numerous textbooks and popular works, and contributor to countless journals, is, by his own account, obsessed with science and technology. More so, however, he is fascinated by their philosophical, psychological, historical, political, economic, and sociological components.